T0333772

Reinventing the Propeller

An international community of specialists reinvented the propeller during the Aeronautical Revolution, a vibrant period of innovation in North America and Europe from World War I to the end of World War II. They experienced both success and failure as they created competing designs that enabled increasingly sophisticated and "modern" commercial and military aircraft to climb quicker and cruise faster using less power. *Reinventing the Propeller* nimbly moves from the minds of these inventors to their drawing boards, workshops, research and development facilities, and factories, and then shows us how their work performed in the air, both commercially and militarily. *Reinventing the Propeller* documents this story of a forgotten technology to reveal new perspectives on engineering, research and development, design, and the multilayered social, cultural, financial, commercial, industrial, and military infrastructure of aviation.

Jeremy R. Kinney is a curator in the Aeronautics Department at the Smithsonian National Air and Space Museum.

Cambridge Centennial of Flight

General Editors

The series presents new titles dealing with the drama and historical impact of human flight. The Air Age began on December 17, 1903, with the epic powered and controlled flight by the Wright brothers at Kitty Hawk. The airplane rapidly developed into an efficient means of global travel and a lethal weapon of war. Modern rocketry has allowed heirs of the Wrights to orbit the Earth and to land on the Moon, inaugurating a new era of exploration of the solar system by humans and robotic machines. The Centennial of Flight series offers pioneering studies with fresh interpretive insights and broad appeal on key themes, events, and personalities that shaped the evolution of aerospace technology.

Also published in this series

Scott W. Palmer, *Dictatorship of the Air: Aviation Culture and the Fate of Modern Russia*

Michael B. Petersen, *Missiles for the Fatherland: Peenemünde, National Socialism, and the V-2 Missile*

Von Hardesty, *Camera Aloft: Edward Steichen in the Great War*

Asif A. Siddiqi, *The Red Rockets' Glare: Spaceflight and the Russian Imagination, 1857–1957*

Reinventing the Propeller

Aeronautical Specialty and the Triumph of the Modern Airplane

JEREMY R. KINNEY

Smithsonian National Air and Space Museum

CAMBRIDGE
UNIVERSITY PRESS

University Printing House, Cambridge CB2 8BS, United Kingdom
One Liberty Plaza, 20th Floor, New York, NY 10006, USA
477 Williamstown Road, Port Melbourne, VIC 3207, Australia
4843/24, 2nd Floor, Ansari Road, Daryaganj, Delhi – 110002, India
79 Anson Road, #06-04/06, Singapore 079906

Cambridge University Press is part of the University of Cambridge.

It furthers the University's mission by disseminating knowledge in the pursuit of education, learning, and research at the highest international levels of excellence.

www.cambridge.org
Information on this title: www.cambridge.org/9781107142862
10.1017/9781316529744

First published 2017

Printed in the United States of America by Sheridan Books, Inc.

A catalogue record for this publication is available from the British Library.

Library of Congress Cataloging in Publication Data
Names: Kinney, Jeremy R.
Title: Reinventing the propeller: aeronautical specialty and the triumph of the modern airplane / Jeremy R. Kinney, Smithsonian National Air and Space Museum.
Description: New York: Cambridge University Press, 2017. | Series: Cambridge centennial of flight | Includes bibliographical references and index.
Identifiers: LCCN 2016041201 | ISBN 9781107142862 (hard back)
Subjects: LCSH: Propellers, Aerial – History. | Aeronautics – History.
Classification: LCC TL705.K56 2016 | DDC 629.134/36–dc23
LC record available at https://lccn.loc.gov/2016041201

ISBN 978-1-107-14286-2 Hardback

To Mom and Dad, for never hesitating to buy their little boy books about airplanes

Contents

List of Figures — *page* x
List of Tables — xii
Preface — xiii
Acknowledgments — xv
List of Abbreviations — xix
A Note on Terms — xxi

1 Introduction: The Propeller and the Modern Airplane — 1
2 "The Best Propeller for Starting Is Not the Best for Flying" — 16
3 "Engineering of a Pioneer Character" — 45
4 A "New Type Adjustable-Pitch Propeller" — 75
5 "The Propeller That Took Lindbergh Across" — 116
6 "The Ultimate Solution of Our Propeller Problem" — 146
7 No. 1 Propeller Company — 180
8 A Gear Shift for the Airplane — 204
9 Constant-Speed — 235
10 "The Spitfire Now 'Is an Aeroplane'" — 274
11 A Propeller for the Air Age — 305
12 Conclusion: The Triumph and Decline of the Propeller — 345

Essay on Sources — 351
Index — 361

Figures

1. George Grogan's United Air Lines *Mainliner* over Manhattan. *page* 2
2. Orville and Wilbur Wright and their second *Flyer* at Huffman Prairie. 22
3. A wood, fixed-pitch propeller installed on an Army Air Service Curtiss biplane. 38
4. The electric propeller whirl testing rig at the Westinghouse East Pittsburgh factory. 54
5. The Engineering Division whirl rig at McCook Field. 61
6. The Wright Field Propeller Laboratory. 69
7. Earl Daugherty with the Hart and Eustis propeller installed on his biplane. 76
8. The US Army Air Service's 1920 Hart reversible and variable-pitch propeller. 89
9. Wallace R. Turnbull with his electric propeller installed on an Avro biplane. 114
10. Charles Lindbergh, the *Spirit of St. Louis*, and the Standard Steel ground-adjustable-pitch propeller. 117
11. The split hub and clamping ring design of the Standard Steel ground-adjustable-pitch propeller. 130
12. The double shoulder retention system found in the hub and blade root of the Standard Steel ground-adjustable-pitch propeller. 131
13. Lt. David Rittenhouse stands on the float of his Curtiss CR-3 Racer. 147
14. Albert Reed, Casey Jones, and Curtiss employees commemorate the flight of the D-1 propeller. 151
15. The UATC Technical Advisory Committee. 196

16. Bernt Balchen and the early model Hamilton Standard controllable-counterweight propeller. 200
17. Dutch KLM's DC-2 *Uiver* arriving in Melbourne. 205
18. Triumph of the propeller specialists: Hamilton Standard employees with the Collier Trophy. 230
19. The Hamilton Standard Hydromatic Propeller. 243
20. Vickers Wellesleys of the RAF Long Range Development Unit. 269
21. RAF 19 Squadron received Spitfire fighters equipped with two-blade fixed-pitch propellers. 279
22. Spitfires of 65 Squadron taking off from Hornchurch in August 1940. 301
23. Feathered propeller on Boeing B-17 Flying Fortress over northern Italy. 309
24. Curtiss Propeller Division workers in 1943. 312
25. Lockheed-Martin C-130J Hercules transport. 342

Tables

1. Famous flights made with Standard Steel ground-adjustable propellers *page* 141
2. Reed propeller production 174
3. United States and British constant-speed propeller production during World War II 319
4. German and Japanese propeller production, 1934–1945 324

Preface

When I was a small boy, my parents gave me a little wood airplane with a big red plastic propeller. Holding the fuselage in one hand, I turned the propeller to wind up the rubber band "engine." The power of the rubber band spun the propeller and I could feel a breeze flow over my airplane until I let it go. If I did not wind enough, the airplane would jump and skid along the ground, unable to take off. If I wound too much, the propeller would "race" and the airplane would vibrate and careen out of control. I spent hours in our backyard learning how to wind the propeller so my airplane would fly straight and level.

During one of our many family visits to museums and air shows during my childhood, I got close to some of my favorite airplanes. One of those was the Curtiss Jenny, a fabric-covered biplane that the wandering barnstormers flew from town to town across America in the 1920s selling rides to brave and curious folks. As I looked at the Jenny's wood propeller, I could see a long, flat curve that moved all the way along its length, from the tip at one end, through the hub in the center, and on to the tip at the other end. This aerodynamic twist, called pitch, gave the propeller its shape and allowed it to turn the engine's power into thrust – the "breeze" created by my toy airplane – to propel the Jenny forward.

Another favorite was the Douglas DC-3 airliner from the 1930s. The DC-3 was a very different airplane than the Jenny. It was a sleek twin-engine monoplane capable of carrying twenty-one passengers. Unlike my toy airplane and the Jenny, each of its two propellers had three shiny metal blades that could change pitch in flight. When needed, the propellers generated a lot of thrust for takeoff and prevented "racing" as the

DC-3 cruised through the sky. The technical transformation, or reinvention, of the airplane propeller – from the one found on the Jenny to the advanced design installed on the DC-3 – by a community of specialists and what it had to do with increasing the performance of aircraft over the course of the twentieth century is the focus of this book.

Acknowledgments

I began this project as a graduate student studying the history of technology at Auburn University. Stephen L. McFarland served as my major professor and dissertation chair. James R. Hansen provided an unprecedented opportunity to conduct research at various archives throughout the United States through my involvement in the NASA-sponsored multivolume documentary history of aerodynamics, *The Wind and Beyond*. William F. Trimble's skill as an editor improved the text and the ideas expressed in it immeasurably. All three served as thoughtful and patient guides throughout the dissertation and manuscript preparation process. Special mention must be made of a valued colleague, friend, and recent addition to the Auburn history faculty, Alan D. Meyer, who valiantly read and critiqued the final manuscript draft. I would also like to thank Donna J. Bohanan, Lindy Biggs, John Burkhalter, my fellow graduate students, and the staff of the Auburn University libraries for their assistance.

This work concluded at the Smithsonian National Air and Space Museum where I curate the world's largest collection of propeller artifacts, approximately 400 in all. I am especially indebted to past and present colleagues in the Aeronautics Department: John Anderson, Hal Andrews, Dorothy Cochrane, Roger Connor, Tom Crouch, Ron Davies, Roland Foley, Von Hardesty, Peter Jakab, Michael Kern, Russell Lee, Christopher Moore, Tom Paone, Dominick Pisano, Nicholas Renner, Herbert Rochen, Bob van der Linden, Howard Wesoky, and Collette Williams. Within the Archives and Library, Amanda Buel, Elizabeth Borja, Chris Cottrill, Phil Edwards, Dan Hagedorn, Marilyn Graskowiak, Kate Igoe, Mark Kahn, Melissa Keiser, Jessamyn Lloyd, Brian Niklas, Mark Taylor, David Schwartz, Leah Smith, and Larry Wilson provided invaluable support.

The staff of the Collections Department, including Tony Carp, Matthew Nazzaro, and Scott Wood, made available the technological artifacts discussed in this book and shared their knowledge of them. Finally, Jennifer Carlton of the Exhibits Department assisted with the processing of photographs.

Historians working for federal agencies also proved crucial to the completion of this project. Rick Peuser and Don Giordano at the National Archives and Records Administration and James Aldridge, Archie DiFante, Jeff Duford, Wes Henry, Dave Menard, Laura Romesburg, Brett Stolle, and James Young of the United States Air Force History and Museums Program facilitated access to collections central to this study. At the National Aeronautics and Space Administration History Office, I would like to thank Jane Odom and Colin Fries.

The respective staffs of universities, museums, libraries, and archives went out of their way to help: Lois Beattie and Deborah G. Douglas of the Massachusetts Institute of Technology, David White at Kettering University, Janice Baker at the Museum of Flight, the staff of the Chattanooga-Hamilton County Bicentennial Library, Glenda Thornton, St. John Free Public Library, and Stephen Payne and Marcia Mordfield of the Canada Aviation and Space Museum.

Corporate legacy and history programs facilitated this project. Mike Lombardi, Tom Lubbesmeyer, and Pat McGinnis of the Boeing Company allowed me to uncover a veritable treasure trove of textual and photographic materials concerning the history of the propeller. Daniel Coulom, Marilyn Kozynoski, John Mayo, John Misselwitz, Nancy Mottes, and Robert Scheckman of United Technologies Corporation Aerospace Systems, Jamison Roseberry of General Electric Aviation Systems, John Leonard of the Rolls-Royce Heritage Trust, Allison Branch, and Suzanne Petre of the General Motors Corporation Business Research Library also provided valuable assistance.

Among the larger community of engineers, enthusiasts, historians, and participants, Robert Ash, Douglas B. Ayer, Roger Bilstein, Squire and Freda Brown, Frank B. Estabrook, Tom Fey, Bruce and Carol Frazer, Ray Hegy, Charles W. Hitz, Elmo F. Huston, Peter and Joanne Law, Spencer Heath MacCallum, Kimble D. McCutcheon, Edward T. Reilly, R. K. Smith, and Walter G. Vincenti shared their valuable resources, stories, and time.

I would like to recognize the following individuals in Great Britain: aerospace engineer and historian Patrick Hassell; Keith Moore of the Institution of Mechanical Engineers; Spitfire and Hurricane expert Mike

Williams; Peter Elliot, Nina Burls, and Gordon Leith at the Royal Air Force Museum; Andrew Nahum and Rory Cook of the Science Museum; archivist William Spencer of the National Archives of the United Kingdom; and Christiaan van Schaardenburgh, Head of Collections at the Tank Museum.

Special thanks are in order for two individuals closely related to the story of the airplane propeller, Walter H. Caldwell and George Rosen. Caldwell shared his personal knowledge of his father, Frank W. Caldwell, a central figure in the technical development of the propeller. Rosen paved the way for my research with his book, *Thrusting Forward: A History of the Propeller*, and proved to be of great help when the author had questions about sources. Both Walter and George passed away before this project's completion, and it is unfortunate they were unable to see the final product in print.

At Cambridge University Press, I must recognize Deborah Gershenowitz, Dana Bricken, Kristina Deusch, Amanda George, and Centennial of Flight series editors John Anderson and Von Hardesty. John, one of the pioneers of the historical study of aeronautical engineering, was an especially cheerful and encouraging advocate for this study. I also appreciate greatly the contribution of the peer readers, Richard P. Hallion and David Zimmerman.

Financial support came from the American Historical Association, Auburn University, the National Aeronautics and Space Administration, the Royal Air Force Museum American Foundation, and the Smithsonian.

The research and writing of history, like the creation of technology, is a communal effort. This book would not have been possible without the gracious help of these and many more individuals and institutions. While that is true, all errors of fact, interpretation, or omission are solely my own.

Final thanks are due to family members Stanton, Margaret, Fannie, Kristopher, Saki, Cassie, and a little white dog named Sneezy for their inspiration and support during the course of this project. My wife, Cheryl, has been a devoted believer in this story from the day we first met and we welcomed our beautiful daughter, Piper, as writing the book drew to a close. I dedicate this book to my parents, Jerry and Susan, because they never hesitated to encourage their youngest son's enthusiasm for the history of flight.

Abbreviations

ASME	*American Society of Mechanical Engineers*
AVCO	*Aviation Corporation of Delaware*
BAP	*Bureau of Aircraft Production*
BATC	*Boeing Airplane and Transport Corporation*
BCA	*Boeing Company Archives*
BEF	*British Expeditionary Force*
BF	*Biography Files*
BMW	*Bavarian Motor Works*
BuAer	*Bureau of Aeronautics*
CAA	*Civil Aeronautics Authority*
CAMC	*Curtiss Aeroplane and Motor Company*
CATD	*German/Japanese Captured Air Technical Documents*
CE	*Claire Egtvedt Files*
CHC	*Chattanooga-Hamilton County Bicentennial Library*
CKC	*Charles Kettering Collection*
CMK	*Clement M. Keys Papers*
CO	*Commanding Officer*
COAC	*Chief of the Air Corps*
COAS	*Chief of the Air Service*
CWC	*Curtiss-Wright Corporation Records*
DAD	*D. Adam Dickey Papers*
DVL	*Deutsche Versuchsanstalt für Luftfahrt*
FWP	*Fred E. Weick Papers*
GE	*General Electric Company*
GHQ	*General Headquarters Air Force*
GM	*General Motors Corporation*
HFM	*Henry Ford Museum and Greenfield Village Research Center*
HRA	*United States Air Force Historical Research Agency*

HSCRD	*Hamilton Sundstrand Community Relations Division*
IAS	*Institute of the Aeronautical Sciences*
JCH	*Jerome C. Hunsaker Papers*
KUA	*Kettering University Archives*
MAA	*Manufacturer's Aircraft Association*
MFRC	*Museum of Flight Resource Center*
MIT	*Massachusetts Institute of Technology*
NAA	*National Aeronautic Association*
NACA	*National Advisory Committee for Aeronautics*
NARA	*National Archives and Records Administration*
NASA	*National Aeronautics and Space Administration*
NASAHO	*National Aeronautics and Space Administration History Office*
NASM	*National Air and Space Museum*
NRC	*National Research Council*
OPEC	*Organization of Petroleum Exporting Countries*
PCF	*Propulsion Curatorial Files*
PTF	*Propulsion Technical Files*
RAE	*Royal Aircraft Establishment*
RAeS	*Royal Aeronautical Society*
RAF	*Royal Air Force*
RCAF	*Royal Canadian Air Force*
RD	*Research and Development File*
RG	*Record Group*
RLM	*Reichsluftfahrtministerium*
RO	*Registrar's Office*
SAE	*Society of Automotive Engineers*
STOL	*Short-takeoff and Landing*
TAC	*Technical Advisory Committee*
TNA	*National Archives of the United Kingdom*
TWA	*Transcontinental and Western Air/Trans World Airlines*
UAC	*United Aircraft Corporation*
UAL	*United Air Lines*
UATC	*United Aircraft and Transport Corporation*
USCC	*United States Court of Claims*
USSBS	*United States Strategic Bombing Survey*
UTC	*United Technologies Corporation*
VDM	*Vereinigte Deutsche Metallwerke*
WFTD	*Wright Field Technical Documents Library*

A Note on Terms

Why do we call this technology a propeller? That is its basic function, to propel an airplane forward through the air by the generation of thrust. Propellers have also been known by other names since the early days of aviation. The Wright brothers first called their creation a "fan screw" in 1902 to express its ability to move air through the fluid medium of the sky in the same way a carpenter's wood screw twists through a pine board. By 1908, they then used "screw propeller" before ultimately settling on "propeller" by 1913, with early American aeronautical enthusiasts following suit.[1] The European aeronautical community chose the English airscrew, French *hélice aérienne*, and German *Luftschraube* since it literally referred to the motion of the technology through the air.

Those different choices stirred a significant debate over whether or not airscrew or propeller was the correct terminology. From the perspective of engineering education, textbooks on both sides of the Atlantic reflected the divide.[2] In 1920, British engineer Dr. Henry C. Watts acknowledged that they were both abbreviations for "air screw-propeller," which described the medium in which the technology operated (air), its specific form

[1] "'Fan Screw Experiments,' December 15, 1902, Wilbur *Wright's Notebook H, 1902–1905*, Appendix III: The Wright Propellers," in*The Papers of Wilbur and Orville Wright, Vol. 1, 1899–1905*, ed. Marvin W. McFarland (New York: McGraw-Hill, 2001), 598; Wilbur and Orville Wright, "The Wright Brothers' Aeroplane," *Century Magazine* (September 1908), and Orville Wright, "How We Made the First Flight," *Flying* (December 1913), both in Peter L. Jakab and Rick Young, eds., *The Published Writings of Wilbur and Orville Wright* (Washington, DC: Smithsonian Institution Press, 2000), 29–30, 41.

[2] Examples include Charles B. Hayward, *Aerial Propeller* (Chicago: American School of Correspondence, 1912); and M. A. S. Riach, *Air-Screws* (London: Crosby, Lockwood, and Son, 1916).

of motion (screw), and its fundamental name (propeller). He admitted his preference for the "more logical" *airscrew*, which was the formal term adopted by the Royal Aeronautical Society (RAeS), but deferred to the "verdict of custom and usage" delivered by aircraft manufacturers and military air services to adopt and use *propeller*.[3] An open letter from an aspiring pilot, W. A. Chase, to the RAeS Technical Terms Committee and to C. G. Grey of the influential British journal *The Aeroplane* reinforced that argument against the use of the impolite and "ill-sounding" "airscrew" in Britain.[4]

Despite Watts's desire for order and Chase's wish to be respectable, the use of both terms persisted. Fred Weick noted in his landmark 1930 engineering textbook, *Aircraft Propeller Design*, that usage continued to reflect international boundaries. The American aeronautical community stuck to propeller while the members of the larger international aeronautical community used their linguistic equivalent of airscrew.[5] Over time, the terms became the synonyms they were intended to be originally, with propeller being the dominant of the two. In fact, while the British aeronautical community remained steadfastly attached to airscrew, at least until the late twentieth century, the rest of the international aeronautical community allowed propeller to become a nonliteral translation.[6] Both propeller and airscrew appear in this book depending on the national locale being discussed.

[3] Henry C. Watts, *The Design of Screw Propellers with Special Reference to Their Adaptation for Aircraft* (London: Longmans, Green, and Company, 1920), v; W. Barnard Faraday, ed., *A Glossary of Aeronautical Terms* (London: Royal Aeronautical Society, 1919), 69.

[4] W. A. Chase, "Terminological Tortuosities," *The Aeroplane* 18 (January 28, 1920): 213.

[5] Fred E. Weick, *Aircraft Propeller Design* (New York: McGraw-Hill, 1930), 2.

[6] John D. Anderson, Jr., *Introduction to Flight*, 7th ed. (New York: McGraw-Hill, 2012), 753. The trend, at least from the perspective of historical writing, in Great Britain is edging toward "propeller." "Airscrew" does not appear in a late twentieth century authoritative work on the technical development of the interwar airplane with an emphasis on developments in Great Britain. See Philip Jarrett, *Biplane to Monoplane: Aircraft Development, 1919–1939* (London: Putnam, 1997).

1

Introduction

The Propeller and the Modern Airplane

On the afternoon of Wednesday, April 6, 1938, a United Air Lines *Mainliner*, a Douglas Sleeper Transport, departed from the Newark, New Jersey, airport with fourteen passengers aboard, a crew of three, and enough fuel to reach Chicago. They rolled on the runway for only a brief fifteen seconds before lifting off. Then pilot George Grogan, sitting in the left seat, pointed the futuristic silver twin-engine airliner northeast toward New York City. As they cruised over Central Park at 6,000 feet and just over 200 mph, the propeller on the right engine stopped turning (Figure 1). To any observer, that was an indication of engine failure, loss of power, and an impending crash. Remarkably, Grogan maneuvered the *Mainliner* through the clouds, climbed, and turned without losing speed or altitude. When he was finished, Grogan restarted the right engine and the motionless propeller began whirling again. He then shut down the left engine, stopped its propeller and set its blades parallel to the wind like he had for the other, and proceeded on with an aerial tour of New York, Connecticut, and New Jersey before returning to Newark on two running engines.[1]

The passengers aboard Grogan's *Mainliner* had experienced the first public demonstration of a revolutionary aeronautical innovation, a practical propeller capable of maximizing the power of an aircraft engine and the performance and safety of an airplane overall. The next day, United Air Lines inaugurated its fifteen-hour coast-to-coast transcontinental service with Douglas Sleeper Transport (DST) and DC-3 airliners capable of

[1] James V. Piersol, "Air Currents," *New York Times*, April 10, 1938, p. 164; "New Propellers Installed," *Aero Digest* 32 (May 1938): 27.

FIGURE 1 George Grogan's United Air Lines *Mainliner* DST with one engine shut off over Manhattan in New York City in April 1938.
Courtesy of Hagley Museum and Library.

flying as high as 20,000 feet at an unprecedented speed of three miles a minute across the United States.[2]

The Douglas airliners and their propellers were "modern" in every sense of the word. Compared to the slow, fabric-covered biplane of the World War I era, they were the state of the art in aeronautical technology; the flying embodiment of the combined aerodynamic, structural, and propulsive innovations that first made flight a global endeavor. These high-speed streamlined metal monoplanes resulted from an Aeronautical Revolution that swept through North America and Europe during the twenty years between the world wars.[3] Parallel and intertwined advances

[2] "Fast Air Liners Go into Service," *Los Angeles Times*, April 8, 1938, p. A2; United Air Lines, "United Announces the Nation's Fastest, Most Powerful, Quietest Large Airliners!" *Chicago Daily Tribune*, April 20, 1938, p. 4; American Society of Mechanical Engineers, *Hamilton Standard Hydromatic Propeller: International Historic Engineering Landmark*, Book No. HH 10 90 (November 8, 1990), 3.

[3] "Aeronautical Revolution" goes beyond the technical emphasis of John B. Rae's "airframe revolution" or Larry Loftin's "design revolution" to fully encompass the interrelated technical and nontechnical changes witnessed in aviation overall during the

in technology, governmental regulation, entrepreneurial growth, and cultural awareness created aviation as we know it today.

At the center of this revolution in the sky was an international community. These designers, engineers, inventors, entrepreneurs, military officers, pilots, and many others represented a myriad of aeronautical specialties. They reinvented existing technology to create new airplanes. Their shared culture of performance inspired, pushed, and enticed competing visions of expanding the capabilities of aeronautical technology in the name of advancing humankind.

Within the aeronautical community, there was a group of specialists dedicated to one critical part of the modern airplane, the propeller. Their work to reinvent that one component resulted in a transformative technology that enabled commercial and military aircraft to climb quicker and cruise faster using less power and, if need be, fly to safety on one engine. When integrated into increasingly sophisticated aircraft, the airlines used them to connect the world by airplane and air forces used them to fight a global war in the air. These modern propellers and their ability to "shift gears in the air" to meet different operating conditions helped make the system of the airplane a world-changing technology. This book is the story of those specialists, their beliefs, their successes and failures, and the propellers they created while standing at the intersection of the technical and cultural forces that shaped the airplane over the course of the twentieth century.

A Culture of Performance and the Reinvention of the Airplane

The first airplane took to the air at Kill Devil Hill on the desolate windswept dunes off the Outer Banks on the coast of North Carolina on Thursday morning, December 17, 1903. At the controls was Orville Wright, who with his older brother, Wilbur, brought their *Flyer* from Dayton, Ohio. This original flying machine had two wings, a horizontal elevator mounted in front, and a vertical tail in the rear constructed from wood, covered in muslin, and joined together with wood struts and wire. Power came from an aluminum reciprocating piston engine connected by chains to two wood propellers. The *Flyer* flew four times that morning.

1920s and 1930s. John B. Rae, *Climb to Greatness: The American Aircraft Industry, 1920–1960* (Cambridge, MA: MIT Press, 1968), 58; Laurence K. Loftin, Jr., *Quest for Performance: The Evolution of Modern Aircraft* (Washington, DC: National Aeronautics and Space Administration, 1985), 77.

The last saw Wilbur pilot the biplane as high as fourteen feet, at speeds approaching 20 miles per hour over a distance of 852 feet. During those 59 seconds, the brothers knew full well they were the first to conceptualize, design, and construct, in other words, invent, an airplane capable of practical, sustained flight. While the achievement is theirs, Wilbur and Orville were also the preeminent members of an emergent community of aeronautical enthusiasts.[4]

This aeronautical community grew as it improved, developed, and used the Wrights' creation through those early days of flight, World War I, and on into the 1920s. The famous flights of Frenchman Louis Blériot, German Manfred von Richthofen, American Charles Lindbergh, and other pioneering pilots represented the cultural acceptance of the airplane from an entertaining novelty flown by gentlemen-aviators at air meets into an instrument of commerce, a weapon of war, and a vehicle for spectacle. In response, a significant portion of Western culture, primarily in the United States, embraced a new form of technological enthusiasm called "airmindedness" that called for the zealous support of aviation to bring about the next great era in human civilization, the Air Age.[5] That reaction revealed a strong cultural undercurrent, characterized by the rapid and continual technical development of time- and distance-shattering technologies. The popular passion for the fast-sailing clipper, steamship, railroad locomotive, automobile, and airplane as well as their perceived contributions to society reflected the Enlightenment ideal of "Progress" where "Technology" autonomously drove humankind forward through history. For airminded America and Europe, the concept of the airplane as a vehicle to utopia was easy to grasp.

New communities of aeronautical specialists in North America and Europe created the technical foundation that energized the growth of airmindedness in the popular imagination. Unlike the Wright brothers,

[4] Tom D. Crouch, *A Dream of Wings: Americans and the Airplane, 1875–1905* (New York: Norton, 1981; reprint, Washington, DC: Smithsonian Institution Press, 1989), 19.

[5] Joseph J. Corn, *The Winged Gospel: America's Romance with Aviation, 1900–1950* (New York: Oxford University Press, 1983), 12, 31, 135. Studies focusing on European and transnational aeronautical enthusiasm that complement Corn's work include: Peter Fritzsche, *A Nation of Fliers: German Aviation and the Popular Imagination* (Cambridge, MA: Harvard University Press, 1992); Robert Wohl, *A Passion for Wings: Aviation and the Western Imagination, 1908–1918* (New Haven, CT: Yale University Press, 1994), and *The Spectacle of Flight: Aviation and the Western Imagination, 1920–1950* (New Haven, CT: Yale University Press, 2005); Guillaume de Syon, *Zeppelin! Germany and the Airship, 1900–1939* (Baltimore, MD: Johns Hopkins University Press, 2002).

these designers, engineers, and inventors focused on creating, innovating, and introducing improved components and systems of the airplane, from innovative aerodynamic shapes, structures, and engines to cockpit instruments, landing lights, and structural fasteners.[6] These largely anonymous technical personnel worked either for themselves, small companies, government research institutions, or octopus-like corporations. They shared the historical stage with executives, entrepreneurs, government and military officials, pilots, journalists, factory workers, and even family members in the larger aeronautical community. To one observer, it was the "passion of the specialists" and the "fervor of those individuals" that created a heritage of research, development, and foundational work in aeronautics.[7] There were many lifetimes of work, investment, and dedication to these distinct areas within aeronautics.

The specialists saw themselves as pioneers of a technological frontier in the sky as they toiled to make airplanes fly higher, farther, and faster than ever before.[8] Adopted by historians almost to the point of cliché today, the phrase "higher, faster, and farther" reflected an interrelated technical and cultural pursuit by the aeronautical community since the days of the Wright brothers.[9] To the specialists, a seemingly limitless increase in

[6] Walter G. Vincenti's pioneering study of engineering knowledge documented "communities of practitioners" acting as "social agents for the production of knowledge" during the first half century of flight. Walter G. Vincenti, *What Engineers Know and How They Know It: Analytical Studies from Aeronautical History* (Baltimore, MD: Johns Hopkins University Press, 1990), 238–239.

[7] H. M. "Jack" Horner, in *We Saw It Happen*, United Aircraft Corporation, 1953, motion picture.

[8] David T. Courtwright, *Sky as Frontier: Adventure, Aviation, and Empire* (College Station, TX: Texas A&M University Press, 2004), 14–15, 97–105.

[9] Dominick Pisano, "The Literature of Aviation," in *Milestones of Aviation*, ed. John T. Greenwood (New York: Hugh Lauter Levin Associates, 1995), 311. Greenwood's entire volume discusses the development of the airplane through chapters entitled "farther," "higher and faster," "bigger," and "better." For works that also convey the "higher, faster, and farther" theme, see Mark P. Friedlander, Jr., and Gene Gurney, *Higher, Faster, and Farther* (New York: Morrow, 1973); Terry Gwynn-Jones, *Farther and Faster: Aviation's Adventuring Years, 1909–1939* (Washington, DC: Smithsonian Institution Press, 1991); and Stephen L. McFarland, "Higher, Faster, and Farther: Fueling the Aeronautical Revolution, 1919–1945," in *Innovation and the Development of Flight*, ed. Roger D. Launius (College Station, TX: Texas A&M University Press, 1999), 100–131. The first significant appearance of the phrase was in a popular history of the United States, which celebrated the fact that "planes [*sic*] steadily flew higher, faster, and farther" as they circumnavigated the earth, raced the sun across the continent in one day, and soared over the North Pole among other aeronautical achievements. Walter S. Hayward and Dorothy A. Hamilton, *The American People: A Popular History of the United States, 1865–1941* (New York: Sheridan House, 1943), 222–223.

aircraft altitude, speed, and range as the measure of performance served as a clear indicator of two things. First, ever more sophisticated aircraft increasingly expanded the commercial and military capabilities of aviation. Second, that performance validated their culture's celebration of the airplane as a vehicle of "Progress." The influential editor of the British aeronautical journal, *The Aeroplane*, C. G. Grey, reflected in 1924 on the first twenty years of flight and observed this relationship as culture and technology merged into the "conquest of the air." He suggested what might follow in the future of aviation – bigger, more efficient, and safer aircraft flying along established airways – and that they would "all come in due course."[10] This airminded culture of performance saw inevitability in the continually progressive evolution of the airplane and what the result would do for the world.[11]

During the 1920s and 1930s, the aeronautical community came together collectively and created new technology as it expressed its enthusiasm for flight. Through innovation in aerodynamics, structures, and propulsion, the specialists and their nontechnical partners reinvented the airplane.[12] They took the system invented by the Wright brothers, the wood, strut-and-wire braced biplane, and refined further by an international aeronautical community during the 1910s and World War I, and made it better. The end result was the airplane in its first "modern"

[10] C. G. Grey, "On the Coming of Age of Aviation," *The Aeroplane* 27 (December 17, 1924): 565–566.

[11] Historians have debunked the idea of the inevitability of technology shaping history, specifically referred to as technological determinism, and the temptation to oversimplify historical trends in their lexicon in favor of more complex and nuanced interpretive frameworks. Merritt Roe Smith and Leo Marx, *Does Technology Drive History? The Dilemma of Technological Determinism* (Cambridge, MA: MIT Press, 1994).

[12] Robert T. Jones, a discoverer of the swept wing, used the term "reinvented" to describe his experience working with pioneering aerodynamics researcher Fred E. Weick at the Langley Memorial Aeronautical Laboratory of the National Advisory Committee for Aeronautics (NACA) in the 1930s. Fred E. Weick and James R. Hansen, *From the Ground Up: The Autobiography of an Aeronautical Engineer* (Washington, DC: Smithsonian Institution Press, 1988), vii. James R. Hansen provided the historiographical framework for interpreting the technical development of aircraft in the 1920s and 1930s as a process of reinvention beginning with the Wright *Flyer* and ending with aircraft like the Douglas DC-3. James R. Hansen, "Flight and Technology: An Overview," in *National Aerospace Conference Proceedings: The Meaning of Flight in the Twentieth Century, October, 1-3 1998* (Dayton, OH: Wright State University, 1999), 156–158; and "Introduction to Volume II: Reinventing the Airplane," in *The Wind and Beyond: Journey into the History of Aerodynamics in America; Volume II: Reinventing the Airplane*, NASA SP-2007-4409, ed. James R. Hansen (Washington, DC: Government Printing Office, 2007), xxi–xxv.

form – the high-speed, metal, streamlined, cantilever-wing monoplane with retractable landing gear and advanced engines and propellers. On both sides of the Atlantic, the introduction of revolutionary aircraft in the mid-1930s astonished the international aeronautical community and a largely airminded mainstream society with their ability to transport people, cargo, and weapons higher, farther, and faster over countries, continents, and oceans.

The emergence of the modern airplane was a major event in the history of Western civilization. For many, Grey's predictions were coming to fruition. On Thursday evening, May 30, 1935, American aeronautical engineer and aviation entrepreneur, Donald W. Douglas, delivered the prestigious Wilbur Wright Memorial Lecture before the Royal Aeronautical Society of Great Britain. His talk, given in the shadow of the Wright brothers' *Flyer* on display at the Science Museum in London, addressed the development of the "modern" airliner in the United States. The products of American manufacturers, built to fly high and fast over the long distances of North America, epitomized the state-of-the-art in aeronautical technology. To Douglas, the "rapid technological progress" that culminated in his successful DC-series airliners reflected the existence of a "golden age" of aeronautics.[13] In making that distinction, he placed the development of the modern airplane squarely within a longstanding cultural mind-set. Beginning with the Greek poet Hesiod in 800 BCE, observers and historians described a particular moment of extraordinary achievement in the history of a culture as an idealized period of great happiness and prosperity.[14]

Douglas believed there were many groups responsible for the success of the modern airplanes that were about to help change the world. The airlines, research organizations, and airframe, engine, instrument, and radio manufacturers all had their part in the process as members of the aeronautical community. Of those contributors, he saw propeller makers and their creations as fundamental to that achievement. Without them, the "glorious future" Douglas predicted for aviation was impossible.[15]

[13] Donald W. Douglas, "The Development and Reliability of the Modern Multi-Engine Air Liner," *The Journal of the Royal Aeronautical Society* 40 (November 1935): 1042–1043, 1046.

[14] H. C. Baldry, "Who Invented the Golden Age?" *The Classical Quarterly* 2 (January–April 1952): 83–92.

[15] Douglas, "The Development and Reliability of the Modern Multi-Engine Air Liner," 1043.

Lessons from an "Invention of a Smaller Nature"

In the ten years following Douglas's speech, world society embraced the airplane and excited, amazed, shocked, and terrified itself with the promise and power of aeronautical technology. It was, in the words of the prominent American sociologist and pioneering historian of technology, William Fielding Ogburn, a "big invention" that enabled individuals, institutions, nations, and communities to express their hopes, dreams, and ambitions for a myriad of economic, political, military, and technological reasons.[16] The exploration of those themes and the inclusion of social and cultural factors have expanded our understanding of flight and technology as a major force in society into the twenty-first century.[17]

Understanding how that "big invention" came to be as a technical system from the perspective of designers and manufacturers of complete aircraft is fascinating and important.[18] Another way to comprehend the origins of the modern airplane is to consider it, once again in Ogburn's words, as a "cluster of inventions of a smaller nature."[19] Unraveling and explaining the evolution of the different technologies making up the airplane offers different perspectives on engineering, research and development, design, and the multilayered social, cultural, financial, commercial, industrial, and military infrastructure of aviation during the Aeronautical Revolution.[20] To understand fully the modern airplane and the community that created it during the 1920s and 1930s, so much depends

[16] William Fielding Ogburn, *The Social Effects of Aviation* (Boston, MA: Houghton Mifflin, 1946), 58.

[17] The writing of aviation history has grown with the larger discipline of the history of technology from a focused look at famous engineers and the "nuts-and-bolts" of great machines to investigations of nontechnical issues involving society and culture. To learn more about the crucial turning point in the discipline, see James R. Hansen, "Review Essay: Aviation History in the Wider View," *Technology and Culture* 30 (July 1989): 643–656. Roger Launius recognized that many of those trends were coming to fruition in the form of a "New Aviation History" that in the case of innovation, cast the technical development of aeronautical technology as an integral part of the human experience. Roger D. Launius, "Patterns of Innovation in Aeronautical Development," in *Innovation and the Development of Flight*, 14–15.

[18] See Philip Jarrett, ed., *Biplane to Monoplane: Aircraft Development, 1919–1939* (London: Putnam, 1997); and John D. Anderson, Jr., *The Airplane: A History of Its Technology* (Reston, VA: AIAA Press, 2002).

[19] Ogburn's main purpose in *The Social Effects of Aviation* was to predict future trends in aviation at all levels of civilian life, but his acknowledgment of the airplane's place in American society up to the 1940s as well as its technical makeup was especially prophetic. Ogburn, *The Social Effects of Aviation*, 58.

[20] James R. Hansen asserted that a focus on these topics would contribute to a "demystification of the aeronautical enterprise." James R. Hansen, "Aviation History in the Wider View," 649; and "Demystifying the History of Aeronautics," in *A Spacefaring*

upon an understanding of its component parts and the specialists that produced them.[21]

An airplane is a synergistic collection of technologies, with the propeller as the crucial link among them during the Aeronautical Revolution. It bridged the gap between the two major technological advances that made higher performance possible: innovations that reduced aerodynamic drag, and improvements in propulsion technology that increased power output. Streamlined design, which included enclosed cockpits, cantilever monoplane wings (that is, wings without external, drag-producing struts or braces), and retractable landing gear, reflected the latest in aerodynamic knowledge. Sophisticated engines, fuels, and supercharging increased power. The propeller, through its spinning helical motion, converted that energy into thrust the same way a wing generated lift to propel an airplane forward. Reinventing propellers to "shift gears in the air" for maximum efficiency during takeoff, climb, and cruise resulted in higher performance at crucial moments in the technical development of the airplane and the history of the national aeronautical cultures that used them.

The "propeller people," as one executive called them, represented a distinct specialist community within aviation.[22] At its core were airminded inventors and engineers that were either celebrated leaders in their fields or largely anonymous in the historical record. Their educational backgrounds ranged from being self-taught to extended formal training in

Nation: Perspectives on American Space History and Policy, ed. Martin J. Collins and Sylvia D. Fries (Washington, DC: Smithsonian Institution Press, 1991), 153–166.

[21] Early American historians have provided important models for studying material culture to gain a broader understanding of technology and the cultures that created them. Judith A. McGaw advocated a "direct treatment of the 'thing'," specifically "small things," as she explored everyday technologies of colonial American life. Judith A. McGaw, "'So Much Depends upon a Red Wheelbarrow': Agricultural Tool Ownership in the Eighteenth-Century Mid-Atlantic," in *Early American Technology: Making and Doing Things from the Colonial Era to 1850*, ed. Judith A. McGaw (Chapel Hill, NC: University of North Carolina Press, 1994), 328–357. Focusing on insights gleaned from the study of "ordinary" household goods, Laurel T. Ulrich sought to reveal how nineteenth century New Englanders found meaning in a world "cross-snarled and twined" together. Laurel T. Ulrich, *The Age of Homespun: Objects and Stories in the Creation of an American Myth* (New York: Alfred A. Knopf, 2001), 6–8. For broader discussions of the topic, see Steven Lubar and W. David Kingery, eds., *History from Things: Essays on Material Culture* (Washington, DC: Smithsonian Institution Press, 1993).

[22] George J. Mead, in Report of the Second Meeting of the Technical Advisory Committee of the United Aircraft and Transport Corporation at Seattle, Washington, December 2–6, 1929, Claire Egtvedt Files (hereafter cited as CE), Boeing Company Archives (hereafter cited as BCA), 379.

industry or at a university. At the personal level, they yearned for the same things other technologists desired. They wanted to reap financial rewards, achieve professional recognition, satisfy their intellectual curiosity, and to experience the satisfaction of a career in a new and exciting industry. To the spouse of one of the specialists, "all they wanted was a drawing board and a sharp pencil and a chance" to reinvent their one part of the airplane.[23]

For this specialist community, the development of propeller technology generated a dynamic environment that facilitated coexistence, collaboration, and conflict. A few worked as individuals in the mode of the celebrated independent inventor of the late nineteenth and early twentieth centuries.[24] Most found employment during World War I in newly created government and industrial engineering organizations. Those establishments served as collective "inventive institutions" that provided an equally nurturing and challenging environment in which experimenters could cultivate, explore, and evaluate their novel ideas for propellers.[25] In response, the members of the propeller community built coalitions throughout the 1920s and 1930s in the United States and Europe to complete the research and development process.[26] Inventors and engineers and government research institutions and manufacturers all needed each other. Individuals needed resources to support further development of their ideas while organizations sought the latest innovations to increase aircraft performance.

Cooperation was never universal. Individuals, teams, groups, and institutions did not triumphantly move forward in lockstep toward the common and singular goal of developing improved propellers. Depending on their place within the community, whether it was a lone inventor,

[23] Juliet Blanchard, *A Man Wants Wings: Werner J. Blanchard, Adventures in Aviation*, 1986, www.margaretpoethig.com/family_friends/pete/wings_bio/index.html (Accessed May 24, 2012), 14.

[24] Thomas P. Hughes, *American Genesis: A Century of Invention and Technological Enthusiasm, 1870–1970* (New York: Penguin, 1989), 7–8.

[25] David Edgerton, *The Shock of the Old: Technology and Global History since 1900* (New York: Oxford University Press, 2007), 192, 197. The concept of "inventive institutions" has been explored in more detail in regard to aircraft propulsion technology by Hermione S. Giffard, in "The Development and Production of Turbojet Aero-Engines in Britain, Germany, and the United States, 1936–1945" (Ph.D. diss., Imperial College, 2011), 20–21, 177–242.

[26] Called "heterogeneous engineering," the process is a central tenet within the history of technology. Launius, "Patterns of Innovation in Aeronautical Development," 13; John M. Staudenmaier, "Rationality, Agency, and Contingency: Recent Trends in the History of Technology," *Reviews in American History* 30 (March 2002): 173.

a government engineer, or an expert member of a professional society, the propeller people identified with the needs of their respective corporate, institutional, and personal networks first. The resultant competing viewpoints and methodologies reveal the existence of rival visions of what innovation represented to the propeller specialists. The creation of improved propellers was not a foregone conclusion, a predestined idea whose "time had come," reflecting the unavoidable and implicit triumph of technology.[27] A multitude of technical, cultural, economic, social, and even personal factors, ranging from large to small in consequence, and independent to competing in importance, shaped their technical mindset.[28] The weight of those factors varied at different times and levels of intensity depending on the individuals and groups facing the challenge.

Achieving the future by reinventing the propeller took a lot of trial-and-error. The need for an improved propeller offered a straightforward technical challenge, even though no one could visualize a practical design or the materials used in its construction when the process of reinventing the airplane began during World War I.[29] As a result, the specialists offered practical rather than abstract theoretical solutions. What made a methodology, material, and the resultant design workable and ultimately perceived as a measure of "Progress" was entirely subjective and based on the individual and institutional experience of practitioners within the aeronautical community. The intertwined experience of successes and failures inherent in innovation tempered many views of the world-changing

[27] The desire to synthesize the broader trends of technical development can often convey a belief in the inevitability and glorious destiny for aeronautical technology while losing the nuanced and often arduous details that make up the fabric of history, especially regarding the multifaceted nature of innovation. See Peter W. Brooks, *The Modern Airliner: Its Origins and Development* (Manhattan, KS: Sunflower University Press, 1961); Rae, *Climb to Greatness*; Ronald Miller and David Sawers, *The Technical Development of Modern Aviation* (New York: Praeger, 1970).

[28] Walter G. Vincenti, "The Retractable Landing Gear and the Northrop 'Anomaly': Variation-Selection and the Shaping of Technology," *Technology and Culture* 35 (January 1994): 8–10; Deborah G. Douglas, "Three-Miles-A-Minute: The National Advisory Committee for Aeronautics and the Development of the Modern Airliner," in *Innovation and the Development of Flight*, 159.

[29] Vincenti identified the concept of identifying a future engineering solution without knowing how to go about achieving it as the "perception of a new technological possibility." Vincenti, *What Engineers Know and How They Know It*, 203. Edward Constant defined a community's recognition that a technical system would fail or would be best replaced in the future by radically different system as a "presumptive anomaly." His definition makes the distinction that scientific and not engineering knowledge provided the basis for that realization. Edward W. Constant, *The Origins of the Turbojet Revolution* (Baltimore, MD: Johns Hopkins University Press, 1980), 15.

power of aircraft performance and the modernity it represented. The pursuit of practicality as an engineering ideology and the choices that it generated was both a technical and cultural construct.

The propeller specialists' pursuit of practicality in their designs drove them to consider a variety of materials. It has been argued that the aeronautical community's adoption of aluminum aircraft structures during the interwar period resulted from a "progress ideology of metal" that created the perception that metal symbolized modernity while wood represented the backward and traditional past. Consequently, that belief overrode the absence of real technical data that proved metal was the superior material.[30] While it is debatable that a singular ideology took precedence over test results and other factors in the introduction of metal aircraft structures, the same is not found in the propeller community. During the 1920s and 1930s, there were concurrent wood, plastic, and a variety of metal "ages" that excited humankind about what those materials symbolized in terms of "Progress."[31] The cultural impetus to make airplanes fly higher, faster, and farther, drove the propeller specialists to identify their work as a necessary part of aviation's future, but they did not emphasize one material over another until they had a practical reason to do so. At important moments of both success and failure in the history of the propeller, they embraced different types of materials and their related construction methods in their pursuit of both performance and practicality. When justified, the propeller specialists certainly celebrated the use of wood, plastic, aluminum alloy, or steel in their designs, but those materials were a means to their ends, not the motivation for making the choices they made no matter how symbolic. Their expression of modernity was in making their component of the airplane durable, reliable, efficient, and safe.

The important transition from creation to use offers important perspectives on the place of a technology in society broadly defined. The propeller specialists provided a crucial voice in the shaping of the modern airplane. Their work paralleled advances in aerodynamics, structures,

[30] Eric M. Schatzberg, *Wings of Wood, Wings of Metal: Culture and Technical Choice in American Airplane Materials, 1914–1945* (Princeton, NJ: Princeton University Press, 1998), 3–6, 15–19, 21.

[31] See Brooke Hindle, ed., *America's Wooden Age: Aspects of its Early Technology* (Tarrytown, NY: Sleepy Hollow Press, 1975); Jeffrey L. Meikle, *American Plastic: A Cultural History* (New Brunswick, NJ: Rutgers University Press, 1995); Thomas J. Misa, *A Nation of Steel: The Making of Modern America, 1865–1925* (Baltimore, MD: Johns Hopkins University Press, 1995); and Mimi Sheller, *Aluminum Dreams: The Making of Light Modernity* (Cambridge, MA: MIT Press, 2014).

and power plants that resulted in an indeterminate synergy of invention, innovation, and use. They fought for their place in their own particular networks as well as the larger hierarchy of the aeronautical community.[32] As they negotiated for the increased use of their designs and reinforced the role of the specialist in engineering organizations, they faced acceptance, ambivalence, resistance, and rejection. The exploration of that story allows the propeller community to rejoin the ranks of its contemporaries in the more well-known aerodynamics, airframe, and engine communities to recast a fuller view of aeronautical innovation during the interwar period.

The larger aeronautical community, represented by aircraft manufacturers, airlines, and military air services, either accepted and used or rejected reinvented propeller technology. The clarion call in this greater culture of performance was higher, faster, and farther, but it, too, possessed competing visions of what constituted true innovation. The adoption and use of propellers on increasingly "modern" aircraft was not just the logical incorporation of a technology based solely on technical merits. The identification of the "best" technology that reflected the "right choice" on the part of an operator, a concept that is easy to embrace with the advantage of hindsight today, was not obvious at the time.[33] The existence of multiple propeller designs illustrated that there was more than one way to attempt to improve the performance of the airplane. Additionally, nontechnical flashpoints, rooted in factors such as the needs of business, politics, and war, influenced many moments of adoption during the 1920s and 1930s. Those moments could result in spectacular demonstrations of the airplane's potential. Even then, the aeronautical community's reactions to improved propeller designs varied due to their specific needs based on time and place.

When those members of the aeronautical community traditionally responsible for choosing new innovations failed to adopt them for either technical or nontechnical reasons, members from the broader community became involved. Pilots, the ultimate users in aviation, expressed their own distinctive style of technological practice and became zealous advocates for the incorporation of more advanced propellers into their

[32] Vincenti, *What Engineers Know and How They Know It*, 9.

[33] Judith McGaw asserted that the belief that there was "one best way" to perform a task in colonial America was more an historical artifact of an industrial society interpreting its preindustrial past rather than an accurate depiction of history. McGaw, "'So Much Depends upon a Red Wheelbarrow'," 355.

aircraft.[34] For a group that had little or no voice in the process of creation, adoption, and application, these users found a way to advance their own visions of aircraft performance.

Once launched, North American and European propeller designs were competing technologies in a modern aeronautical world. These rival designs produced and adopted by different manufacturers and used by various operators ebbed and flowed in importance as they moved forward, persevered, and came and went in competition with each other. Their role in the aerial drama of the 1920s, 1930s, and 1940s highlights a contextual interpretation of the development of technology where the pathways that led to success or failure reveal that there was no seamless linear transition from idea to innovation and reinvention to use for any given design. Overall, the process of creating and using a new aeronautical design, as seen through this case study of the propeller specialists, was ambiguous, complicated, and indeterminate. The propeller in the form presented in the skies over New York City during the spring of 1938, as well as the modern airplane that it thrust through the sky, was not preordained nor was its development and adoption easy.

As the modern airplane reached the pinnacle of its use in the 1950s, the turbojet engine, a revolutionary aeronautical technology with its own story, became the main form of high performance and long-distance aircraft propulsion.[35] Commercial and military operators rejected the slower propeller and piston engine in the name of higher, faster, and farther. In an ironic twist of history, they reshaped the cultural perception of the airplane propeller. No longer was it "modern," but a crude relic of a past age of aviation that was neither exceptional nor novel and perhaps, above all else, justifiably forgotten in the progress-driven history and memory of flight. The popular reaction to the propeller in the Jet Age invites a reconsideration of how a culture perceives technology at different times in its history. The specialists and propeller-driven aircraft persisted as an integral part of aviation in an interconnected world of technology where "old" and "new" coexisted.[36]

[34] Ronald Kline and Trevor Pinch, "Users as Agents of Technological Change: The Social Construction of the Automobile in the Rural United States," *Technology and Culture* (October 1996): 764–765.

[35] The revolutionary "paradigm shift" caused by the introduction of the turbojet either displaced members of the community or forced them to adopt the new technology to survive. Constant, *Origins of the Turbojet Revolution*, 117, 120.

[36] Edgerton, *Shock of the Old*, 28–29.

Propellers became central components of one of the great inventions of the twentieth century, the modern airplane that appeared during the 1930s and 1940s. They enabled those aerial expressions of technological modernity to "shift gears in the air" as the world became interconnected through airborne networks of commerce and war. The story of their invention, reinvention, and use reveals the dynamic and nonlinear nature of aeronautical innovation as a fundamental part of the human experience during the first half of the twentieth century. From the engineers who first imagined a propeller, to the workers who made it, the organizations that ordered it, and the operators who took it into the air and maintained it on the ground, an airminded community of specialists shaped and reshaped the technology to reflect their perceptions of shared needs and values. While these individuals and groups were not all successful in the end, their communal efforts made the airplane even better than before. In the process, they helped to reinvent the airplane.

2

"The Best Propeller for Starting Is Not the Best for Flying"

Aviation journalist Carl Dienstbach traveled to Dayton in July 1905 to interview the world's leading aeronautical technologists. Beginning in 1896, Wilbur and Orville worked to invent the first airplane through a synergy of the technical systems of lift, control, structures, and propulsion, which culminated in the December 17, 1903 flight of their *Flyer* at Kitty Hawk, North Carolina. In the process, Wilbur and Orville created the first practical, efficient, and purpose-built aerial propellers and a design theory to support them. Acknowledging their ability to design an efficient propeller according to their aerodynamic principles, Wilbur still recognized the limitations of their invention. He informed Dienstbach that "the best propeller for starting is not the best for flying."[1] The elder Wright's statement highlighted the fact that the technology that turned the engine's power into thrust and enabled aircraft to take to the air was far from meeting the needs of aviation.

The Wrights were the leading members of a distinct community of American and European technologists that created the airplane in the late nineteenth century.[2] Their efforts provided the basis from which modern flight evolved. They chose the wood, fixed-pitch propeller as its standard through their adherence to the already established design paradigms of light weight and simplicity. At the same time, this early flight community took

[1] Carl Dienstbach, "Can an Airplane Jump Straight Up?: Recent Progress in Propellers," *Popular Science Monthly* 93 (July 1918): 62.

[2] Tom D. Crouch, *A Dream of Wings: Americans and the Airplane, 1875–1905* (New York: Norton, 1981; reprint, Washington, DC: Smithsonian Institution Press, 1989), 19.

the Wrights' original creation and attempted to change its composition as well as its ability to alter pitch.

The Early History of the Propeller

The idea of how an airplane propeller might work existed for centuries in human history. The Greek mathematician and engineer Archimedes described a method of transporting water uphill through the helical motion of his "water screw" during the third century BCE. The emergence of propeller-like windmills in Western Europe in the twelfth century CE indicated a reverse understanding of the principles and capabilities of thrust. Their builders used wind-driven wood-framed and fabric-covered blades, called sails, to create power for agricultural purposes. The inventor-philosopher-artist Leonardo da Vinci sketched an "aerial screw" in the late fifteenth century that was a precursor of the helicopter. Though mainly conceptual, these ideas influenced the development of both marine and aerial propellers. Seagoing ships, lighter-than-air balloons and airships, and the fanciful flying machines from the late eighteenth and nineteenth centuries all featured various forms of propellers.

Propellers appeared first on lighter-than-air aircraft. Aeronaut Jean-Pierre Francois Blanchard was the first person to employ an airborne propeller in 1784. His passenger, John Sheldon, turned the ineffective propeller device called a "moulinet," or little windmill, by hand during a balloon ascent over the Military Academy in the Chelsea neighborhood of London. That same year, Jean Baptiste Meusnier employed human-powered "turning oars" in his dirigible. Meusnier's crew cranked the propellers to reach a speed of three miles an hour.[3] In the nineteenth century, French railway engineer Henri Giffard was the first to apply mechanical power successfully to propellers. He fitted a small, three-horsepower steam engine with an eleven-foot-diameter three-blade propeller to a gondola platform under an elliptical gas bag. Giffard made a spectacular seventeen-mile flight from Paris to Elancourt on September 24, 1852. He intended to return to Paris, but the airship's heavy and inefficient engine made it an impractical flying machine.[4]

[3] Charles B. Hayward, *Practical Aeronautics: An Understandable Presentation of Interesting and Essential Facts in Aeronautical Science* (Chicago, IL: American School of Correspondence, 1912), 411; Basil Collier, *The Airship: A History* (New York: G.P. Putnam's Sons, 1974), 21–22, 32; George Rosen, *Thrusting Forward: A History of the Propeller* (Windsor Locks, CT: United Technologies Corporation, 1984), 13, 19.

[4] Rosen, *Thrusting Forward*, 16, 19.

The use of propellers for heavier-than-air craft had its precedents in the eighteenth century. French mathematician Alexis-Jean-Pierre Paucton described in his 1768 treatise, *Theorie de la vis d'Archimede*, a flying machine fitted with two propellers, one for propulsion and the other for lift. After briefly experimenting with paddle propulsion on his pioneering 1799 design, Sir George Cayley became a proponent of the propeller as the main form of propulsion for aerial vehicles in 1804. His 1837 airship design incorporated flapping wings and propellers. Cayley's 1843 convertiplane design featured both propellers and horizontal spinning discs, which were prototypical helicopter rotors. His research influenced later experimenters to reject flapping wings, oars, and paddle wheels as power transmission devices and to adopt propellers instead. Thomas Moy's large-scale model, the *Aerial Steamer*, appeared at the 1875 Crystal Palace Exhibition and featured twelve-foot-diameter six-blade propellers. Each blade consisted of eight slats that could be adjusted on the ground to produce maximum thrust for specific conditions.[5]

The idea of varying the angle of propeller blades first appeared on ships in the early nineteenth century. The earliest patent for a variable-pitch propeller dates from Great Britain in 1816. The HMS *Aurora* went to sea with one in 1853. These early attempts responded to the need to remove the drag-producing effect of the blades when the propeller was not in use. Moving the blades to a position parallel to the water flow enabled a ship to operate more efficiently while under sail, which was the primary source of marine power up to the first half of the nineteenth century.[6]

For an airplane, a variable-pitch propeller changed the angle, or pitch, at which the propeller blades rotated through the air. That ability maximized the power of the engine at different speeds and operating conditions. An easy way to explain how that works is to see its operation in the same way a transmission makes that possible in a car. Airplanes have two primary operating ranges – takeoff and cruise – that require entirely different propellers. Much like the low gear in an automobile transmission, a low pitch propeller spins very fast to produce a lot of thrust to allow the airplane to accelerate during takeoff and climb quickly into the sky.

[5] Ibid., 10, 16–18.

[6] K. M. Molson, "Some Historical Notes on the Development of the Variable Pitch Propeller," *Canadian Aeronautics and Space Journal* 11 (June 1965): 177; Rosen, *Thrusting Forward*, 10; J. L. Nayler and Ernest Ower, *Aviation: Its Technical Development* (London: Peter Owen/Vision Press, 1965), 156.

Like that low gear, however, a low pitch propeller works best when the airplane is traveling slowly, and it quickly becomes inefficient at higher speeds. A high pitch propeller, like the high gear in your car, turns slower at first but, once up to speed, covers more ground and enables an airplane to fly farther and faster at cruising altitude. A "gearshift of the air" combined low-pitch for takeoffs, high-pitch for cruising, and any pitch in between for that matter, to increase the capability and performance of airplanes in flight.

Researchers in France and Germany recognized the advantages of variable pitch for flying machines in the late nineteenth century. While working on an early helicopter design from 1868 to 1871, the Italian-born Joseph Croce-Spinelli developed a collective pitch control that he applied to the aerial propeller to increase efficiency at takeoff. He designed a large steam-driven rotorcraft, the *Navire Aerien*, which featured two twenty-three-foot lifting screws and two six-foot-diameter propellers. His hub mechanism featured a hydraulic piston that exerted pressure against a spring to change blade angle. The hydraulic actuation mechanism was an indicator of the future, but the inclusion of the spring was impractical. Overall, the hub mechanism was too fragile and failed to offset blade flutter. Moreover, it could not retain a selected pitch position setting without being locked in place on the ground.[7]

In France, Alphonse Pénaud, a brilliant young engineer with a world-class education exhibited a fundamental understanding of what design elements future aircraft would have. The eleven-second flight of his small, rubber-powered airplane, the *Planophore*, relied upon a propeller for propulsion during its dramatic flight at the Tuileries Gardens in Paris in August 1871. Pénaud's pioneering airplane patent of February 18, 1876, filed with mechanic Paul Gauchot, described a futuristic two-seat monoplane with elliptical wings, a control system consisting of a control column connected to elevators and a rudder, retractable landing gear, and a variable-pitch propeller with metal blades.[8] Unfortunately, circumstances prevented Pénaud from ever building the bold new design. Disheartened, ill, and facing ridicule for his efforts – it was a common

[7] Prewitt Aircraft Company, *History of the Helicopter from Its Beginnings to the Year 1900*, Vol. 1 (United States Air Force Air Materiel Command, March 1, 1951), 25; Ronald Miller and David Sawers, *The Technical Development of Modern Aviation* (New York: Praeger, 1970), 71; E. K. Liberatore, *Helicopters Before Helicopters* (Malabar, FL: Krieger Publishing Company, 1998), 46–47, 247.

[8] Alphonse Pénaud and Paul Gauchot, "Airplane," French Patent No. 111,574, February 18, 1876; "1876 Plane Was Modern," *New York Times*, June 21, 1936, section 10, p. 9.

belief that powered flight was impossible and any individual attempting it was insane – Pénaud committed suicide in 1880 at the age of thirty.[9]

The Swiss artist and aeronautical experimenter, Carl von Steiger-Kirchofer, proposed a design for mechanically actuated variable-pitch propeller as part of his surprisingly advanced and futuristic flying machine in his 1891 *Vogelflug und Flugmaschine*, or *Bird Flight and the Flying Machine*. He created the design in reaction to his recognition of the relationship between blade angle and engine speed and its effect on propeller efficiency. Von Steiger's design featured the actuation control directly connected to the engine throttle. His sketches, surprisingly modern in appearance, demonstrated an understanding of the aerodynamic theory that illustrated the value of the variable-pitch propeller.[10]

Despite this consensus that propellers – rather than, say, mechanical flapping wings and reciprocal oars, paddle wheels, or even rockets – represented the best way to propel an aircraft through the sky, none of the early aeronautical enthusiasts knew how to conceptualize, design, and fabricate an effective airplane propeller. Samuel P. Langley, Secretary of the Smithsonian and widely regarded as America's leading scientist at the end of the nineteenth century, was one of those experimenters who took on the challenge of solving the problem of heavier-than-air powered flight. His *Great Aerodrome* had two triangular wood and canvas propellers that resembled windmill blades and could barely transform the engine's power into thrust. Langley's spectacular public failures on the Potomac River near Quantico, Virginia, in October and December 1903, where the *Great Aerodrome* crashed into the river like "a handful of mortar," indicated to a skeptical public that airplanes, and the propellers intended to push or pull them through the air, were a silly dream.[11]

The Wrights theorized, designed, and constructed the first practical aerial propeller and the aerodynamic theory to calculate its performance during the process of creating their historic 1903 *Flyer*. Wilbur and Orville created the airframe first, followed by the engine, and set about designing a propeller as the last major phase of their development

[9] Octave Chanute, *Progress in Flying Machines* (New York: *American Engineer and Railroad Journal*, 1894; reprint, Mineola, NY: Dover, 1997), 119–122; Rosen, *Thrusting Forward*, 18.

[10] Carl Steiger, *Vogelflug und Flugmaschine* (Munich: G. Franzische h.b. Hofbuchhandlung, 1891), 84, 90. Passage translated by Deborah R. Behrend, February 15, 2002; Molson, "Some Historical Notes on the Development of the Variable Pitch Propeller," 177; Charles Harvard Gibbs-Smith, *Aviation: An Historical Survey from its Origins to the End of World War II* (London: Her Majesty's Stationery Office, 1970), 85–86.

[11] "Buzzard a Wreck: Langley's Hopes Dashed," *Washington Post*, October 8, 1903, p. 1.

program during the winter of 1902 and 1903. They learned from a basic literature search that marine-propeller theory lacked a sound foundation and they formulated a new approach.[12] Conceptualizing the propeller as a rotating, twisted wing moving forward through the air in a helical path, the brothers used airfoil data calculated from their wind tunnel to design blades able to convert the energy of their twelve-horsepower engine into thrust.[13] Believing that "the propeller should in every case be designed to meet the particular conditions of the machine to which it is applied," Wilbur and Orville designed and fabricated their propeller for one specific, or fixed, performance regime.[14] Using a drawknife and hatchets, they shaped two long and thin propellers from two-ply spruce, covered them in linen, and sealed them with aluminum powder suspended in varnish (Figure 2). The Wright propellers produced 66 percent efficiency, which was enough to get the *Flyer* off the ground at Kitty Hawk.[15]

The importance of the Wrights' invention of the aerial propeller cannot be overestimated. Historian Peter L. Jakab asserted that "before the Wright propeller there were none like it, and after it there were none that were different."[16] The twisted, rotating wing of the Wrights was a far cry from the flat plates of previous propeller experimenters. Even though they recognized the overall performance limitations of their fixed propellers as early as 1905 when they met with Carl Dienstbach, Wilbur

[12] Unknown to the Wrights, their propeller theory already existed. In France, Russian-born Stefan Drzewiecki devised in 1885 a theory that considered the propeller a twisted airfoil where each of its segments represented an ordinary wing as they traveled in a helical path. Drzewiecki was the first to calculate the forces on blade segments to find the thrust and torque output for the entire propeller. He also pioneered the use of airfoil data to determine propeller efficiency. Drzewiecki published various papers and texts culminating with *Théorie Générale de l'Hélice Propulsive* in 1920. Harvey H. Lippincott, "Propulsion Systems of the Wright Brothers," in *The Wright Flyer: An Engineering Perspective*, ed. Howard S. Wolko (Washington, DC: National Air and Space Museum, 1987), 80; Walter G. Vincenti, "Air-Propeller Tests of W.F. Durand and E.P. Lesley: A Case Study in Technological Methodology," *Technology and Culture* 20 (1979): 718–719; Fred E. Weick, *Aircraft Propeller Design* (New York: McGraw-Hill, 1930), 37–38.

[13] Orville and Wilbur Wright, "How We Made the Flight," *Flying* 2 (December 1913): 10–12, 35–36.

[14] Orville and Wilbur Wright, "The Wright Brothers' Aeroplane," *The Century Magazine* 76 (September 1908): 648–649.

[15] Crouch, *A Dream of Wings*, 33, 294; Peter L. Jakab, *Visions of a Flying Machine: The Wright Brothers and the Process of Invention* (Washington, DC: Smithsonian Institution Press, 1990), 184, 194–195; John D. Anderson, Jr., *A History of Aerodynamics and Its Impact on Flying Machines* (New York: Cambridge University Press, 1997), 237–238.

[16] Jakab, *Visions of a Flying Machine*, 195.

FIGURE 2 Orville and Wilbur Wright and their second *Flyer*, complete with its two fixed-pitch wood propellers, at Huffman Prairie outside of Dayton, Ohio, in May 1904. Smithsonian National Air and Space Museum (NASM 2002–16610).

and Orville never experimented with propellers capable of changing their pitch in flight.[17] They simply did not need them.

A New Technological Community: The Propeller Specialists

As the Wrights initiated the use of the wood, fixed-pitch propeller by the early flight community, a new industry emerged in Europe and the United States. Lucien Chauviére became the first specialty propeller manufacturer. He constructed the propeller on Louis Blériot's channel-crossing monoplane in July 1909.[18] A Chauviére "Integrale" propeller was instantly recognizable due to the distinctive curved leading edges of the blades that resembled two scimitars laid opposite to each other and

[17] Dienstbach, "Can An Airplane Jump Straight Up?," 62.

[18] "Lucien Chauviére," File CC-231900-001, Biography Files (hereafter cited as BF), National Air and Space Museum (hereafter cited as NASM); Rosen, *Thrusting Forward*, 28.

joined at the grips to form a long, flat "S." Chauviére's design was not rooted in any sophisticated aerodynamic analysis, but it would nonetheless influence an entire generation of propeller manufacturers.

In the United States, Hugo C. Gibson, an active member of the Aeronautic Society of New York, founded the first American propeller manufacturer in New York City in 1909. The Requa-Gibson Company began crafting copies of Chauviére designs, but it then pioneered distinctive designs by E. W. Bonson. A prospective client had two choices. They could have Bonson design a propeller specifically built for their engine and airframe. If frugality was a concern, they could select a standard propeller in six, seven, and eight foot diameters that cost $50, $60, or $70 and weighed six and one-half, nine, and twelve pounds, respectively. By the fall of 1910, Requa-Gibson boasted in advertising that for "men who fly every day," the company's propellers were capable of generating 58 percent more power and worth the extra expense compared to other designs. The success of this pioneer propeller manufacturer was short-lived and the small company went bankrupt in June 1911.[19]

Shortly after Requa-Gibson started operations, James Lee Simmons began designing, experimenting with, and making propellers at his Washington Aeroplane Company factory located on Water Street in the southwest area of the District of Columbia. By 1913, the company produced a line of "Columbia" monoplanes, biplanes, and flying boats based on European and American designs. Like other early manufacturers, Simmons fabricated variations of Wright and Chauviére propellers in two-, three-, and four-blade configurations.[20]

The predominant manufacturer in the United States, Spencer Heath's American Propeller and Manufacturing Company, opened in Baltimore in 1909. Born in 1867 in Vienna, Virginia, near Washington, DC, Heath went to the Corcoran Scientific School and worked in Chicago as an electrical and mechanical engineer. After his marriage to suffragette and Susan B. Anthony–confidant, Johanna Maria Holm, in 1898, he returned to Washington. By day, he designed coaling stations for the Navy Department and studied law by night. After graduation, Heath became

[19] Aeronautic Society of New York, *An Epitome of the Work of the Aeronautic Society from July 1908 to December 1909*, 2002, www.earlyaviators.com/aso1.htm (Accessed January 6, 2006), 35, 37–38; "Propellers: Requa-Gibson Company," *Aeronautics* 6 (April 1910): 2; "Requa-Gibson Propellers," *Aero* 1 (October 15, 1910): 1; "Bankruptcy Notices," *New York Times*, June 20, 1911, p. 17; Rosen, *Thrusting Forward*, 26, 28.

[20] Washington Aeroplane Company, "Simmons Propellers," July 1913, File B5-800000–20, Propulsion Technical Files (hereafter cited as PTF), NASM.

a patent lawyer and engineering consultant to some of America's leading inventors. One of his clients was Emile Berliner, who not only developed the gramophone, the flat-disk record, and the microphone, but also was an early pioneer in helicopter design. From 1908 to 1909, Heath designed and built the rotor blades for Berliner's experimental craft, which kindled his interest in aeronautics.[21]

Heath started American Propeller and began producing propellers emblazoned with the "Paragon" brand name. He quickly became known as the "propeller expert" within the early aviation community. He first displayed his products – "high" and "low" speed wood, fixed-pitch propellers – in the aeronautical exhibit of the Washington Auto Show at the district's Convention Hall in January 1910. The high speed propellers, designed for engines developing thirty-five to fifty-horsepower, featured spruce construction, weighed fourteen pounds, and measured seven feet in diameter. He served often as a consultant to would-be aircraft designers, who, in turn, displayed their aircraft at the Auto Show. One associate, Robert Moore, who was Emile Berliner's chief mechanic, attempted to demonstrate his lightweight tractor monoplane before military officers and the general public at Fort Meyer across the Potomac in Virginia in March 1910. Ironically, Moore blamed the failure of his aircraft to get into air on the propeller.[22] Other, more successful early aeronautical enthusiasts selected Paragon propellers and found them perfectly suitable as they introduced the airplane to adoring crowds across Europe and North America.

American Propeller was also a contractor to the US Army and Navy. The Army's first aircraft, Signal Corps Dirigible No. 1, relied upon a Paragon propeller and a Curtiss engine. In April 1914, naval aviator Lt. Patrick N. L. Bellinger made the nation's first aerial combat mission, a reconnaissance flight searching for underwater mines, during the six-month American occupation of Veracruz, Mexico. Heath made much of the fact that Bellinger's Curtiss AB-3 flying boat, operating from the battleship USS *Mississippi*, and the six other naval aircraft used for the expedition utilized his propellers.[23] Governments around the world also

[21] Spencer Heath MacCallum, telephone interview by author, March 6–7, 2001; Alvin Lowi, Jr., "The Legacy of Spencer Heath: Conclusions," *Journal of Socioeconomics*, 1998, www.logan.com/afi/spencer3.html (Accessed April 16, 2003).

[22] "Build Private Aeros," *Washington Post*, January 27, 1910, p. 11; "Washington Auto Show Has Aero Division," *Aeronautics* 6 (March 1910): 98; "Monoplane is Tested," *Washington Post*, March 10, 1910, p. 5.

[23] "Equips Air Navy Here," *Baltimore Sun*, August 5, 1914, p. 8.

placed orders for their nascent air organizations. Crates of propellers were leaving Baltimore for Brazil, China, France, and Japan. As America's military and industrial presence in the world expanded, so did the fortunes of the American Propeller and Manufacturing Company.

The Burgess Company of Marblehead, Massachusetts, supplied the largest American aircraft and engine manufacturer, the Curtiss Company of Hammondsport, New York, with wood propellers beginning in 1910. Even though Burgess delivered at a rate of two per week, Curtiss started producing its own propellers at its new Buffalo, New York, factory in 1916.[24] Curtiss aimed to consolidate aircraft, engine, and propeller manufacturing under one roof.

Materials and Mechanisms

The use of wood for the construction of airplane propellers was intuitive as well as illustrative of the technological culture of late nineteenth and early twentieth century America and Europe. For the United States, it was the predominant construction material because of its abundant availability and low cost. Those two factors dictated engineering development and symbolized a national culture rooted in wood. Traditionally, engineers used wood as a static architectural material in bridges and buildings or as part of large mechanical devices such as water wheels that did not require high strength or precision of manufacture. Generations of experienced woodworkers further ensured the material's dominance by creating a vast wood products industry that ranged from crude railroad ties to highly ornate furniture for the American home.[25] While Europeans had to be more mindful of their natural resources, both were cultures defined by wood in many ways.

A wood propeller did not consist of a single piece of lumber. Fred T. Jane, editor of *All the World's Airships* (the present day *Jane's*), noted in 1909 it was "next to impossible to obtain perfect blocks of wood in

[24] "The Third Part of a Plane: Curtiss Electric Propellers," n.d. [1942], File B5-260140-01, PTF, NASM.

[25] Brooke Hindle, ed., *America's Wooden Age: Aspects of its Early Technology*, 2nd ed. (Tarrytown, NY: Sleepy Hollow Press, 1985), 3–4. The author based all discussions regarding the general nature of wood products and woodworking from Jerry L. Kinney, interviews by the author, Trinity, North Carolina, September 1995–April 2004, and Jeremy R. Kinney, "'We Hold the Merchandising Idea As Paramount': The Virtues of Flexible Mass Production in the 1920s American Furniture Industry," *Business and Economic History* 28 (Winter 1999): 1–11.

the required lengths of from seven to twenty feet" to make propellers.[26] A wood propeller was laminated together, meaning a fabricator glued alternating layers of wood boards together, shaped them to form, and coated the final product with varnish or paint. The primary methods of laminated construction in the United States were the Wright, American Propeller, and the Curtiss techniques while the French followed the Chauviére and de Grandeville systems. The methods differed in the number of layers and the exact placement of the laminates in the finished product.[27] Any wood species could be used to make a propeller. Early manufacturers favored hardwoods, specifically mahogany, black walnut, or birch, but they also used white oak, maple, poplar, and the softwood spruce.[28]

Besides the cultural familiarity with the material and the ease of construction, a laminated wood propeller was also ideally suited for flight. A typical two-blade nine-foot-diameter propeller weighed approximately forty-five pounds, which could be handled by one mechanic and did not offset the balance of the aircraft in the air. A running aircraft engine vibrated in the form of harmonic oscillations that made propeller blades flex in flight. Wood was a logical choice for that environment due to its high resistance to fatigue.

Despite the arguments in favor of wood, the aeronautical community also regarded metal as an acceptable propeller construction material. A common fixed-pitch metal propeller used by French aeronautical enthusiasts during the early flight period was the R.E.P. propeller developed by Robert Esnault-Pelterie in 1909. This four-blade propeller featured a steel hub and blades connected with steel tubing. Early in 1910, Antoinette, Blériot, and Voisin manufactured two-blade propellers from aluminum sheets riveted to steel stems radiating from a steel hub. Blériot was so sure of the potential success of his propeller that he called it his "Type Populaire." In Great Britain, the Hollands propeller of 1912 featured thin steel plates brazed together and shaped into an airfoil.[29]

[26] Fred T. Jane, *All the World's Airships* (London: Sampson Low, Marston and Company, 1909), 354.

[27] Hayward, *Practical Aeronautics*, 428–430.

[28] Frank W. Caldwell, "The Construction of Airplane Propellers," *Aviation* 2 (May 1, 1917): 300–301.

[29] Jane, *All the World's Airships*, 355; Hayward, *Practical Aeronautics*, 410, 430; Grover C. Loening, "Comparison of Successful Types of Aeroplanes," *Aeronautics* 6 (February 1910): 44; S. Albert Reed, "Technical Development of the Reed Metal Propeller," *Transactions of the ASME* 48 (1928): 55.

There was an initial effort to construct variable-pitch propellers while the early propeller community experimented with both wood and metal fixed-pitch propellers. Early propeller designers fabricated propellers that automatically changed pitch under varying thrust loads. Louis Breguet constructed a hinged three-blade propeller in France that used springs to keep the blade at the appropriate pitch during flight. Spencer Heath's American Propeller and Manufacturing Company introduced a 1913 design with thin, flexible blades. These so-called flexible type propellers were popular, but empirical experiments over time revealed that they were aerodynamically inefficient and structurally unsound.[30]

Finding flexible propellers to be inadequate, the early flight pioneers learned that the type of experimental metal construction facilitated controlled pitch alteration, albeit on the ground. Aeronautical enthusiasts had been flying metal two-blade ground-adjustable diameter and pitch propellers as early as 1909. These early multipiece propellers featured blades consisting of curved sheet metal fastened to a bar and a hub that clamped the blades to the engine shaft. Mounting clips behind the blades or a clamp-type hub allowed adjustment of the pitch and diameter of the blades according to the anticipated performance regime.

Besides the French propeller makers, Geoffrey de Havilland in Great Britain experimented with ground-adjustable-pitch propellers. In a letter to the editors of *Flight* magazine, de Havilland shared his reasons for constructing a new type of propeller for his first airplane beginning in 1908. "Conventional" wood propellers impaired effective experimentation because they were expensive, fragile, and fixed in pitch. His two seven-foot-six-inch metal ground-adjustable propellers could be set at any angle or pitch. De Havilland found the ideal setting by running them on an engine and measuring the thrust with a spring balance. The budding aeronautical engineer and his partner, Frank T. Hearle, found the adjustable propellers to possess "a very great advantage" for designing new aircraft through trial and error. The No. 1, as it came to be called, was a dismal failure with the left wings collapsing seconds into its first flight in April 1910. The successful second airplane, the F.E.1, used a five-foot spruce propeller since de Havilland and Hearle "knew more of the

[30] Hayward, *Practical Aeronautics*, 425; Weick, *Aircraft Propeller Design*, 180; Molson, "Some Historical Notes on the Development of the Variable Pitch Propeller," 178, 182.

game."[31] They did not comprehend that an adjustable propeller could be retained for use on an operational airplane.

The first metal ground-adjustable-pitch propeller developed in the United States appeared in mid-1910 on the dirigible, *Rhode Island*. The airship resulted from a partnership between Stuart Bastow, a Pawtucket, Rhode Island, electrical sign contractor, and Victor W. Pagé, a mechanical engineer who worked for the *New England Automobile Journal*. While Bastow designed the gas bag and keel of the *Rhode Island*, he bought the propulsion system from Thomas Scott Baldwin's *California Arrow*, a two-cylinder, eight-horsepower Curtiss A-2 engine and a crude bamboo and fabric propeller. Pagé, unsatisfied with what he believed was no better than a "windmill sail," suggested constructing a propeller from aluminum, then a relatively new, but proven, material, because of its high strength and light weight. The new propeller consisted of aluminum blades riveted to steel tubing that connected through a threaded joint to a central aluminum hub. It was six-foot nine-inches in diameter and weighed approximately thirty pounds. The design was a compromise between the dual goals of maximum strength and minimum weight. The curved aluminum blades retained an aerodynamic shape while the steel tubing flattened toward their ends provided the required strength to offset centrifugal force.[32]

While the use of aluminum as the primary material represents a response to the ever-present desire to balance weight versus strength, the adjustable blades reflected Bastow and Pagé's desire for flexibility in operation. They simply did not know which pitch setting would be best for the dirigible except through trial and error. Pagé recalled almost twenty-five years after the flight of the *Rhode Island* that, "we did not know the pitch that would be necessary to pull that big bag through the air with this small [Curtiss A-2] motor ... so it seemed to me that a good way of solving that problem without much mathematics would be to make the blades so that you could twist them around and try them, which we did." Flight tests revealed that the most desirable setting allowed a traveling distance of five feet per revolution. Pagé's empiricism resulted in

[31] Geoffrey de Havilland, "Correspondence: Propellers," *Flight* 2 (March 12, 1910): 197; "Enterprise in Airscrews: First Details of a Mighty New de Havilland Airscrew and the Story of 21 Years of Achievement," *Flight* 69 (March 2, 1956): 237.

[32] "With the Aviators," *New England Automobile Journal* 29 (November 1910): 38–40; "Bastow-Pagé Propeller," File A19420042000, Registrar's Office (hereafter cited as RO), NASM; "Brief," February 13, 1936, *Reed Propeller Company v. United States*, United States Court of Claims (hereafter cited as USCC) No. 42,133, 1281, 1288, 1289–1292, in "Bastow-Pagé Propeller," File A19420042000, RO, NASM.

the earliest American attempt at multipiece construction that facilitated deliberate pitch variation. The metal propeller emerged unscathed during several forced landings, which indicated their inherent performance and durability benefits.[33]

Another important precursor to the metal, variable-pitch propeller in the United States was a ground-adjustable-pitch propeller constructed by the Aeromarine Plane and Motor Company of Nutley, New Jersey, from cast duralumin. Duralumin was a new aluminum alloy developed in Germany in 1909, which would figure prominently as an aircraft structural material over the next five decades. The president of Aeromarine, Inglis M. Uppercu, assisted by Henry Ivan Stengel, began work on the propeller during the summer of 1914. Whereas the Bastow-Pagé propeller utilized a threaded joint to alter pitch, the Aeromarine propeller featured a split bronze hub that bolted together in two pieces. The blades fit within each hub where the pitch angle could be positioned in relationship to a small notch on the front of the hub. Overall, the cast aluminum design proved to be too weak for the stresses inflicted upon aerial propellers in flight. The design exhibited low fatigue strength, meaning it could not withstand the rigors of constant flexing and operation under both aerodynamic and physical loads.[34]

Bastow, Pagé, Uppercu, and Stengel all saw that metal, specifically aluminum, offered advantages that influenced their experimentation and improved the performance of aircraft. They did not envision that the propeller would be able to change its pitch in flight. Other aeronautical enthusiasts recognized the potential value of the variable-pitch propeller to the synergistic technical system of the airplane. From 1909 to 1913, they experimented with various types of mechanical linkages, centrifugal governors, and electrical regulators to actuate and control pitch changes while airborne. In Italy, Giovanni Caproni designed a pair of experimental mechanically actuated propellers for his first biplane, the Ca 1, in 1910. He used concentric shafts, a screw mechanism, and levers to change pitch in flight.[35]

[33] "With the Aviators," 38–40; "Bastow-Pagé Propeller"; "Brief," *Reed Propeller Company v. United States*, USCC No. 42,133, Folder GJ42133(3), Box 4504, Record Group (hereafter cited as RG) 123, National Archives and Records Administration (hereafter cited as NARA), 1292.

[34] O. Westover to H. W. Holden, December 11, 1935, File 452.8, "Curtiss-Reed and All-Metal Propellers," Box 1083, RG 18, NARA; P. C. I. Stengel to the United States National Museum, November 19, 1942, "Aeromarine 1914 Propeller," File A19330044000, RO, NASM; Rosen, *Thrusting Forward*, 37–38, 40.

[35] Molson, "Some Historical Notes on the Development of the Variable Pitch Propeller," 177–179.

In the United States, James B. Lund and R. D. Dwight of Chicago, Illinois, designed and constructed a mammoth tandem quadruplane that was eighteen feet tall, thirty-three feet long, and weighed seventeen hundred pounds during the summer of 1911. For thrust, they created a twelve-foot-diameter variable-pitch propeller that featured separate drive and control mechanisms and blades consisting of steel tubes joined to wood tips. Lund and Dwight blamed the small fifty-horsepower motor for being unable to get the giant aircraft, nicknamed the "Flying Bridge," off the ground at the Aero Club of Illinois' Cicero Field during the spring and summer of 1912. Unable to gain continued funding, Lund and Dwight abandoned their project soon after.[36]

Pioneer aircraft manufacturer, Edson F. Gallaudet of Norwich, Connecticut, conceptualized an experimental duralumin, three-blade variable-pitch propeller in December 1912. This propeller was both controllable and reversible, reflecting Gallaudet's belief that the propeller would enhance overall aircraft performance in addition to the better handling of aircraft and land, water, and in the case of dirigibles, in the air. The propeller control system was part of an engine-propeller drive mechanism intended for his Model B flying boat and Hydro-Monoplane No. 3 designs. The engine, mounted longitudinally in the fuselage, connected to two ninety-degree gearboxes that turned the propeller shafts. Within that linkage system, the rotary motion of the pilot's control lever in the cockpit converted to linear motion to move the hollow propeller drive-shaft back and forth, which operated the adjustable propeller hub. Gallaudet received a United States patent for his design in October 1916, but he never built a full-size example for use on any of his aircraft.[37]

The Integral Propeller Company of Great Britain claimed that it introduced a practical, mechanically actuated variable-pitch propeller for airships in 1916. The propeller featured two detachable wood blades joined to a central metal hub where a system of wheels and gears provided pitch control. According to *Aerial Age Weekly*, the propeller had two "readily obvious" advantages for airships. First, the ability to reverse the

[36] "A Novel Multiplane," *Ignition and Accessories* 1 (April 1912): 209–210; James B. Lund to Director, Smithsonian Institute, July 6, 1931; and James B. Lund to A. Wetmore, October 3, 1931, "Lund-Dwight Propeller," File A19320025000, RO, NASM; William H. Smith to Frank A. Taylor, August 9, 1942, File B5-332000–01, PTF, NASM

[37] E. F. Gallaudet, "Propeller," U.S. Patent No. 1,203,557, October 31, 1916; J. M. Gwinn, Jr., to Materiel Division, December 2, 1940, Folder "Propellers–Edson F. Gallaudet, 1940," Box 6589, Research and Development File (hereafter cited as RD) 3403, RG 342, NARA; Robert A. Gordon, "The Gallaudet Story, Part 5 – 1913–1914 Designs and Patents," *World War I Aero* (November 2003): 24–37.

propeller's pitch assisted in maneuvering the airship. Second, the ability to alter pitch enabled maximum efficiency at high altitudes. Despite the apparent advantages of the Integral variable-pitch propeller over the "ordinary" fixed-pitch propeller, the aeronautical community did not adopt it for widespread use.[38]

Despite those initial attempts at pitch variation and control, the early aviation community reverted to the wood, fixed-pitch propeller as the standard component on aircraft. Fred Jane believed that wood was the superior material. Wood propellers, especially laminated propellers, offered great strength in relation to their light weight. Chauviére's highly successful Integrale wood propeller series soon became the propeller of choice for early pilots. Adhering to the dominant design philosophy emphasizing light weight and strength, the aeronautical community correctly believed that laminated wood propellers possessed a considerable advantage over metal propellers. Any metal propeller reduced in weight would be weaker structurally and would fail before its wood counterpart. Wood was "much safer" than metal. As for metal ground-adjustable and variable-pitch propellers, Jane noted that they were acceptable for "experimental work," but they were the "worst possible form for use as a permanent screw," because the overall design was heavy and inefficient compared to wood, fixed-pitch propellers.[39]

Many aeronautical pioneers believed that metal was not durable enough to survive in the extreme physical environment in which a propeller operated. Echoing the metallurgical knowledge of the times, their fears centered on crystallization, a form of metal fatigue that led to cracking and potentially catastrophic failure during use. Jane held the widely held perception that a metal propeller, when connected directly to an engine shaft, would be subject to "internal crystallization due to the inherent hammering of the engine." Orville Wright acknowledged that metal was unsuitable because it "quickly crystallizes." Inglis Uppercu recalled that when he began work on the duralumin propeller, his peers criticized him for his "lack of metallurgical knowledge in attempting to produce an aeroplane propeller." The general belief was that "there was no metal that would stand the constant vibration without crystallization." Due to the inability to design and construct a strong metal propeller, Charles B. Hayward observed in *Practical Aeronautics* that "very few

[38] "The Integral Variable-Pitch Propeller," *Aerial Age Weekly* 3 (June 26, 1916): 443, 458.
[39] Jane, *All the World's Airships*, 354–355.

metal propellers of any kind" were in use in 1912.[40] In other words, they believed metal was too brittle. The aeronautical community had placed its faith in the wood, fixed-pitch propeller.

A Specialist Works to Improve the Propeller

During the period 1908–1912, pilot-designers determined the form and composition of the aerial propeller through empirical trial-and-error. At the same time, the enthusiasm for flight reached into American universities to inspire professionally trained engineers to enter aeronautics. One of those early pioneers was Frank W. Caldwell, who was about to embark upon a thirty-year process of consistent innovation in propeller design.

Frank Walker Caldwell was born on December 20, 1889, at Lookout Mountain near Chattanooga, Tennessee. Caldwell's father, Frank Hollis, came to Chattanooga from Georgia in 1869 and married Mary Ella Walker. Frank Hollis worked for the Cahill Iron Works, a manufacturer of structural building materials, bath tubs, sinks, and fireplace grates, and eventually became the company's president. The Caldwell family settled in nearby Lookout Mountain and was the first to build a permanent year-round home there. They were instrumental in building the first Christian church on the mountain, the Lookout Mountain Presbyterian Church. As a prominent entrepreneur and both a civic and religious leader, Caldwell also served as the mayor of Lookout Mountain.[41]

The Caldwells were privileged in terms of social position and enjoyed the benefits of being an upper-middle-class American family in a prospering southern industrial city. There were no limits or obstacles for young and privileged southern men from this class in terms of education and upward mobility. In his teens, he left Lookout Mountain to begin his

[40] Jane, *All the World's Airships*, 354; Orville Wright, "Sporting Future of the Airplane," *U.S. Air Services* 1 (February 1919): 5; Inglis M. Uppercu to J.E. Graf, March 22, 1933, "Aeromarine 1914 Propeller," File A19330044000, RO, NASM; Hayward, *Practical Aeronautics*, 430.

[41] "Frank Walker Caldwell," in Lester D. Gardner, *Who's Who in American Aeronautics* (New York: Gardner, Moffat Company, 1922), 31; "Letter From Founding Member Frank H. Caldwell" [Frank H. Caldwell to J. E. Smartt, September 27, 1939], in T. Cartter Frierson, *Lookout Mountain Presbyterian Church Centennial History – The First 100 Years, October 16, 1892–1992*, 1999, www.tcfgroup.com/LMPCHistory/Caldwell.html (Accessed November 5, 1999); "Exhibit by Cahill Iron Works, 1906," File 318.1916, Paul A. Hiener Collection, Chattanooga-Hamilton County Bicentennial Library (hereafter cited as CHC); "Frank H. Caldwell, 1922," p. 96, vol. 2, Chattanooga Albums, CHC.

early higher education at the Tome Preparatory School in North East, Maryland, and then studied at the University of Virginia from 1907 to 1908.[42]

Caldwell entered the mechanical engineering program at the Massachusetts Institute of Technology (MIT) in 1908. Exhibiting a strong interest in aviation, he studied all of the available reference books on the newly emerging field of aeronautics and adapted his course of study and extracurricular activities to fit his interests. Caldwell was a founding member of the Tech Aero Club. He and fellow student, Hans Frank Lehmann, designed and flew gliders from 1910 to 1912, including one that won a contest. They also collaborated on a bachelor's degree thesis in 1912 entitled, "Investigation of Air Propellers," which was one of the earliest attempts at a comprehensive propeller-testing program in the world. Caldwell and Lehmann used the whirling arm apparatus created by Professor David L. Gallup at Worcester Polytechnic Institute to undertake the program. At the completion of the project, Caldwell had a clear understanding of the state of the art in the aerodynamic design of propellers and the methods employed to evaluate them. Caldwell graduated from MIT in 1912 with a bachelor's of science degree in mechanical engineering.[43]

MIT would not offer a dedicated aeronautical engineering curriculum until 1914. Caldwell graduated five years before the first generation of MIT-trained aeronautical engineers – Donald W. Douglas, Virginius E. Clark, and Alexander Klemin – entered leading positions in the community. Jerome C. Hunsaker, one of Caldwell's former classmates, was their teacher.[44]

The young mechanical engineer's first professional experience was not in aeronautics. He worked with his father at the Cahill Iron Works as a production engineer where he focused on improving methods of manufacturing porcelain enamel wares. The sudden departure from an

[42] "Class from Lookout Mountain Public School, 1897," p. 98, vol. 2, Chattanooga Albums, CHC; "Frank Walker Caldwell," *Who's Who in American Aeronautics*, 1922, 31; "Frank Walker Caldwell," *Collier's Encyclopedia*, 4 (New York: Collier, 1957), 322–323; "F.W. Caldwell Dies at 85," *West Hartford News*, January 2, 1975, p. 4.

[43] "Testimony of Mr. Frank W. Caldwell," October 8–9, 1928, *Reed Propeller Company v. United States*, USCC No. E-544, Box 4509, RG 123, NARA, 2–3; Frank W. Caldwell and Hans F. Lehmann, "Investigation of Air Propellers" (B. S. thesis, Massachusetts Institute of Technology, June 1912); "Frank Walker Caldwell," *Collier's Encyclopedia*, 322–323; "F.W. Caldwell Dies," *West Hartford News*, 4.

[44] William F. Trimble, *Jerome C. Hunsaker and the Rise of American Aeronautics* (Washington, DC: Smithsonian Institution Press, 2002), 33–34.

intensive university engineering program that had exposed him to the new century's most exciting new technology to the relatively mundane practice of manufacturing the perfect iron or brass casting in industrial Chattanooga was dramatic. He read what aviation materials were available and corresponded with individuals within the industry. Unable to stay out of aviation, Caldwell soon joined the leading American aeronautical company, the Curtiss Aeroplane and Motor Company of Buffalo, New York, as the foreman and production engineer for its newly formed propeller department in 1916.[45]

As Caldwell settled down in Buffalo, the embryonic American air arm learned firsthand the limitations of the wood propeller during the failed Punitive Expedition into Mexico from March 1916 to February 1917. Intending to cause an international incident between the United States and the army of Mexican President Venustiano Carranza, the bandit-revolutionary Pancho Villa staged a raid on Columbus, New Mexico, which resulted in the deaths of seventeen Americans. In retaliation, Brig. Gen. John J. Pershing led two columns of American troops into Mexico to capture Villa. Accompanying the troops was the First Aero Squadron, which was the first American Army air unit to conduct operations in the field. Pershing intended to use the squadron's Curtiss JN-3 aircraft in reconnaissance and search operations as well as a rapid courier service between American units operating in the high mountains of northern Mexico. The general, interested in the use of the airplane as a military weapon since 1908, acted in accordance with a 1915 War Department directive that aircraft should provide the main form of strategic intelligence for military operations.[46]

The Curtiss JN-3 was woefully inadequate for the job at hand. Its underpowered ninety-horsepower Curtiss OX-5 engine did not provide the margin of safety for operations in a mountainous and hot environment. Many Army aviators believed the JN-3 was ill-suited for its designed purpose – training – much less anything else. The JN-3 was of questionable quality in terms of the workmanship and materials used in its construction. The major problem with the biplane quickly centered on its propeller. Curtiss manufactured its propellers from mahogany, walnut, and birch in the cool and more humid American northeast. After three to four hours of

[45] "Testimony of Mr. Frank W. Caldwell," 3; "Frank Walker Caldwell," *Who's Who in American Aeronautics*, 1922, 31; "Frank Walker Caldwell," *Collier's Encyclopedia*, 322–323.

[46] Herbert A. Johnson, *Wingless Eagle: U.S. Army Aviation through World War I* (Chapel Hill, NC: University of North Carolina Press, 2001), 161–163, 167.

operation at temperatures exceeding 130 degree Fahrenheit in the southwestern desert, the JN-3 propellers dried, warped, and split, rendering the aircraft useless. The immediate solution was to store all propellers in humidors to preserve their structural integrity when not being used. The removal and re-attachment of propellers on aircraft for individual sorties was time-consuming and interfered with the Army's daily operations.[47]

The wood propeller problems forced Pershing to contact Curtiss directly by telegraph. Caldwell and two assistants arrived in Columbus in June 1916. They studied local furniture-making practices and found that the use of native woods, primarily poplar, and improved glues that could withstand the heat, was the solution. Caldwell and his team established a small factory and constructed eighty propellers by hand with spoke shaves, scrapers, and sand paper for the expedition. The *New York Times* reported that their efforts alleviated the propeller problem and helped ensure that the Army fliers kept up with Pershing's campaign.[48]

The oppressive climatic conditions found on the Mexican border by Army aviators during the Pershing expedition forced the nascent American aeronautical community to evaluate the practicality of the wood propeller. At the meeting of the Executive Committee of the National Advisory Committee for Aeronautics (NACA – the present day NASA) in Washington, DC, on July 13, 1916, the Army's representative, Lt. Col. George Owen Squier, recommended a thorough investigation into materials for aircraft propellers. The committee members discussed the issue and possible remedies, but did not reach a conclusion.[49] That

[47] Johnson, *Wingless Eagle*, 164–166; Maj. Gen. Howard C. Davidson, Interview by Brig. Gen. George W. Goddard, August 20, 1966, interview K239.0512-996 C.1, transcript, United States Air Force Oral History Program, United States Air Force Historical Research Agency (hereafter cited as HRA), 11; "Army Fliers Have Learned to Conserve Moisture in Propellers in Arid Regions – Metal Being Used," *Washington Post*, August 20, 1916, evening section, p. 2; Juliette A. Hennessy, *The United States Army Air Arm: April 1861 to April 1917* (Washington, DC: USAF Historical Division, 1958; reprint, Washington, DC: Office of Air Force History, 1985), 168–169.

[48] "Aero Propeller Factory on Border to be Erected for Army Machines," *Washington Post*, June 25, 1916, p. 11; Caldwell, "The Construction of Airplane Propellers," 300; Walter H. Caldwell, telephone interview by author, July 23, 1999; "Frank Walker Caldwell," *Who's Who in American Aeronautics* (New York: Aviation Publishing Corporation, 1928), 17–18; "Frank Walker Caldwell," *Collier's Encyclopedia*, 322–323; "F.W. Caldwell Dies," *Hartford Courant*, 4; Rosen, *Thrusting Forward*, 31; "Aerial Patrol on Border," *New York Times*, June 26, 1916, p. 5; "Fourth Annual Dinner," *Journal of the Aeronautical Sciences* 4 (February 1936): 114.

[49] Minutes of the Meeting of the NACA Executive Committee, July 13, 1916, NACA Files, National Aeronautics and Space Administration History Office (hereafter cited as NASAHO).

same month, the *Washington Post* reported that the only remedy to the "aeroplane-propeller situation" was the "development of a metal propeller that will not be susceptible to climatic conditions."[50]

World War I and the Push for Performance

When the United States entered World War I in 1917, it committed itself to an ambitious aviation production program. An unprecedented $640 million government appropriation on July 24, 1917, charged the American military and industry to produce 22,625 aircraft and 45,625 engines. The board divided the production figure between 12,000 aircraft for service in France, about 700 for defense of the continental United States, and over 10,000 trainers for the instruction of American and foreign flight cadets in the United States.[51] If each of those aircraft required one operational and one spare propeller, then a virtually nonexistent industry had to manufacture almost 50,000 propellers to meet the demand.

Part of that program involved the establishment of a dedicated military aeronautical research and development facility to be operated by the newly formed Engineering Division of the United States Army Air Service at McCook Field in Dayton.[52] The division's specialized subdivisions cooperated in modifying existing aircraft to increase performance and in designing, testing, and manufacturing new aircraft. The experimental factory was responsible for the construction of aircraft at McCook, and its engineers worked on developing advanced aircraft and power plant production procedures. The engine assembly and testing department conducted exhaustive tests on aircraft power plants and related systems. The machine shop assumed the responsibility of performing specialized fabrication duties for all the departments. The chemical and physical material laboratory devoted most of its energy to evaluating differing types of metals, alloys, and aviation engine oils and fuels. The mechanical research department developed an improved engine cooling system and

[50] "Army Fliers Have Learned to Conserve Moisture in Propellers," 2.

[51] Benedict Crowell, *America's Munitions, 1917–1918* (Washington, DC: Government Printing Office, 1919), 239–241.

[52] Before the creation of the United States Air Force in 1947, the Army's air arm was known as the Air Service (1918 to 1925), Air Corps (1926–1941), and the Air Forces (1941–1947). The Army's aeronautical research component also changed its name several times during the 1920s and 1930s. For the purposes of clarity, the term "Engineering Division" will refer to the organization from 1918 to 1926. The term "Materiel Division" will refer to it after the creation of the Air Corps in 1926.

other intricate design solutions. The aeronautical research department concentrated on flight testing, wind tunnel research, and structures testing. The armament department performed the purest military function by developing aircraft weapons.[53]

The propeller unit was the first Engineering Division branch to open at McCook in December 1917. It dealt with enhancing the efficiency and durability of propellers for military aircraft. The Army began work on improving propellers for military aircraft within the Signal Corps Plane Design Section in Washington, DC, earlier during the spring of 1917 in response to its experience during the Pershing expedition. Already recognized as a leading authority on propeller design in the United States due to his work at MIT and Curtiss, Frank W. Caldwell became the civilian chief engineer of the propeller unit at McCook. He was responsible for the research, design, and testing of all aircraft propellers used by the Army and Navy in World War I.[54] Caldwell shaped the direction and tone of America's propeller program from the standpoint of design and construction as well as the methods used for successful engineering evaluation.

Despite the lessons learned on the Mexican border, wood remained the dominant construction material for propellers in the World War I aviation production program, which meant they would be fixed-pitch following the general design knowledge of the time. To the American military, the advantages of wood outweighed the disadvantages (Figure 3). The materials were plentiful and the strength limitations in terms of fatigue were understood. The wood, fixed-pitch propeller was a known quantity. A metal propeller was not.

Unfortunately, other serious disadvantages came with the use of the wood propeller during a new age of mechanized warfare and industrial mobilization. No single wood species offered a balanced combination of light weight, high strength, resistance to climatic conditions, and ease of manufacture. None of the major propeller woods offered a clear-cut advantage. Black walnut was the most durable, but its heavy weight and

[53] Lois E. Walker and Shelby E. Wickam, *From Huffman Prairie to the Moon: The History of Wright-Patterson Air Force Base* (Washington, DC: Government Printing Office, 1986), 178–181; Terence M. Dean, "The History of McCook Field, Dayton, Ohio, 1917–1927" (Master's thesis, University of Dayton, 1969), 50.

[54] "Frank Walker Caldwell," *Collier's Encyclopedia*, 322–323; Walker and Wickam, *From Huffman Prairie to the Moon*, 178–181; Dean, "The History of McCook Field," 50, 51–52; Daniel Adam Dickey, Interview by Lois E. Walker, September 2, 16, 30, 1983, interview K239.0512-1712, transcript, United States Air Force Oral History Program, HRA, 13.

FIGURE 3 By the end of World War I, the wood, fixed-pitch propeller had become the standard for aeronautical propulsion as seen on this Army Air Service Curtiss biplane at Langley Field, Virginia. Smithsonian National Air and Space Museum (NASM USAF-8088AC).

low tensile strength made it unattractive. Oak offered strength with a significant weight penalty. Honduras mahogany was light, durable, and easy to glue together, but it was not strong. Birch was strong and light, but hard to manufacture and not durable in operation.[55] The choice of a particular wood for a propeller was a compromise and the aeronautical community knew that from the outset.

Despite those limitations, an American propeller industry of approximately forty different government contractors located in fifteen different states emerged as a result.[56] Members of the wood products

[55] Caldwell, "The Construction of Airplane Propellers," 300–301.

[56] Propeller Memoranda, File 216.21052–216.21053, Propeller Memoranda, Bureau of Aircraft Production (hereafter cited as BAP), HRA; "Photographs of Propellers Whirl-Tested by the Westinghouse Electric and Manufacturing Company for the United States Government," October 24, 1918, Folder 168.7329-12, Part 1, D. Adam Dickey Papers (hereafter cited as DAD), HRA; G. Rosen to R. B. Meyer, Jr., January 19, 1978, Propulsion Curatorial Files (hereafter cited as PCF), NASM; Folder "Propellers–1918," Box 5469, RD 3085, RG 342, NARA.

manufacturing industry enlisted directly in the nationwide war effort and entered into competition against the pioneer companies, the American Propeller and Manufacturing Company and the Curtiss Aeroplane and Motor Company. Instead of chairs, tables, and pianos, these manufacturers converted to the production of airplane propellers.[57] The Southern Aircraft Company of High Point, North Carolina, better known as the Giant Furniture Company, used its woodworking machinery to make propellers from local hardwoods, primarily white oak, for use on United States Navy flying boats, training aircraft, and airships for service during the war.[58]

The Matthews Brothers Manufacturing Company of Milwaukee, Wisconsin, was another newcomer. Like other woodworking companies, the maker of fine bar and dining room furniture – the company made the barrel chairs for houses designed by Frank Lloyd Wright in the Chicago area – became a primary propeller contractor to the United States government. The general manager of the aircraft department was Thomas F. "Tom" Hamilton. Born in Seattle, he flew his first airplane, a Voisin-type biplane, in 1909 earning him membership in the Early Birds, the exclusive group of aviators who flew prior to 1916. He quickly offered his own line of biplanes and wood propellers. Hamilton trained British pilots in Vancouver, British Columbia before American entry into World War I. After joining Matthews Brothers in 1917, Hamilton oversaw the production of two- and four-blade mahogany propellers for the Navy's Curtiss HS2L, H16, NC, and F-5-L flying boats.[59]

While not a trained engineer or even known for getting his hands dirty as a builder, Hamilton had what could be called a "technical appreciation" for propeller technology. A born entrepreneur, he found the right people to make his business dealings a success.[60] His workers at Matthews Brothers had a simple solution for constructing a four-blade propeller. Other manufacturers spliced two blades perpendicularly into the hub of

[57] "Building Industries to Coordinate for War Work," *Manufacturers' Record* 74 (July 11, 1918): 63.

[58] David N. Thomas, "Early History of the North Carolina Furniture Industry, 1880–1921" (PhD diss., University of North Carolina at Chapel Hill, 1964), 270.

[59] "Thomas Foster Hamilton," *Who's Who in American Aeronautics* (New York: Aviation Publishing Corp., 1928), 48; "Thomas F. Hamilton: Early Plane Builder – Aviator – Propeller Manufacturer," from the *Flying Pioneers Biographies of Harold E. Morehouse*, n.d., Museum of Flight Resource Center (hereafter cited as MFRC).

[60] Fred E. Weick, Interview by Harvey H. Lippincott, March 4, 1981, Folder "Interview: United Aircraft and Hamilton, 1929-1930," Box 3, Fred E. Weick Papers (hereafter cited as FWP), NASM, 97.

an existing design, which was an expensive process requiring time and skill and ultimately proved to be weak and failure-prone. Hamilton's employees bolted two two-blade propellers together to form an "X."

American Propeller was, by far, the leading wartime manufacturer. The company had been producing propellers for the aircraft of Great Britain's Royal Flying Corps and Royal Naval Air Service for two years before the United States entered the conflict.[61] By 1918, American's facilities consisted of three manufacturing plants, centered around the original factory on East Hamburg Street in south Baltimore, and employed 450 workers. The newest building on Jackson and Gitting streets was the largest and most advanced propeller factory in the country, built out of concrete and steel at a cost of $100,000. American Propeller's innovative use of electric power reflected a growing trend in industry. Small electric motors enabled the use of compact and safer woodworking machinery and overhead lighting facilitated two shifts a day.[62] While there are no clear numbers regarding the overall number of propellers manufactured for the Allies, Heath claimed his company had produced 75 percent of them by war's end.[63]

The standard method of manufacturing a World War I–era propeller was a complicated and time-consuming process. Workers glued thin kiln-dried sheets of solid wood together to form a laminated rectangular block called a "blank." Care had to be taken where the grain of each sheet opposed each other to neutralize warping. The next step involved removing excess wood from the blank to create the rough outline of a propeller. Depending on the size of the company, a worker assigned the task of "roughing out" a propeller would go about it in two different ways. Smaller concerns completed that task by hand with spoke shaves one wood chip at a time like the Wright brothers. Larger manufacturers employed labor-saving machinery, specifically the mechanical copy lathe used by furniture and firearms manufacturers to form irregular wood shapes such as table and chair legs and gunstocks. Long established woodworking machinery manufacturers like Defiance and Mattison produced rebranded "propeller turning lathes" for the war effort.[64] After sanding and finishing, workers drilled installation holes into the hub,

[61] "Propellers Made Here," *Baltimore Sun*, April 26, 1916, p. 6.

[62] "Large New Propeller Factory for Baltimore," *Manufacturers' Record* 73 (January 3, 1918): 79.

[63] Lowi, "The Legacy of Spencer Heath: Conclusions."

[64] Defiance Machine Works, "A Big Output of Aeroplane Propellers," *Aviation* 3 (September 1, 1917): 200; "Mattison Machine Works," *Aviation* 3 (December 15, 1917): 722.

balanced the blades, and added a protective layer of fabric or metal on the leading edge.[65]

Besides being time-consuming and complicated in construction, the use of wood also highlighted American military's reliance upon both domestic and foreign sources for its propeller materials. In the United States, black walnut was in short supply and could not be grown fast enough to meet the demand. Mahogany had to be imported from South America or Africa. The *Wall Street Journal* reported in September 1917 that the United States government required an unprecedented 7,500,000 board feet of quarter-sawn oak just for propellers.[66]

Real and perceived wood shortages during the war influenced a concerted effort to find a new construction material for both propellers and aircraft. Considering the many variables in the material, construction, and use of the wood propeller, the key to manufacturing the device was to use lumber that had an adequate level of moisture in it to ensure stability. The ideal way to prepare wood was to store it in an open-air lumber yard for five years before placing it in a dry kiln for three weeks. The kiln ensured that the propeller lumber retained approximately 7 to 9 percent of its moisture without totally drying out. The lumber remained for another three weeks in the workshop before laminating and shaping. Ideally, from forest to flying field, it took almost six years to manufacture one wood propeller. The pressures and realities of producing propellers for the growing aeronautical community, especially as demand surged with American's entry into World War I, shortened the process to two years. Not surprisingly, accelerating the process significantly lessened the quality of finished product.[67]

Once installed on an airplane, extreme variations in temperature and humidity as well as any water, sand, dirt, gravel, or tall grass kicked up while taxiing before takeoff easily damaged the aerodynamic and structural integrity of wood propellers. Waging an aerial war on a world-scale exacerbated the need for a more consistent material. Even with brass or copper guards, or "tips," installed for protection along the leading edges of the blades, an average wood propeller lasted only six months in service in Europe. In addition, wood propellers often deteriorated under

[65] Caldwell, "The Construction of Airplane Propellers," 300–301; Frank W. Caldwell, "The Shaping of Airplane Propellers," *Aviation* 3 (October 1, 1917): 308; and "Finishing Up and Balancing Airplane Propellers," *Aviation* 3 (November 1, 1917): 462.

[66] "Black Walnut for Propellers," *New York Times*, June 2, 1918, p. 58; "Government Wants Lumber," *Wall Street Journal*, September 1, 1917, p. 8; Dickey Interview, 1.

[67] Caldwell, "The Construction of Airplane Propellers," 300–301.

the stress of high output engines such as the 400-horsepower Liberty V-12 engine, making them even more unsuitable for use on new higher-performance aircraft.[68]

Caldwell worked to make the propeller a standard component of the airplane to enable the American military to go to the war in the air. Regarding production, each propeller carried standardized markings denoting drawing and part number identification, engine and aircraft application, and designated operational revolutions per minute (rpm) and pitch.[69] Caldwell created a body of technical literature on propeller design, manufacturing, and operational maintenance in the official bulletin of the Engineering Division.[70] Nevertheless, as he had learned early on, only so much could be done to improve the wood propeller such as devising an improved method of tipping wood propellers with a new scalloped design.[71] Caldwell, an engineer well-versed in using both wood and metal as a material, was beginning to see the problems with the former as soon-to-be insurmountable.

Identifying a Novel Possibility

The experience on the Mexican border and with the World War I aviation production program influenced the American aeronautical community's call for the development of a new type of propeller, specifically a metal

[68] Dickey Interview, 1; Rosen, *Thrusting Forward*, 32, 36.

[69] Pitch, in this case, referred to the distance in feet an airplane traveled forward during one revolution of the propeller. "Conference on Propellers," June 8, 1918, Folder "Propellers, May-June 1918," Box 11, RD 3085, RG 342, NARA.

[70] See "List of Propellers and Propeller Clubs for Service and Training Planes and Engines," *Bulletin of the Experimental Department, Airplane Engineering Division, United States Army* 2 (December 1918): 112–113; "Outline of Propeller Design," *Bulletin of the Experimental Department, Airplane Engineering Division, United States Army* 2 (January 1919): 91–106; "Propeller Hub Dowel and Key Location," *Bulletin of the Experimental Department, Airplane Engineering Division, United States Army* 2 (December 1918): 113; "Propeller Storage Room Conditioning Apparatus," *Bulletin of the Experimental Department, Airplane Engineering Division, United States Army* 2 (December 1918): 74–83; "Protecting Propellers Against Abrasion," *Bulletin of the Experimental Department, Airplane Engineering Division, United States Army* 2 (October 1918): 83–89; "Woods for Propeller Manufacture," *Bulletin of the Experimental Department, Airplane Engineering Division, United States Army* 2 (January 1919): 119–127. Much of that information and knowledge would be organized into a single volume in 1920 that addressed design, manufacturing, inspection, storage, testing, and materials. Propeller Section, Engineering Division, U. S. Army Air Service, *The Airplane Propeller* (Washington, DC: Government Printing Office, 1921).

[71] "Metal Tipping of Aircraft Propellers," *Air Service Information Circular* No. 246, 3 (July 15, 1921): 28.

variable-pitch propeller. During the spring of 1918, the NACA identified the construction of a practical metal propeller as one of the "very important problems now confronting the air services of the nation." Specifically, the NACA issued a call for assistance in developing a steel propeller that would be "coincident" with the introduction of a variable-pitch propeller, one that could change its blade angle, or pitch, in flight to meet all operating conditions.[72]

Later that year, the chairman of the NACA, William F. Durand, announced before the Royal Aeronautical Society (RAeS) that the invention of such a device was "of the highest order of importance" and "outstanding as one of the appliances for which the art of navigation is definitely in wanting."[73] He was well-qualified to speak on the matter. An 1876 graduate of the United States Naval Academy and professor emeritus from Cornell University's prestigious Sibley College of Engineering, Durand served as the head of Stanford University's mechanical engineering department beginning in 1904. A noted authority on marine propellers, he became interested in aeronautics, specifically propellers, in 1914. His influential article of the same year, "The Screw Propeller: With Special Reference to Aeroplane Propulsion," that appeared in the *Journal of the Franklin Institute*, secured his charter membership in the NACA.[74]

The NACA's request reflected the pursuit of two important and intertwined propeller design trends that began in the United States during World War I. The first trend – the search for new materials and construction methods to facilitate changing blade angle – provided the foundation for implementing variable pitch. Engineers and designers had to develop a type of propeller that offered controlled pitch variation. The second, the perfection of the mechanism for changing blade pitch, involved a myriad of design possibilities. Each required investigation before a suitable system could be identified. A practical method of pitch actuation at the propeller hub together with a reliable system of controlling pitch from the cockpit had yet to be found.

[72] "Problems in Propeller Design," *Aviation and Aeronautical Engineering* 4 (June 1, 1918): 108.

[73] William F. Durand, "Some Outstanding Problems in Aeronautics," Wilbur Wright Memorial Lecture, *Annual Report NACA, 1918* (Washington, DC: Government Printing Office, 1919), 40–41.

[74] William F. Durand, "The Screw Propeller: With Special Reference to Aeroplane Propulsion," *Journal of the Franklin Institute* 178 (September 1914): 259–286; Frederick E. Terman, "William Frederick Durand (1859–1958)" in *Aeronautics and Astronautics: Proceedings of the Durand Centennial Conference Held at Stanford University, 5–8 August 1959*, ed. Nicholas J. Huff and Walter G. Vincenti (New York: Pergamon Press, 1960), 4–7.

Besides these new trends, propeller designers had to address the ever-present issue of weight in aircraft design, which was the dominant criterion in the design of new aircraft from World War I on into the 1920s and early 1930s. Pioneer American aeronautical engineer Grover C. Loening summarized the design community's attitude toward weight in his 1918 textbook *Military Aeroplanes*:

> While great attention must be paid to streamlining and aerodynamic efficiency, the newcomer does not get the proper appreciation of aviation without realizing that weight is the deadly enemy of flying. After all is said and done, the first thing to seek in a flying machine is light weight. Every extra pound weighs like lead in the hands of the flyer, uses up power and materials, and limits maneuvering.

Loening recognized that pilots and designers measured the performance of aircraft through the proportional figure "pounds per horsepower." The term expressed the relationship between the airplane's weight and the output of the engine. The fewer the pounds to one horsepower meant the airplane could fly faster, that is, more efficiently, and that was the ultimate goal for any new aircraft design.[75] The adherence to keeping aircraft light was a constant in the search for improving the performance of the airplane. Within that context, even with its inherent limitations, the lightweight wood, fixed-pitch propeller was the most logical design choice.

As the head of the United States government's propeller research and development program, Frank Caldwell was ready to face those obstacles. He was on his way to developing a propeller that was strong enough and capable of being efficient whether it was "starting" or "flying." It was becoming increasingly clear to him that the new propeller would not be fixed-pitch in configuration and more than likely not made from wood either. In the process, Caldwell would become the leading member of a new specialist propeller community that blossomed and grew with the airplane over the next four decades.

[75] Grover C. Loening, *Military Airplanes: An Explanatory Consideration of Their Characteristics, Performances, Construction, Maintenance and Operation*, 2nd ed. (Boston, MA: W. S. Best Printing Company, 1918), 191, 192.

3

"Engineering of a Pioneer Character"

On October 1, 1918, the Propeller Testing Laboratory of the United States Army Air Service's Engineering Division began operations at McCook Field in Dayton, Ohio. The core structure within the facility was a streamlined propeller whirl-test rig that was a functional indicator of the promising future of aeronautics. In a confidential 1918 publication, the Army claimed that the collection of what was the largest and most powerful propeller testing equipment in the world was the result of "engineering of a pioneer character."[1] Under the leadership of civilian specialist engineers, primarily Frank W. Caldwell, the Engineering Division developed propeller testing techniques and facilities that were major contributions to the field of aeronautical engineering. The laboratory was a clear indication of the Army's cutting-edge work in aeronautics and reflected the US government's intent to wage a modern industrial and aerial war in Europe.

The story of the development of propeller whirl-testing facilities and techniques by the Army Air Service from World War I to the early 1930s highlights the almost singular role that the US government played in nurturing a new industry and its engineering community. Engineers used their experience from different disciplines to shape a new one, aeronautical engineering, as they reacted to dynamic design requirements. In the process, they created the fundamental infrastructure from which the modern airplane evolved.[2] The pioneering propeller testing procedures

[1] "New Propeller Testing Laboratory," Bulletin of the Experimental Department, Airplane Engineering Division 2 (October 1918): 31.

[2] James R. Hansen, "Review Essay: Aviation History in the Wider View," *Technology and Culture* 30 (July 1989): 649; and ""Demystifying the History of Aeronautics," in *A Spacefaring Nation: Perspectives on American Space History and Policy*, ed. Martin

and practices at the Westinghouse Electric and Manufacturing Company of East Pittsburgh, Pennsylvania, from 1917 to 1918, McCook Field from 1918 to 1928, and Wright Field in Dayton, from 1928 to 1931, demonstrated the evolution of this new professional field as it pertained to the unique and highly-sophisticated "Army style" of engineering centered on development and innovation.[3]

Building a Foundation for Engineering Development

World War I was the catalyst for increased aeronautical development in Europe and the United States and it directly stimulated the development of the propeller. The wood, fixed-pitch propeller, which was most efficient for only one predetermined flight condition, gave satisfactory performance for aircraft that operated at low speeds and altitudes. As the war raged on, the aeronautical industries of the warring nations produced increasingly faster, more powerful, and deadlier aircraft capable of flying in excess of 100 miles per hour and ever higher altitudes. New propellers had to be stronger and able to withstand those ever-expanding performance thresholds.

Frank Caldwell faced an enormous challenge. The knowledge required to design an efficient and durable propeller was scant when he joined the Engineering Division in 1917. Caldwell knew that acquiring an acceptable level of design expertise required a considerable amount of time, experimentation, testing, and investment. The design constants of weight and size limitations persistently shaped this development. Almost overnight, propeller engineers required a more complete understanding of the forces that acted upon a propeller and the characteristics of the materials that could withstand them.[4]

The primary American aeronautical technology to emerge during the war was the 400-horsepower United States Army Standardized V-12, or "Liberty," engine. Four times more powerful than the ninety-horsepower Curtiss OX-5 used in JN-4 aircraft, the Liberty offered Caldwell and the engineers at McCook a significant challenge. Not only was there an

J. Collins and Sylvia D. Fries (Washington, DC: Smithsonian Institution Press, 1991), 153–166.

[3] Peter L. Jakab, "Aerospace in Adolescence: McCook Field and the Beginnings of Modern Flight Research," in *Atmospheric Flight in the Twentieth Century*, ed. Peter Galison and Alex Roland (Boston, MA: Kluwer Academic Publishers, 2000), 47.

[4] Daniel Adam Dickey, Interview by Lois E. Walker, September 2, 16, 30, 1983, interview K239.0512-1712, transcript, United States Air Force Oral History Program, HRA, 163.

increase in power, but there were also plans to supercharge the Liberty engine to increase high-altitude performance. These factors required that the new propellers be able to alter their pitch to perform efficiently at different altitudes.[5] If the American military required aircraft with higher performance at higher speeds and altitudes, then it needed a practical variable-pitch propeller.

A fixed-pitch propeller was designed in response to three specific performance parameters: airplane speed, engine horsepower, and engine rpm. The airplane designer first estimated the desired operating speed in miles per hour (mph). The same designer next determined the available engine horsepower based on the power plant chosen for the intended airframe. Using this information, the engine designer determined the rpm for the engine at the desired airspeed. The propeller designer used these three factors – mph, horsepower, and rpm – to design a propeller for that specific desired performance/engine/airframe combination.[6]

Before Caldwell and the propeller unit could proceed with experimentation, they had to create the basic tools and practices necessary for engineering development. Increased engine horsepower required heavier, larger, and more complicated propellers, which pushed the boundaries of engineering design knowledge. From his background in mechanical engineering, Caldwell knew he had to ensure that a potential design was structurally sound before production could proceed. He devised a way to evaluate the strength and durability of new designs, materials, and types of construction to guarantee that no unsafe propellers were accepted into military service. Caldwell believed the best way to replicate an actual flight condition was to use testing apparatus that rotated, or whirled, propellers in excess of their designed speed and power ratings for extended periods of time. Any material could withstand static loads from a press testing machine for a dangerously deceptive period of time. A fluctuating load created by a vibrating or fluttering propeller on a whirl rig could cause immediate and catastrophic failure.[7] Caldwell's concept of the whirl-testing rig was a fundamental contribution to the development of the propeller.

[5] Dickey Interview, 6, 163; Ronald Miller and David Sawers, *The Technical Development of Modern Aviation* (New York: Praeger, 1970), 71–72, 163.

[6] Gardner W. Carr, "Organization and Activities of Engineering Division of the Army Air Service," *U.S. Air Services* 7 (February 1922): 25.

[7] Dickey Interview, 20, 163; D. A. Dickey to R. C. Gazley, n.d. [March 23, 1931], Folder 168.7329-3, DAD, HRA.

When working to make his idea into a reality, Caldwell investigated the challenge of designing a whirl-test rig thoroughly by addressing the myriad of factors involved in such an undertaking. First, he had to choose between gasoline aircraft engines, steam turbines, and electric motors as the power source. Caldwell and the propeller unit originally intended to test propellers only using the service engines that they were designed for, but this approach had proven both time consuming and expensive. The goal was to develop an effective, efficient, and economical mode of testing. The destruction of an expensive, precision-built aircraft engine as part of normal testing operations was not an option.[8]

Caldwell's decision to use electric motors reflected widespread changes in American industrial power. During the decade between 1909 and 1919, the widespread introduction of electric-powered manufacturing equipment rapidly transformed the American factory. Prior to this, the primary source of power for a factory was water (using waterwheels or turbines) or the steam engine. Steam power was inherently dangerous with the threat of boiler explosions commonplace. A complex system of belts, shafting, pulleys, and clutches transmitted the energy from this single power source to machinery located all around the shop floor. By contrast, smaller electric motors could be used to drive each individual machine, thus eliminating the inefficient, complicated, and dangerous web of belts and pulleys. These motors, primarily manufactured by the General Electric Company (GE) and Westinghouse, produced more power, were cheaper and easier to use, took up less space on the factory floor, and could operate for extended periods beyond their designed capacity.[9] The electrification of American industry was a revolution in progress when Caldwell chose to electrify his whirl rig.

Whirl testing was an empirical process that reflected Caldwell's and the Army's desire to "over-engineer" aeronautical equipment for the rigors of aerial operations. An Engineering Division publication confirmed this attitude when it stated, "it is not considered sufficient that ... [propellers] be constructed according to design, of the best materials, but in addition they are subjected to a whirling test ... at speeds ranging up to 200 percent in excess of that required in service."[10] Over-engineering, over-building, and over-testing a propeller indicated the common engineering practice

[8] Dickey Interview, 163.

[9] Richard B. DuBoff, *Electric Power in American Manufacturing, 1889–1958* (New York: Arno Press, 1979), 62, 66–67, 86, 140.

[10] *A Little Journey to the Home of the Engineering Division, Army Air Service* (Dayton, OH: Thompson Print Company, n.d. [1924]), 23, 25.

of incorporating a "factor of safety" into a structural design. This factor of safety took into account unknown performance parameters to ensure that a structure survived under all foreseeable conditions.[11]

The procedure for whirl-testing propellers was an extensive three-level process. The first level consisted of basic testing at various speeds for calibration purposes and to test the operation of control mechanisms. This also included a ten-plus-hour endurance run at 100 to 200 percent overload on the electrically driven test stand. The second level saw the propeller installed on a high-horsepower aeronautical engine and run for twenty to fifty hours at full throttle. The third and final level, destructive testing, involved placing the propeller back on the electric test stand and whirling it at higher speeds and horsepower with the intention of destroying the propeller to determine its overall structural strength. The order and procedure of each level was a deliberate effort to acquire as much data as possible before destruction.[12]

The main emphasis of whirl testing was to establish the strength and endurance of a specific propeller design. Thrust, torque, centrifugal force, and gyrostatic moments all acted upon the propeller in flight.[13] According to Caldwell, the theoretical calculation of propeller stresses was tedious and misleading unless carried through in an exacting manner. He emphasized that a ten-hour destructive whirling test at 50 percent in excess of the propeller's designed operating regime was more than satisfactory in determining its strength. The unbalanced centrifugal force and resultant rotating leverage could exceed hundreds of thousands of pounds. Thus, if one or more of a propeller's blades failed, it could, at the least, break the engine shaft. At worst, it could rip the engine from its mountings, resulting in a crash. Even if the pilot, crew, and passengers survived a propeller failure, the resultant damage was expensive.[14]

While whirl testing concentrated on evaluating the structural integrity of propeller designs, the Army was concerned also with the aerodynamic

[11] Aubrey F. Burstall, *A History of Mechanical Engineering* (Cambridge, MA: MIT Press, 1965), 369.

[12] Dickey to Gazley, n.d. [March 23, 1931].

[13] Experts in the history of modern engineering recognize that "centrifugal force" – the perceived force that pulls a propeller blade away from the hub as it rotates – does not exist. Centripetal force acts upon a propeller in the same way gravitational force acts upon a satellite in earth's orbit where physical forces pull the blades toward the hub. This study will use the contemporary term "centrifugal force" for the purposes of continuity. J. Lawrence Lee, interview by author, Auburn, Alabama, May 2000.

[14] F. W. Caldwell, "Propellers," Lecture Number 8, February 23, 1921, Box 1, Wright Field Propeller Test Reports, NASM, 3–4; Dickey Interview, 36, 81, 82.

efficiency of its propeller designs. Flight testing a new propeller design was the final check for ensuring that aerodynamic calculations derived from wind tunnel testing research were accurate. The difference in scale between a model propeller in the wind tunnel and the full-size propeller mounted on an airplane (a phenomenon known as "scale effect") could result in an unsuitable design. The Aircraft Branch at McCook performed full-scale propeller flight tests with a thrust meter, which allowed engineers to evaluate wind tunnel designs and to ascertain their relationship to full-scale propellers. Designed by Caldwell, the new device fit on any airplane equipped with a Liberty V-12 engine. The thrust meter allowed flight test personnel to measure the actual thrust of the propeller under true operating conditions. The device consisted of an integrated external and internal hub that secured the propeller to the engine shaft while giving it the mobility to move forward under thrust. As the propeller produced forward thrust, it pulled the external hub forward against a calibrated spring, which it contacted via a roller. While the spring compressed under load (just like the spring on a scale), a circular drum recorded the deflection of the spring.[15]

Flight-testing allowed the evaluation of the propeller during its two main performance conditions: takeoff and cruise. The test pilot and technical personnel aboard the plane and on the ground took calculations of the aircraft's performance at varying altitudes and speeds. The most basic of these tests was to time these periods with a stopwatch. The overall process involved the empirical comparison of designs that differed in airfoil shape and blade angle, but were for the same engine and airframe combination. Test propellers could include modifications of standardized designs or represent departures from established practice. A standard propeller, such as the Signal Corps drawing number 8–45 for the Liberty engine as installed in the American variant of the de Havilland DH-4 "Liberty Plane," was often used as a constant in these tests. It was ideal to use the same aircraft and engine combination in each of these tests to ensure a proper evaluation.[16]

[15] F. W. Caldwell, "Propellers," Lecture Number 8, 8; F. W. Caldwell, "Thrustmeter for Liberty 12 Engine," July 15, 1918, Propeller Memorandum Number 16, File 21052-16, BAP, HRA; "Recording Propeller Thrustmeter," *Bulletin of the Experimental Department, Airplane Engineering Division* 1 (September 1918): 77–82.

[16] F. J. Patchell, "Summary of Information Contained in Reports of Tests on Propellers," July 12, 1918, File D52.43/71, Wright Field Technical Documents Library (hereafter cited as WFTD), NASM; Dickey Interview, 36.

The Army engineers also made measurements of a propeller's thrust versus power and rpm to correlate the data with engine specifications operating under specific flight conditions. Whirl-testing calibrations corresponded to propeller thrust on the ground when there was no forward velocity before the start of takeoff. Flight-testing ascertained the propeller's performance as it traveled through the air at the aircraft's normal flight speed, something that could not be replicated during the whirl test. The propeller engineers then used this data to deduce normal operating flight speed.[17]

The Army's work to determine the aerodynamic efficiency of a propeller blade was not for use by the aeronautical community. As early as 1915, the executive committee of the NACA recognized that the lack of consistent aerodynamic propeller design data was one of the general problems facing American aeronautics. The need for "more efficient air-propellers," able to retain their efficiency over a variety of flight conditions, was a primary concern.[18] They sponsored a comprehensive propeller aerodynamics research program at Stanford University near Palo Alto, California, under the direction of William F. Durand and Everett P. Lesley beginning in 1916. They used their expertise in marine propeller design to conduct a systematic broad-based study entitled "Experimental Research on Air-Propellers."[19] By 1922, the groundbreaking series of experiments combining mathematical calculation and wind tunnel studies established a standard table of propeller blade coefficients available to designers.[20]

At the Langley Memorial Aeronautical Laboratory in Virginia, NACA researchers under the leadership of propeller specialist Fred E. Weick opened a new facility to research full-size propellers and aircraft

[17] Dickey Interview, 37.

[18] "Problems," *Annual Report NACA, 1915* (Washington, DC: Government Printing Office, 1916), 13, 15.

[19] Lesley received a master's degree in naval architecture from Cornell University and served for two years at the Navy's Experimental Towing Tank. He came to Stanford's mechanical engineering department in 1907 with a considerable knowledge (like Durand) of marine-propellers. Walter G. Vincenti, "Air-Propeller Tests of W.F. Durand and E.P. Lesley: A Case Study in Technological Methodology," *Technology and Culture* 20 (1979): 720, 722. Vincenti asserted that there were inherent similarities in marine propeller and airplane propeller research, specifically regarding methodology. For a contemporary view of the relation between aeronautical and marine engineering, see Jerome C. Hunsaker, "Aeronautics in Naval Architecture," *Transactions of the Society of Naval Architects and Marine Engineers* 32 (1924): 1–25.

[20] William F. Durand and E. P. Lesley, "Experimental Research on Air-Propellers – V," NACA Technical Report No. 141 (1922), 169; Walter G. Vincenti, *What Engineers Know and How They Know It: Analytical Studies from Aeronautical History* (Baltimore, MD: Johns Hopkins University Press, 1990), 154, 158.

components in July 1927 that was in response to and complemented the work of Durand and Lesley at Stanford.[21] The Propeller Research Tunnel came to be an important tool in the development of the NACA radial engine cowling and the integration of the propeller, radial engine, and the nacelle into the overall aerodynamic configuration of the airplane in the late 1920s and early 1930s.[22] Both the Stanford and Langley programs became models of engineering methodology that reflected the NACA's desire to address basic problems, meaning the new knowledge was applicable to the design of blades regardless of whether they were wood or metal or to be used on fixed- or variable-pitch propellers.

Mathematical aerodynamic analyses were useful from the standpoint of design layout, but not for the majority of the work conducted by the propeller engineers at McCook and Wright fields. Whirl testing was the only way to evaluate the most critical structural factors in a propeller design. The Army's propeller engineers believed the "safety of the propeller was much more important than its variation in [aerodynamic] performance."[23] They even went so far as to regard the propeller as a simple cantilever being subjected to thrust and torque loads alone without considering the restoring moment of centrifugal force on the propeller during testing. That omission was a matter of convenience and simplification in the testing process.[24]

The reasons for testing propellers to destruction were clearly outlined in 1918. The justifications reflected a practical emphasis on efficient contracting and manufacturing processes and the harder-to-define creation of pilot confidence in the technology. An Airplane Engineering Division bulletin stated:

> The purposes [*sic*] of propeller testing are both engineering and phychological [*sic*] in their character. A pilot who knows exactly what his propeller will stand and has withstood under terrific overload, will have a self-confidence impossible for him, did he not possess test records prepared for each propeller type and sub-type.[25]

[21] Fred E. Weick and Donald H. Wood, "The Twenty-Foot Propeller Research Tunnel of the National Advisory Committee for Aeronautics," NACA Technical Report No. 300 (1928), 3-16.

[22] James R. Hansen, *Engineer in Charge: A History of the Langley Aeronautical Laboratory, 1917–1958* (Washington, DC: National Aeronautics and Space Administration, 1987), 87–90, 444.

[23] Dickey Interview, 36, 81, 82.

[24] Dickey to Gazley, n.d. [March 23, 1931].

[25] "New Propeller Testing Laboratory," 45.

Because the primary goal for the Army's propeller testing program was to gain approval for service use, whirl testing ensured that government contractors manufactured their products according to predetermined specifications. The type of construction, materials including wood laminates and glue, and the finish could be quickly evaluated during the whirl test and the results could be startlingly clear. The Army knew a standardized design could withstand the whirl test, but it had to ensure that the versions of those designs built by the propeller manufacturers followed those specifications exactly.[26] This particular style of developmental type testing was at the foundation of the transition from wood to metal propellers.

Under Caldwell's influence, the Army began to define its requirements for a permanent destructive propeller whirl rig in September 1917. The original intent was to construct a destructive propeller testing facility at Langley Field in Hampton, Virginia, where the Army intended to center its experimental engineering program. Those plans were short-lived, however, because the Army quickly decided to concentrate its engineering and test facilities at McCook Field in Dayton. Army planners believed it was better that the new site be strategically placed between Washington, DC and industrial Detroit rather than remote southeastern Virginia. The Chief Signal Officer, Brig. Gen. George Owen Squier, requested an estimate from Westinghouse Electric for the cost and delivery of an electric motor to whirl propellers, generating equipment to power the motors, and for the related structures of the facility. Specifically, the Signal Corps required a minimum of 500-horsepower and 2,500 rpm generated by a 220-volt, four-pole, squirrel cage induction motor. The Army envisioned routine tests at speeds up to 1,480 rpm.[27]

Early Work at East Pittsburgh

Caldwell's initial attempt at designing a propeller whirl-testing rig occurred at Westinghouse's East Pittsburgh plant. The Army paid for the setup and installation of the whirl rig at Westinghouse, but did not own the equipment or rights to the use of the facility.[28] Created to meet the unprecedented wartime demand for propellers, the Westinghouse

[26] Dickey Interview, 38.

[27] G. O. Squier to W. Sykes, "Equipment for Making Destructive Propeller Tests," September 5, 1917, Folder 168.7329-2, "Pre-1920 Correspondence," DAD, HRA.

[28] Dickey Interview, 13, 19, 163–164; "Frank Walker Caldwell," *Collier's Encyclopedia*, 4 (New York: Collier, 1957), 322–323; "New Propeller Testing Laboratory," 31.

FIGURE 4 The world's first electric propeller whirl testing rig at the Westinghouse East Pittsburgh factory. National Archives and Records Administration.

facility was nothing more than a temporary measure that resulted from a cooperative effort between the Army and private industry. Nevertheless, the Westinghouse equipment was the first whirl rig in the world. Its constantly evolving design reflected the continual learning process of how to design the necessary testing apparatus needed to evaluate propellers.

Westinghouse installed the propeller whirl-testing apparatus in aisle "D" of the East Pittsburgh plant's electric motor assembly and testing area (Figure 4). The area featured a large test floor for assembling and testing machinery such as generators, motors, and dynamometers. The industrial location offered two other advantages. First, its location offset the anticipated high volume of noise and its negative effect on the surrounding area outside the factory. Second, the industrial setting ensured that a large amount of electrical power was readily available. Due to the horizontal orientation of the electrical motor, the test propellers rotated in the vertical plane, just as they would on an actual airplane.[29]

[29] Dickey Interview, 13–14, 163–164; Caption for photograph of Propeller Whirl Test Stand at the East Pittsburgh, Pennsylvania, Plant of the Westinghouse Electric and Manufacturing Company, 1917, Folder 168.7329-12 Part 1, DAD, HRA.

According to D. Adam Dickey, a young and newly hired electrical engineer working in the Westinghouse Research Department, workers "knocked a big hole in the wall" at the end of the aisle to divert the resultant air blast and ventilate the overheating interior air that accumulated during a whirling test. The inside of the factory would have been uninhabitable otherwise. Unfortunately, on a cold day the Westinghouse whirl rig rapidly depleted the factory building of its warm air during testing. As a result, the propeller engineers had to conduct their tests when the factory was closed. The routine schedule for propeller whirl testing at Westinghouse began at 12 noon on a Saturday and continued until early Monday morning, right before the factory workers returned to work. The propeller engineers used a desk in the poorly heated shop area to sleep in between individual tests.[30]

The power unit for the Westinghouse whirl-testing rig was a 500-kilowatt shunt-wound direct current motor that generated 670 horsepower at a maximum rotational speed of 300 rpm. The motor's top speed was well below the Army's requirement of 1,500 and 2,000 rpm for adequately testing new designs. Caldwell and Westinghouse engineers separated the motor from the test shaft to increase the rotational speed of the whirl rig. They mounted a sixty-six-inch diameter pulley on the motor and fourteen- and twelve-inch diameter pulleys on the propeller shaft and connected them with a thirty-six-inch wide thick leather belt. The combination of the large-diameter pulley gear on the motor turning one of the small diameter pulleys on the test shaft generated the requisite rpms. The pressure to get the testing rig up and running negated the efforts to design a dedicated "step-up" gearbox that would have eliminated such a complicated, but typical for the time, arrangement.[31] The entire arrangement was similar to the belts, shafting, and pulleys found in different types of industrial applications across the country.

Caldwell incorporated into the Westinghouse whirl rig several original safety features proposed in the original Army whirl rig specifications of September 1917. The rig featured an extension to the test shaft designed intentionally to break off with the propeller in the event of a blade failure and to prevent damage to the test shaft and drive system from the

[30] Dickey Interview, 13–14, 163–164; Caption for photograph of Propeller Whirl Test Stand at the East Pittsburgh, Pennsylvania, Plant of the Westinghouse Electric and Manufacturing Company; "Adam Dickey," File CD-330500-01, BF, NASM.

[31] Dickey Interview, 16, 21, 163–164; "Description of Westinghouse Propeller Whirl Equipment," n.d. [1918], Folder 168.7329-12 Part 1, DAD, HRA; "New Propeller Testing Laboratory," 32.

resultant unbalance. Originally, propellers were attached directly to the test shaft in the same manner they were connected to an aircraft engine. One of the eight-inch extensions broke during routine testing, thus validating the Army's 1917 proposal. The extension also acted as an adapter that facilitated the quick exchange of propellers during testing.[32]

In addition, Caldwell installed a barrier to protect the propeller engineers, test equipment, and factory machinery called a "bombproof." The substantial structures built within military forts of the nineteenth century to defend people, ammunition, and important supplies from bombardment influenced the new safety feature. Caldwell's structure was a twelve-foot long archway in the factory wall that surrounded the whirling propeller's plane of rotation. Inside the archway, wood planks, each measuring six-inches thick, absorbed the impact from fragments of disintegrated propellers. Wood was the best material overall because it absorbed propeller fragments and debris rather than deflect them as metal boilerplate or concrete would in the dynamic and lethal environment of whirl testing.[33]

One of the major operating parameters the McCook engineers wanted to evaluate was propeller deflection, or the amount the blades distorted under aerodynamic load. The propeller engineers measured deflection using an optical sighting device in the side of the Westinghouse bombproof. The device moved along stationary guides mounted in two slots arranged perpendicular to the rotation of the propeller. It was the first time that engineers were able to observe and record the extent to which a propeller blade flexed, distorted, and changed in pitch while in use.[34]

The Westinghouse whirl rig began operations in November 1917. Caldwell created the precedent for documenting testing activities when he ordered W. A. Hale, the Engineering Division's representative at Westinghouse's East Pittsburgh facility, to prepare a complete report with text and photographs for each propeller test at East Pittsburgh. This format became the standard for disseminating and preserving testing data until the 1950s. Caldwell wanted to correlate the results of the tests with

[32] Dickey Interview, 16, 19; Squier to Sykes, "Equipment for Making Destructive Propeller Tests."

[33] Dickey Interview, 19, 163–164; Frank W. Caldwell, "Destructive Propeller Tests," Propeller Memorandum No. 1, December 21, 1917, File 216.21052-1, HRA, 15; C. F. Taylor to G. E. A. Hallett, "Rig for Testing Propellers on Engines," February 6, 1922, Folder "Propeller Test Rig, 1922," Box 112, RD 3112, RG 342, NARA.

[34] Caldwell, "Destructive Propeller Tests," 12; Dickey Interview, 20, 163–164.

the current methods of stress calculation, which facilitated the design of strong and durable propellers for the Army and Navy.[35]

The engineers of the propeller unit whirl tested more than 130 sample propellers at the Westinghouse facility from December 1917 to October 1918. The majority of those propellers were slated for immediate production and operational use. Some were wood fixed-pitch propellers made from cherry and oak, but most were constructed using mahogany, black walnut, or birch. The facility also evaluated a few experimental types, including seven steel, six composite, and two variable-pitch propellers.[36]

The detail design data acquired from a whirl test informed a propeller maker of required improvements to the design. If they were a possible government contractor, the information indicated how they could improve their next submission. If it were an Army-designed and manufactured propeller, modifications were made right at the whirl rig if possible. The data also provided the basis for the rejection of a submitted propeller. Several of the experimental designs that arrived at Westinghouse reflected an ignorance of the types of stresses that developed in a whirling propeller. As a result, the young engineers operating the equipment had to be mindful of flying propeller parts at the East Pittsburgh plant.[37]

Even before the Westinghouse facility went into operation the limitations of its ad hoc propeller whirl-testing equipment were well apparent. By far, the inability to generate ever higher rpms was the equipment's most obvious disadvantage. The new standard test for a typical propeller was ten hours at 2,500 rpm. Westinghouse recognized that it was dangerous to attempt to rotate the propeller at that speed because it required the large leather belt to travel at a rate of 82 mph. A heavy, thick belt that could withstand the force could not be adequately shaped around a pulley less than ten inches in diameter. The highest rotational speed they had achieved previously was 68 mph. As a result, W. A. Hale doubted that the required 2,500 rpm was available from the Westinghouse equipment. He recommended that if the Signal Corps required higher speeds of rotation, a direct-drive test shaft and purpose-built motor would be necessary.[38]

[35] F. W. Caldwell to W. A. Hale, November 14, 1917, Folder 168.7329-2, "Pre-1920 Correspondence," DAD, HRA.

[36] "Photographs of Propellers Whirl-Tested by the Westinghouse Electric and Manufacturing Company for the United States Government," October 24, 1918, Folder 168.7329-12 Part 1, DAD, HRA.

[37] Dickey Interview, 19.

[38] W. A. Hale to Airplane Engineering Department, "Destructive Whirling Test of Airplane Propellers," November 9, 1917, Folder 168.7329-2, "Pre-1920 Correspondence," DAD, HRA.

Additionally, the equipment offered limited protection from propeller debris and was unable to measure thrust. Another limitation of the equipment was the weakness of its component parts. C. E. Skinner, one of the propeller engineers detached from McCook Field, reported to Caldwell that despite the use of the breakable extension, the test shaft had been bent during the test of the experimental Ingells steel propeller in April 1918. The requisite repair took weeks to accomplish.[39]

The McCook Field Whirl-testing Facility

It was obvious to Caldwell from the beginning that the Westinghouse facility was only a stopgap and temporary measure in meeting the goals of the wartime production program. He campaigned for the construction of a new propeller whirl-testing facility at the new Engineering Division facility at McCook Field. Caldwell felt it was even more imperative to have new propeller testing equipment constructed as quickly as possible in Dayton to meet the demands of the aviation production program in November 1917.[40]

The propeller unit reported to Lt. Col. Jesse G. Vincent, chief engineer at McCook Field, its justification for whirling and destructive tests in March 1918. The Air Service did not have the capability at Westinghouse to whirl test propellers according to specifications provided by the Production Engineering Division. Primarily, there was an immediate need to test propellers at 15 percent in excess of the rate of their designed engine speed, which could be up to 1,500 rpm. For example, the Forest Products Laboratory at the University of Wisconsin at Madison had been conducting a series of experiments with wood propeller construction based on varying moisture content, density, and tree species. It was imperative for the Air Service to test these experimental propellers to destruction and to whirl them at their rated speed for a long period to gauge their expected suitability for service. The ability to perform regular destructive and whirling tests at McCook Field enhanced the process of propeller design. It kept both the factor of safety and the weight of the propeller down.[41]

[39] Dickey Interview, 163–164; C. E. Skinner to F. W. Caldwell, "Test of Micarta Propeller Z-11," April 8, 1918, Folder 168.7329-2, "Pre-1920 Correspondence," DAD, HRA.

[40] Dickey Interview, 164; Caldwell to Hale, November 14, 1917.

[41] H. C. Marmon to J. G. Vincent, "Whirling and Destructive Stand for Propellers," March 9, 1918, Folder "Propeller Testing Equipment, 1918," RD 3085, RG 342, NARA.

In the long term, it would be more economical to invest the money and effort into permanent testing facilities. The proposed electric whirl-test facility would provide the capability to calibrate propellers and engine "test clubs" – propeller-like devices that allowed an engine to rotate at its designed rpm without generating thrust – to match the performance of an engine. Electric whirl testing also could subject propellers to stresses far beyond normal operating conditions so they developed fatigue weaknesses quickly, which facilitated a rapid testing schedule.[42]

Another argument for the construction of a purpose-built and dedicated whirl-testing facility at McCook Field was its potential value to the ambitious American aviation production program. Before a quantity production order of propellers manufactured according to government specifications could be initiated, the design had to be whirl-tested by the Army for approval. Until then, the manufacturer had to wait for the test of a sample propeller to be completed. In April 1918, it was apparent that the Westinghouse facility did not meet the demand. The lack of propellers for service use was a potential bottleneck in the entire American aviation production program. The Army's inspector in Buffalo, New York, a major center of propeller manufacturing for the war effort due to the large number of woodworking companies located there, believed that the backlog at the Westinghouse facility would stymie production goals for Liberty and Hispano-Suiza propellers.[43]

The Army decided to comply with Caldwell's and the Engineering Division's wishes and construct an electric whirl-testing facility at McCook Field. The estimated cost of the proposed propeller testing equipment was $80,000. The capacity of the propeller testing equipment was 1,200-horsepower with the capability of being overloaded to 2,400-horsepower at 3,000 rpm. Until the completion of the new facility, Westinghouse provided the facilities and location for the government's propeller testing program. The Engineering Division administered and funded the overall program.[44]

The Engineering Division's Propeller Testing Laboratory began operations at McCook on October 1, 1918. At the time, it was the largest

[42] Marmon to Vincent, "Whirling and Destructive Stand for Propellers"; Dickey to Gazley, n.d. [March 23, 1931].

[43] J. G. Perrin to H. C. Marmon, "Spinning Tests for Propellers," n.d. [April 1918], Folder "Propeller Testing Equipment, 1918," RD 3085, RG 342, NARA.

[44] Marmon to Vincent, "Whirling and Destructive Stand for Propellers"; H. E. Blood to J. G. Perrin, "Propeller Testing Equipment," April 29, 1918, Folder "Propeller Test Equipment, 1918," Box 12, RD 3085, RG 342, NARA.

and most powerful whirl-testing facility in the world and a definite leap forward in the capabilities that had been available at Westinghouse. The official report introducing the new facility noted:

> In planning the dynamometer and calibration apparatus and their foundations, it was necessary to perform engineering of a pioneer character...The possibilities and usefulness of the new propeller testing laboratory at McCook Field are enormous, and great satisfaction exists now that it has been placed in commission. It should not only lead to material advancement in the scientific design of this important feature of aircraft, but should enable the testing of a great many more propellers in a given period than has been possible heretofore.[45]

The propeller engineers at McCook had created a new form of engineering testing equipment. Dickey recalled that their work was "so far out in front that you almost had to design your testing methods and your equipment and whatever you needed."[46]

Caldwell, a mechanical engineer, worked with electrical engineers, Dickey and Mascom A. Smith, Jr., to design the whirl rig. They took readily available equipment and used it creatively to suit their purposes. Dickey grew up in Germantown, Ohio, a town just southwest of Dayton. He received his electrical engineering degree from Ohio State University in 1916 and worked at Westinghouse before joining the propeller unit early in 1919. Smith previously worked on the test floor at Westinghouse's East Pittsburgh plant and came to Dayton when the new McCook whirl-testing rig opened. Smith designed the electrical system for the power building as well as for the whirl rig itself. The Army classified Caldwell, Dickey, and Smith as aeronautical engineers due to their propeller work at Westinghouse and they were responsible for establishing specifications and directing the operation of the outdoor substations that powered the entire facility.[47]

The general layout of the new McCook Field facility consisted of two major portions: the equipment for producing the power and the mechanism and devices used by the laboratory to test and evaluate propellers (Figure 5). Constructed out of cream-colored brick, the power building and small blower building housed a motor-generator that converted the 6,600-volt alternating current supplied by the Dayton Power and Light Company into direct current. Another motor-generator powered the drive system at a variety of voltages for speed control. Concrete tunnels

[45] "New Propeller Testing Laboratory," 31, 32, 45.
[46] Dickey Interview, 22.
[47] Dickey Interview, 21–22, 27, 47–48; "Adam Dickey" File, NASM.

FIGURE 5 The Engineering Division whirl rig at McCook Field. Smithsonian National Air and Space Museum (NASM WF-2242).

and conduit connected the power equipment to the whirl rig and provided electricity and ventilation for the air blast.[48] Despite the McCook Field engineers' assertions that, "there is nothing in the power building that would strike an engineer as being very new or original practice," it was original in application.[49]

The Propeller Testing Laboratory's power demands exceeded four times the power used by other units in the Engineering Division. The demand for power was so much that the facility had its own dispatching office to coordinate whirl-testing operations with the Dayton Power and Light Company. The laboratory had to notify the power company of its power requirements, which could exceed 3,000 kilowatts to produce 3,000 to 4,000 horsepower, before a test began. Thus forewarned, the company could then direct another turbogenerator at its plant to accommodate the increased demand for electricity.[50]

[48] Dickey Interview, 27, 164–165.
[49] "New Propeller Testing Laboratory," 32, 33.
[50] Dickey Interview, 47–48.

Located thirty feet away from the power building was the whirl rig. It consisted of four main components – the test housing, propeller drive system, the underground observation/control room, and the bombproof – that in combination resulted in a sophisticated and massive piece of testing equipment. The most recognizable feature was the streamlined housing that contained the propeller drive system. Caldwell designed the structure to resemble an airplane fuselage ensuring even airflow in an attempt to replicate actual flight conditions. The aerodynamically contoured housing created an unobstructed slipstream that guaranteed unbiased thrust measurement of both tractor and pusher propellers.[51]

The sheet iron-skinned structure invoked images of machine-age design and its celebration of high-speed, streamlined shapes that highlighted humankind's progress through the development of technology. It was true that the propeller test rig possessed the potential for increasing the performance of the airplane, but it was also a powerful symbol of the overall progress of aeronautical technology in the immediate post–World War I period. The housing looked more like a streamlined, high-speed train than an esoteric piece of testing equipment and it was certainly more aerodynamically clean than most airplanes of the time.

Underneath the sheet metal covering, which could be removed in sections for maintenance and inspection, was a steel framework that enclosed the rig's power drive system. Caldwell placed the drive system approximately nine feet above the ground to allow clearance for large propellers up to eighteen feet in diameter. A reinforced concrete pedestal, foundation, and trestle supported all of the equipment.[52]

The McCook test stand utilized four inline, direct-drive, engine-testing dynamometers manufactured by the Sprague Electric Works of GE in New York City to eliminate the pulleys and gearboxes used to increase speed. A dynamometer is an electrical device that could be used as a motor or generator in industrial applications and operated at various speed and power ratings. The Propeller Testing Laboratory used them as electric motors to spin the propellers. The whirl rig was capable of rotating an eight-foot propeller at 3,000 rpm, which produced tip speeds of 857 mph, just over the speed of sound. At those speeds, engineers at McCook could conduct destructive whirling tests on propellers designed for engines up to 1,000-horsepower. The dynamometers could measure the power absorbed by a propeller by electrical means. They

[51] Ibid., 31.

[52] "New Propeller Testing Laboratory," 31, 37.

also facilitated testing both right- and left-hand propellers, overcoming another shortcoming of the Westinghouse rig. The ability to whirl propellers in both directions was important regarding the growing use of spur reduction gearing where the engine rotated in one direction and the propeller in another.[53]

A special Francke floating coupling connected the four dynamometers to the propeller shaft, which allowed for small amounts of misalignment between the motors. This floating coupling also protected the motor shaft from the pulling force imposed by a propeller. A thrust bearing installed between the propeller shaft and the dynamometers enabled engineers to measure propeller thrust through a connection to Toledo mechanical scales mounted in a pit below the test rig.[54]

Below the streamlined housing lay the underground observation and control room. From there, the electrical generation and the test equipment could be controlled remotely. The propeller engineers could also observe the plane of rotation for deflection from this protected position. The McCook Field propeller testing facility incorporated a new device for observing and gathering data related to the deflection of the propeller blade. Caldwell believed that a telescope used in a surveyor's transit with vertical and horizontal crosshairs and a focal range of six to fifty feet and mounted on a sliding mechanism provided the best measure of propeller deflections under operating conditions. He requested a price quote and estimated delivery date from the Keuffel and Esser Company of Hoboken, New Jersey, for the equipment in August 1918. Keuffel and Esser responded that their Transit Number 5165, which was "serviceable, shop-worn, and only $45," met the requirements of the Engineering Division because it also had an axis and bubble that made leveling the device easy.[55] The late addition of the telescopic deflection-measuring device was crucial to the success of the whirl rig.

The McCook Field whirl rig bombproof served two purposes. Lined with heavy oak timbers, the structure prevented propeller debris from damaging the facility or injuring McCook Field personnel. Mounted on rails that ran parallel to the test stand, the bombproof also served as a movable gantry-type crane that facilitated the mounting and removal of

[53] F. W. Caldwell, "Propellers," Lecture Number 8, 3–4; Dickey Interview, 21, 31, 164–165; "Sprague Electric Works," Folder 168.7329-2, "Pre-1920 Correspondence," DAD, HRA.

[54] Dickey Interview, 22, 29, 164–165.

[55] F. W. Caldwell to Keuffel and Esser, "Telescope for Sighting Device on Propeller Test Rig," August 7, 1918; Keuffel and Esser Company to AED, August 15, 1918, Folder 168.7329-2, "Pre-1920 Correspondence," DAD, HRA.

test propellers and other equipment with a built-in hoist. Another function of the bombproof was its use in the endurance evaluation of propellers in different climate conditions. A water line mounted in the arch introduced a spray that replicated rain or ice, depending on the time of year, into the airstream of a whirling propeller.[56]

Caldwell believed and successfully argued that his electric whirl rig eliminated the requirement for testing propellers on engines. The Army originally had intended to test propellers only using service engines that they were designed for, but this approach had proven to be inefficient in terms of the time it took to complete a test and the expense of the engines. He eventually realized, however, that it was still necessary to test specific propeller/engine combinations. Engine whirl testing became part of the Army's program at McCook Field in 1922.

The initial tests of propellers mounted on engines could be as spectacular and dangerous as the destructive whirl tests. C. Fayette Taylor, an engineer working in McCook's Power Plant Laboratory, remembered overseeing an early test of an experimental steel propeller on a 300-horsepower Hispano-Suiza inline V-8 engine. During the course of the test, a blade broke off and entered the control room, passing between the heads of the research staff before proceeding through the roof. The rest of the propeller and the test engine lay in shattered pieces strewn about the test area.[57]

The result was the construction of a dedicated propeller/engine-testing stand at McCook Field in 1922 within the Power Plant Laboratory. The structure allowed testing both Liberty and Hispano-Suiza engines in the 300- to 400-horsepower range with propellers up to eleven feet in diameter. Laboratory personnel mounted the engine on a concrete block. The peaked-roof wooden shed was open at two ends with a bombproof-like interior of eight-inch oak planks that covered six feet of the plane of rotation to protect the testing crew as well as nearby facilities and passersby. The new modified propeller/engine test stand was in operation by May 1922 as an integral part of the Army's propeller testing program.[58]

[56] Dickey Interview, 31, 164–165.

[57] Marmon to Vincent, "Whirling and Destructive Stand for Propellers"; C. Fayette Taylor, *Aircraft Propulsion: A Review of the Evolution of Aircraft Piston Engines* (Washington, DC: Smithsonian Institution Press, 1971), 77.

[58] G. A. Hallett to L. W. McIntosh, "Report of Propeller Test," January 27, 1922; Taylor to Hallett, "Rig for Testing Propellers on Engines"; C. F. Taylor to Propeller Branch, "Propeller Testing Stand," May 4, 1922, Folder "Propeller Test Rig, 1922," Box 112, RD 3112, RG 342, NARA.

The roar of overdriven propellers and the shrill overtones of narrow-blade propellers rotating at high speed were unwelcome characteristics of the Army's whirl-testing equipment despite the fact that Caldwell had taken into the effects of noise on the surrounding area. As the speed of a propeller designed for larger and more powerful engines increased under test, the volume of sound became more and more intense. During a routine test on a Good Friday, the field received a call from a church approximately fifteen miles away in Miamisburg, Ohio, requesting that the Army discontinue its tests until the service concluded. The propeller engineers obliged. The Army had to modify the schedule of the whirl-testing rig to accommodate the concerns of people on the field and the surrounding area affected by the high volume of noise, which could make a person physically ill. Basically, the rig operated when people were not around, normally evenings and weekends. A spotlight provided visibility at night and other instances of low-light conditions to enhance the facility's continued operational capability.[59]

Oftentimes, the whirling-test schedule fell behind by weeks due to the failure and subsequent repair of the motors and the floating coupling. The routine test of the large-diameter wood propeller for the Huff-Daland XLB-1, a large, single-engine, three-place bomber, overloaded the equipment resulting in a burned-out motor and an unprecedented three-week delay before testing could resume. The testing of one specific propeller could last as long as a month. Therefore, if the equipment broke down, the government propeller-testing schedule could be in jeopardy while creating a significant backlog.[60]

The limitations of the McCook Field electric propeller testing equipment were apparent by November 1925, reflecting the rapid pace of technological development during the immediate postwar period. Caldwell and his team had designed the equipment seven years earlier when the majority of the propellers were wood and knowledge about their design and construction was limited. Metal propeller designs survived

[59] "Design of Wright Field Propeller Laboratory," n.d. [1931], File 168.7329-3, DAD, HRA; "New Propeller Testing Laboratory," 33; Dickey Interview, 31, 114–115, 164–165.

[60] L. MacDill to Curtiss Aeroplane and Motor Company (hereafter cited as CAMC), "Curtiss Reed Propellers for Consolidated Training Airplane," October 22, 1925, Folder "Heath [Reed], 1922–1925," Box 5582, RD 3143, RG 342, NARA; C. V. Johnson to J. F. Curry, Office of Chief of the Air Service (hereafter cited as COAS), May 21, 1925, Folder "Propeller Test Rig, 1925," Box 5582, RD 3143, RG 342, NARA; F. G. Swanborough, *United States Military Aircraft Since 1909* (New York: Putnam, 1963), 277; L. MacDill to T. P. Wright, December 7, 1925, Folder "Propellers, 1925," Box 5581, RD 3143, RG 342, NARA.

destructive whirl testing only to have their operational versions fail. The dangerous reminder of why there was a need for testing in the first place revealed that the McCook whirl rig simply could not spin the propellers fast enough to destroy them. Caldwell implemented an extra step when he suggested that a propeller run on its designated engine for at least ten hours before receiving approval.[61] The introduction of metal propellers accelerated the demand for new and more powerful testing equipment.

The Propeller Laboratory at Wright Field

The Wright Field propeller test facility was the propeller unit's answer to the increasing demand for better propellers. A new facility gave the Propeller Laboratory an opportunity to correct the shortcomings of the McCook Field whirl rig centered on the need for more power and rotational speed. Since the funds to equip the new facility came from the Army's appropriation for experimental research and development, the propeller unit competed directly for funds with other projects. Those included the experimental Almen eighteen-cylinder barrel engine, contractual work for aircraft with air-cooled engines and supercharging, and the 400-horsepower Curtiss R-1454, the Army's first high-power air-cooled radial.[62] Despite the importance of those projects, the Engineering Division requested a call for proposals be issued to generate bids for the design and construction of the new equipment.[63]

The Chief of the Air Service approved the construction of a new propeller whirl-testing facility at Wright Field in May 1925. The Engineering Division estimated that the equipment would cost $300,000, and Circular E-2575 later that same month calling for proposals for the required electrical equipment. The Allis-Chalmers Manufacturing Company, GE, and Westinghouse submitted proposals. Westinghouse offered the lowest bid at $308,555, which the Engineering Division accepted.[64] Westinghouse estimated complete delivery within nine months. The preliminary drawings of the facility arrived at McCook in October 1925. The excavation

[61] F. W. Caldwell to H. S. Martin, "S.A. Reed Dural Propeller – D-27," March 13, 1923, Folder "Heath [Reed], 1922–1925," Box 5582, RD 3143, RG 342, NARA.

[62] Robert Schlaifer and Samuel D. Heron, *Development of Aircraft Engines and Aviation Fuels* (Boston, MA: Harvard University, 1950), 18.

[63] C. V. Johnson to COAS, "Propeller Test Rig for New Field," May 18, 1925, Folder "Propeller Test Rig, 1925," Box 5582, RD 3143, RG 342, NARA.

[64] Johnson to COAS, "Propeller Test Rig for New Field"; C. V. Johnson to Office of COAS, "Bids on Circular E-2575–Electrical Equipment for Propeller Test Laboratory," June 16, 1925, Folder "Propeller Test Rig, 1925," Box 5582, RD 3143, RG 342, NARA.

of the area at Wright Field took place from March to September 1926.[65] In the end, the new whirl-testing facility would cost $550,000; nearly double its original estimate.[66]

Adam Dickey and Mascom Smith authored the specifications for the electrical power and control equipment and personally oversaw their installation. They designed three large electric motors with different speed and horsepower ranges to evaluate properly the countless propeller designs and applications the military services (and potentially commercial aviation manufacturers) required. They increased the number of test stands from one to three and used longer-lasting synchronous motors capable of maintaining constant speeds similar to turbogenerators instead of the erratic induction motors used at McCook Field. Realizing that the single stand of the McCook rig could not meet the myriad requirements of the Army, Dickey and Smith designed high-, medium-, and slow-speed stands. The high-speed stand whirled at speeds between 1,800 to 4,300 rpm at 2,500-horsepower. The medium-speed stand generated 720 rpm to 1,800 rpm at 3,000-horsepower. The slow-speed stand generated speeds between 300 rpm to 750 rpm at 6,000-horsepower for testing airship propellers up to forty feet in diameter. The high-horsepower rigs operated large propellers, which revolved at slower speeds than small propellers. All three stands were arranged in tandem twenty-five feet above the ground to allow rotational clearance as well as to permit the testing of the propellers in an undisturbed airstream.[67]

The Wright Field bombproof was an arch structure consisted of two layers of six-inch-thick oak with one half-inch-thick steel plate "sandwiched" between them. Each test stand at the new Wright Field facility incorporated an underground observation chamber, which was orientated directly below the plane of rotation of the test propellers. Propeller engineers could sight on the leading and trailing edges of test propellers and measure the linear and angular distortion caused by propeller thrust. The observation window in the chamber's ceiling had steel I-beams spaced twelve inches apart to protect the propeller engineers from high-speed flying debris.[68]

[65] Westinghouse Electrical and Manufacturing Company, "Proposal in Response to War Department Circular E-2575 dated May 25, 1925," June 10, 1925, Folder 168.7329-8 Part 2, DAD, HRA; "Historical Record, Propeller Laboratory, March 1926-July 1943," Folder 168.7329-12 Part 1, DAD, HRA.

[66] Dickey Interview, 52; "Propeller Test Noisiest on Field," *Washington Post*, October 6, 1929, section A, p. 10.

[67] Dickey Interview, 51–52; "Design of Wright Field Propeller Laboratory," n.d. [1931].

[68] Dickey Interview, 57; "Design of Wright Field Propeller Laboratory," n.d. [1931].

Testing different propellers at varying speeds required exact control over the speed and power of each motor being used for the tests. Special generating equipment was installed in the Propeller Laboratory building that could produce 24,000 horsepower. The sophisticated and intricate electrical control system enabled the laboratory staff to have complete control of all three tests from the underground station at the base of each stand or from the main building's main switchboard. To change the speed on the synchronous motors used on the test stands meant their frequency had to be altered. The Wright Field Propeller Laboratory power building could generate from sixty to seventy-two cycles, which could produce the needed rpm for the high-speed stand. The new Propeller Laboratory did not have unlimited power, it had to share its power needs with the newly re-installed McCook Field five-foot and fourteen-inch wind tunnels.[69]

The structural design of the test stands had to compensate for the tremendous forces exerted upon their foundations when a propeller failed. Dickey and Hale incorporated reinforcing steel in the concrete and long, preloaded anchor bolts that were tightened and stretched to secure the stands to their foundations. The preload on the bolts was necessary because they had to have a force exerted on them that was greater than any force exerted during a whirl test. They had to be extra-strength, heavy-duty, and essentially over-engineered. If not, the bolts would "hammer" the foundation and destroy the test stand.[70]

While attempting to ascertain when the new propeller testing facility would open, the Materiel Division still had to contend with the limitations of the McCook facility, which were increasingly restricting the technical development of propellers. It was out of operation due to mechanical difficulties from mid-December 1927 to early January 1928. The Materiel Division somberly informed eager propeller manufacturers that it could not conduct regularly scheduled tests due to the operational breakdown as well as the monumental backlog of other propellers waiting to be tested.[71] The pace of propeller development made the McCook Field facility obsolete, but it had to remain open until October to cover the propeller unit's testing obligations.[72]

[69] "Design of Wright Field Propeller Laboratory," n.d. [1931]; Dickey Interview, 67.

[70] Dickey Interview, 52, 67.

[71] L. MacDill to Standard Steel Propeller Company, "Propeller–Whirl Test," December 24, 1927, Folder "Propeller Test Rig, Testing Equipment, 1927," Box 5664, RD 3164, RG 342, NARA.

[72] L. MacDill to Chief of the Air Corps (hereafter cited as COAC), "Move from McCook Field," December 23, 1927, Folder "Propellers–July through December, 1927," Box 5663, RD 3163, RG 342, NARA.

FIGURE 6 The Wright Field Propeller Laboratory operated the world's most advanced whirl-test rigs when it opened in 1928. Smithsonian National Air and Space Museum (NASM WF-61176).

The new Wright Field propeller laboratory opened at the south end of the field during the fall of 1928. The test stands were the largest structures at the field. The offices were in the nearby Building 16, or "the laboratory building," which also housed the staff of the Aircraft Laboratory (Figure 6).[73] Due to the remote location of the laboratory in relation to the civilian population and the time of day and the week when testing occurred, the propeller unit did not have to worry about designing a system of sound suppression. Brig. Gen. William E. Gillmore, Chief of the

[73] Dickey Interview, 67, 75.

Materiel Division, remarked that the laboratory was "one of the most important and also one of the noisiest" testing facilities at Wright Field.[74]

A Service to American Aeronautics

By 1931, the Materiel Division had fourteen years of experience in propeller whirl testing and evaluating new designs. Elements of the broader American aeronautical community took notice. The Department of Commerce's newly formed Aeronautics Branch awarded approved type certificates to airframe, engine, and propeller designs for use by commercial and private operators. Richard C. Gazely, the chief of the engineering section of the Aeronautics Branch, made an inquiry to the Materiel Division regarding whether or not their type certificate testing of propellers could be conducted at Wright Field.[75] Despite the persistent requests of manufacturers, they could not accommodate the needs of the larger American aeronautical industry. Federal regulations prohibited propeller testing for commercial purposes because of the projected increase in the volume of testing, the need to focus on military needs, and to avoid charges of favoritism toward one company over another.[76] Savvy manufacturers simply marketed the same propeller to both the military and the civilian sectors. Upon the completion of whirl testing, they would request the propeller unit send the results to the Department of Commerce as part of their formal application for an approved type certificate.[77]

The whirl rig was also essential in the Army–inventor relationship. In a response to a would-be propeller inventor who wanted his unspecified innovation to be tested at Wright Field, Maj. Leslie MacDill, chief engineer of the Engineering Division, outlined the purpose of the Army's propeller whirl facility. The Materiel Division's testing equipment was primarily for the routine evaluation of propellers procured on government contracts. New propeller designs that the Propeller Laboratory considered to have technical merit and potential benefit to the government could be integrated into the testing program on a case-by-case basis. The facility, however,

[74] Lois E. Walker and Shelby E. Wickam, *From Huffman Prairie to the Moon: The History of Wright-Patterson Air Force Base* (Washington, DC: Government Printing Office, 1986), 159; W. E. Gillmore, in "Propeller Test Noisiest on Field."

[75] Dickey to Gazley, n.d. [March 23, 1931]; C. H. Howard to R. C. Gazley, n.d. [March 1931], File 168.7329-3, DAD, HRA.

[76] D. A. Dickey to Southern Aircraft Corporation, "Automatic Adjustable Pitch Propeller," June 9, 1930, Folder "Propellers–January through July, 1930," Box 5780, RD 3193, RG 342, NARA.

[77] Erle Martin to D. A. Dickey, January 25, 1932, Folder "Propellers–General, 1932," Box 5867, RD 3216, RG 342, NARA.

was not intended to test any and all unsolicited inventions but to focus on the needs of the American military.[78] In that focused role, the Propeller Laboratory served to foster innovation in a development environment.

The type and extent of testing required for determining the safety of one particular type of propeller was not universal. Propeller engineers believed that a ten-hour test at 50 percent overload was satisfactory for wood propellers. The new aluminum alloy propellers that had been in widespread service since the mid-1920s required ten hours at 100 percent overload. The lack of experience with steel, magnesium, or other propellers of "radically new design" meant the testing requirements were unknown as of 1931.[79]

The one constant during the period was the regulation that approval of a propeller intended for government service was based upon its testing as a complete assembly. The introduction of fillets, additional material that created a smoother physical transition between a hub and blade, or a change in stiffness along the leading edge of the blade, or any minor change in the design of an already-proven propeller, could prove to be catastrophic to the structural integrity of the propeller. The government classified a laminated wood propeller as an example of one-piece construction because any further disassembly rendered it useless. Many metal propellers consisted of several replaceable parts that led to their categorization as multipiece construction. Those components could be tested individually, but the sum of those separate tests could not accurately represent how that propeller would perform in service.[80]

Whirl testing was the key to the development of the metal multipiece propeller and all propellers that followed. This equipment helped engineers learn how to design propellers, which in the end meant that their importance became more of a matter of procedure than of learning new information about the propeller design specifically and propeller testing in general.[81] The equipment also validated their work and told them if they were on the right track.

78 L. MacDill to C. Richard, January 11, 1928, Folder "Propeller Inventions, 1928," Box 5698, RD 3173, RG 342, NARA.

79 Dickey to Gazley, n.d. [March 23, 1931].

80 Ibid.

81 United States Department of Commerce, *Trends In Airplane Design As Indicated By Approved Type Certificates, Bulletin No. 21* (Washington, DC: Government Printing Office, 1931), 14–15; Fred E. Weick, Interview by Harvey H. Lippincott, March 4, 1981, Folder "Interview: United Aircraft and Hamilton, 1929-1930," Box 3, FWP, NASM, 77; "Theoretical Course in Aerodynamics," *Air Service Information Circular* 1 (February 26, 1921): 15–16; Caldwell, "Destructive Propeller Tests," 2; Dickey Interview, 13, 16, 20–21, 36, 79.

Whirl testing was an empirical process that reflected the desire to "over-engineer" aeronautical equipment to withstand the rigors of flight. The creation of that type of testing alone was a major contribution to the development of aeronautical engineering. This success can be attributed to the Army style of engineering that reflected the American aviation industry's emphasis on immediate and practical solutions dictating the path of technological development. The bottom line was strength and safety for American military aircraft. Lauren D. Lyman, a well-known aviation columnist, noted in the *New York Times* that "Wright Field has developed methods of testing so precise that it is now possible to predict exactly the point of failure in a propeller."[82] By 1936, the Materiel Division could boast that "whereas the European props may be lighter in weight than ours, the American safety factor is much greater."[83] The necessary methods and equipment for testing propellers were in existence when World War II began.[84]

The nature of the testing was empirical, but it had to be due to the critical nature of the new propeller technology and its needed application to the airplane. The immediate requirement to produce propellers for the World War I aviation production program stimulated these innovations in propeller technology at Westinghouse and McCook Field. The knowledge gained from those experiences led to the construction of the Wright Field facility, which remained in operation as the United States government's primary propeller whirl-testing facility until 1965. The need for round-the-clock propeller whirl testing during World War II led to the only major change, the construction of massive acoustical enclosures over the test stands to protect the increased number of Wright personnel and the growing community outside the base gates. The air force operated a gas turbine compressor research and test facility at the site from 1975 to 1981. Private contractors to the US government operated the buildings and equipment, virtually unchanged since a second facility upgrade in the 1950s, on into the early twenty-first century.[85]

Caldwell's colleagues in the propeller unit went on into important careers in the propeller community. Smith succeeded Caldwell briefly as

[82] Lauren D. Lyman, "Air Corps Pioneering," *New York Times*, October 6, 1935, p. 173.

[83] T. A. Sims to Chief, Materiel Division, Liaison Section, "Controllable Pitch Propellers," September 16, 1936, Folder "Curtiss Electric Propeller, 1934–1938," Box 6302, RD 3330, RG 342, NARA.

[84] "Proving Ground for Props," *Popular Science* 144 (April 1944): 66–69.

[85] United States Air Force Aeronautical Systems Center History Office, "Historic Buildings and Sites at Wright-Patterson Air Force Base," 1989, www.ascho.wpafb.af.mil/buildings/BUILDINGS.HTM (Accessed February 28, 2003).

head of the propeller unit before taking a position in industry in 1928. Dickey headed the propeller unit from 1928 to 1957. He guided the expansion of the testing facilities during the 1940s and served as an important engineering consultant in the Army's growing inventory of wind tunnels and laboratories. For his leadership during World War II, Dickey received the prestigious Exceptional Civilian Service Commendation from Secretary of War Henry L. Stimson in a ceremony at the Pentagon in 1943.[86]

The Army style grew out of the different backgrounds of the civilian and military specialists who came to work there, and it shaped and directed future aeronautical innovation. Caldwell, Dickey, Hale, and Smith used their knowledge from the established fields of mechanical and electrical engineering to create something entirely new. The basic electrical and mechanical engineering equipment was not new, but the manner in which the propeller specialists used, arranged, and operated this equipment was: to suit the specific purpose of propeller whirl testing within a new discipline called aeronautical engineering.

The existence of a distinctive institutional style also reveals the lesser-known role of the American military in the fundamental development of the airplane. The Army's focus on development and innovation complemented and often acted independently of other organizations, primarily the NACA. The NACA's basic aerodynamic investigations of propellers benefitted from similarly creative, empirical, and pragmatic research methodologies and the requisite testing facilities that expedited the generation of data of use to the aeronautical community. The Army's need for structurally sound propellers meant that its development work needed to take place before the NACA's data could be utilized to its fullest extent.

The propellers for every significant American aircraft built during World War I, from the DH-4 "Liberty Plane" to the Naval Aircraft Factory F-5L flying boat, to the pioneering aircraft of the 1920s, including Charles Lindbergh's *Spirit of St. Louis*, would undergo testing at McCook Field. The Materiel Division at Wright Field evaluated the propellers for the successive generations of "modern" aircraft in the 1930s and the war-winning aircraft of World War II.

Overall, the establishment of permanent propeller whirl-testing equipment by Caldwell and his associates provided the infrastructure for the reinvention of the propeller during the Aeronautical Revolution. Its development illustrates how the US government fostered innovation

86 Dickey Interview, ix–x.

in aeronautical technology, especially the synergistic advances that increased the overall performance of the airplane. The aviation industry did not address the problem of propeller development until after almost a decade of government-sponsored development in the field provided the foundation on which to build. This necessary infrastructure, the result of "engineering of a pioneer character," was at the cornerstone of the development of the modern propeller.

4

A "New Type Adjustable-Pitch Propeller"

The first variable-pitch propeller developed in the United States took to the air over southern California as America entered World War I in April 1917. Pioneer aviator Earl S. Daugherty flew his Daugherty-Stupar Tractor biplane with a variable-pitch propeller designed by Los Angeles inventors Seth Hart and Robert I. Eustis (Figure 7). Starting off with a low pitch setting for high thrust, he gradually increased the pitch of the propeller to generate more speed as the plane leveled off over the airfield. Daugherty was "very much impressed" with the propeller's effect on his airplane's performance. The new propeller was controlled manually by the pilot, meaning that it relied on a series of mechanical linkages and the strength of the pilot to change the pitch.[1] Daugherty's flights were the American propeller community's first steps toward developing a practical variable-pitch mechanism. That work, combined with the efforts of other experimenters in North America and Europe, met varying degrees of success between 1917 and 1927.

World War I served as a catalyst for the reinvention of the airplane in Europe and the United States, but the path and direction that new technology would take, as well as the nature of the organizations that used it, were not clear. In the United States, the newly created Army Air Service was at a crossroads regarding the primary role of military aviation in the postwar period. A tension existed between the old Army leadership, who favored using airplanes in a supplementary tactical role, and the youthful

1 "Tests of an Adjustable Pitch Propeller," *Aerial Age Weekly* 5 (June 4, 1917): 389; "Eustis Adjustable Aeroplane Propeller," September 25, 1917, Folder "Eustis Adjustable Airplane Propeller, 1917," Box 12, RD 3085, RG 342, NARA.

FIGURE 7 Earl Daugherty with Seth Hart and Robert Eustis propeller installed on his biplane at Long Beach during the spring of 1917. National Archives and Records Administration.

Air Service officers who sought to create an entirely new military doctrine based on strategic bombing. As this internal battle played out, the Army Air Service's strategy was to focus on wide-ranging improvements that enhanced the speed, range, load, and maneuverability of observation airplanes, bombers, fighters, transports, trainers, even airships.[2] One of the key technologies for the service's future airplane – no matter what its role – would be the propeller.

The aeronautical community was aware of the potential value of controllable and reversible variable-pitch propellers. Controllable-pitch mechanisms offered enhanced operating performance and fuel economy at the different operating regimes of takeoff and cruise for both single- and multiengine aircraft. The introduction of a practical design

[2] Irving Brinton Holley, Jr., *Ideas and Weapons: Exploitation of the Aerial Weapon by the United States During World War I* (New Haven, CT: Yale University Press, 1953; reprint, Washington, DC: Office of Air Force History, 1983), 82–102, 147–178; Timothy Moy, *War Machines: Transforming Technologies in the U.S. Military, 1920–1940* (College Station, TX: Texas A&M University Press, 2001), 42.

would enhance operations from remote airfields, rough and open seas, and crowded aircraft carrier decks. A propeller that allowed instant pitch adjustment for maximum takeoff efficiency would also allow military aircraft to operate out of heavily damaged and dangerous landing fields.

The most immediate advantage the aeronautical community saw for the variable-pitch propeller was for supercharged engines at high altitudes.[3] By adjusting the propeller's pitch to take a bigger "bite" in the thin, high-altitude atmosphere, the new device allowed a supercharged engine to develop its full power at cruising altitude by helping to reproduce sea-level flying conditions. At the International Air Conference in Brussels in 1925, Whiston A. Bristow, a representative of a British propeller company, stated, "There is every indication that the military aircraft engine of the future will be provided with some type of supercharger necessitating a variable-pitch propeller. This will rule out entirely any propeller of the integral [or fixed] type whether of wood or metal, and the ultimate propeller must have separate metal blades."[4] To achieve that, propeller engineers in the United Kingdom and the United States focused on designing a new multipiece propeller that consisted of separate blades joined to a central hub to facilitate pitch variation.

A reversible propeller, which could change pitch to the extent that it created thrust in the opposite direction of flight even as the propeller continued to spin in its normal direction of rotation, would act as a brake during landing. This feature could help pilots avoid accidents and allow airplanes to operate out of smaller airfields. It also promised to benefit airships. A propeller that provided forward and reverse thrust would enhance the maneuverability of unwieldy lighter-than-air aircraft during takeoff and landing operations. Not only could this improve safety, but it lessened the need for a large number of ground personnel. Those values alone for the two different types of aircraft were reason enough to pursue developing a practical reversible mechanism.

The First Step toward Reinvention

The Hart and Eustis propeller resulted from the work of two individuals from Los Angeles, California. Robert I. Eustis was a career woodworker

[3] David L. Bacon, trans., "Variable-Pitch Propellers," *L'Aeronautique* (September 30, 1920), NACA Technical Memorandum No. 2 (November 1920), 1.

[4] Whiston A. Bristow, "The Design and Construction of Metal Propellers for Aircraft," Paper No. 530, *III° Congrès International de Navigation Aérienne* (Paris: Chiron, 1926), 1.

who found employment at the Glenn L. Martin Company as a propeller maker. While constructing wood, fixed-pitch propellers, he devised an idea for a controllable and reversible propeller in April 1916. Eustis presented his concept to Seth Hart, a successful entrepreneur known for his interest and desire to develop new inventions.[5] After entering the business world in 1898, Hart helped pioneer the sale of insurance in southern California. His mechanical aptitude, abundant source of income, and connections within the southern California business community allowed him to support new inventions both technically and financially.[6]

Eustis finished making the propeller in early 1917. It weighed forty-five pounds and the two blades measured eight-foot two-inches in diameter. The hollow cast aluminum alloy hub was eighteen inches long and four inches in diameter. The key to the design was the use of a "sliding member" that moved operating pins within the hub in a direction parallel to the propeller's axis of rotation. These pins transformed the sliding member's linear movement into the angular motion necessary for a pitch change. From the cockpit, the pilot actuated a push-pull rod connected to a slip ring that in turn moved the sliding members that connected to levers integrated into the root of the propeller blades. The propeller had a pitch range of twelve degrees. Eustis contended that his pitch control mechanism was as delicate and precise in operation as a micrometer.[7]

An innovative feature of the design was the method of attaching the wood blades to the metal hub. The laminated birch blades were set in tapered three and one-eighth-inch diameter metal ferrules. Eustis pressed them onto the end of a blade, and then drove circular steel wedges into the wooden end to make it expand outward and conform to the taper of the ferrule. He drilled a hole directly through the center of the ferrule and the blade and inserted a steel sleeve that contained the operating pin. A thrust collar screwed on the end of the hub held the blades in position.

[5] "Up Nearly a Mile in Two Minutes: New Aircraft Propeller is Highly Praised," *Los Angeles Times*, August 5, 1917, section 2, p. 17; "Adjustable and Reversible Propeller: Robert Eustis' Invention Nears Perfection," *The Ace* 1 (May 1920): 12, 24; "Airship Device Made Here," *Los Angeles Times*, January 27, 1924, p. A12.

[6] "Death Calls Veteran Insurance Man," *Los Angeles Times*, February 18, 1938, section 2, p. 20; Genealogical Data on Seth Hart compiled by Douglas B. Ayer and provided by Frank B. Estabrook, March 2002.

[7] "Description of Mechanism," September 25, 1917, Folder "Eustis Adjustable Airplane Propeller, 1917," Box 12, RD 3085, RG 342, NARA; Walter F. McGinty and Glen T. Lampton, "Characteristics of the Hart Adjustable and Reversible-Pitch Airplane Propeller" (B. S. thesis, University of California, 1923), 2–3; Alvin E. Moore, "The Screw Propeller," *Journal of the Patent Office Society* (December 1941), File B6-000010-01, PTF, NASM.

Ball bearings placed between the thrust collar and the ferrule absorbed the centrifugal load of the spinning propeller and allowed the blades to rotate in flight.[8]

Interested in proving the practicality of the basic design, Eustis first tested his propeller on a sixty-horsepower Hall-Scott engine mounted on a test stand. The propeller yielded an impressive amount of thrust at different pitch settings, which encouraged the inventor to attempt flight tests. Eustis solicited Earl S. Daugherty, an experienced pilot flying out of Long Beach, California, to assist him in his continued experimentation. Daugherty was a pioneer aviator, exhibition flyer, and flight instructor in southern California that had learned to fly at Dominguez Field in June 1911 and was one of the most experienced flyers in the country during the spring of 1917.[9]

From April to August 1917, Daugherty conducted flight tests of the Hart and Eustis propeller at Long Beach with his Daugherty-Stupar Tractor biplane. E. L. Graham, a government inspector assigned to the nearby Wright-Martin aircraft factory in Los Angeles, was one of a number of aeronautical "experts" present.[10] The flight tests yielded impressive results. The *Los Angeles Times* reported on one particular flight in early August 1917:

> [Daugherty's] plane left the ground like a bird in the relatively short distance of forty-three feet. Then, without increasing the speed of the motor, the angle of the blades was changed after an elevation of 6,000 feet had been reached. The speed of the plane immediately jumped from sixty to ninety miles an hour. With increased motor speeds, the new propeller showed an even relatively increased efficiency.

Eustis' propeller increased the biplane's rate of climb by 25 percent to a rate of 2,000 feet per minute.[11] Impressed, Daugherty believed that with a lower pitch setting, he could take off within twenty-five feet by increasing the speed of the engine. Eustis and Daugherty equated the new propeller's value to the airplane in the same way a multigear transmission was important to the smooth operation of an automobile. Using a propeller capable of "starting on low, getting up speed on intermediate, and

[8] McGinty and Lampton, "Characteristics of the Hart Adjustable and Reversible-Pitch Airplane Propeller," 2–4.

[9] "Eustis Adjustable Aeroplane Propeller," September 25, 1917; McGinty and Lampton, "Characteristics of the Hart Adjustable and Reversible-Pitch Airplane Propeller," 3; "Earl S. Daugherty," File CD-045500-01, BF, NASM; "Earl S. Daugherty," *Who's Who in American Aeronautics* (New York: Gardner Publishing Company, 1925), 37.

[10] "Tests of an Adjustable Pitch Propeller," 389.

[11] "Up Nearly a Mile in Two Minutes."

running on high gear" maximized overall airplane performance. Those first flight tests ended when the Army detailed Daugherty to Rockwell Field on North Island in San Diego Bay to serve as a civilian flight instructor during the late summer of 1917.[12]

The new propeller attracted the attention of more Los Angeles entrepreneurs in addition to Seth Hart. They were eager to sell the new device to two primary markets, the United States government and the emergent aviation industry, in the interests of both business and patriotism following America's April 1917 entry into World War I. The American government had committed itself to building an aerial armada that would wage war against the Central Powers and create an aviation industry in the process. Entrepreneurs, hoping to get in on the ground floor of a stunning new business opportunity, searched for new technologies that would help win the war and make them rich in the process.[13] Eustis and Hart filed their patent application for an "adjustable propeller" on September 10, 1917.[14]

Anticipating an opportunity to corner the market for a new and revolutionary invention that could possibly influence the outcome of the war, Eustis, Hart, and others formed a partnership through the Los Angeles Chamber of Commerce. Their representative, stockbroker G. Ray Boggs, sent the Army a package containing a textual description, photographs, and Daugherty's report on the flight tests of the propeller in September 1917. Boggs was anxious to hear if there was interest in the propeller and suggested an official test of the propeller at San Diego.[15]

Interested in new aeronautical innovations, the Army brought an improved Hart and Eustis controllable and reversible propeller to Rockwell Field on January 1, 1918. Eustis fabricated a second hub and ferrules using seamless steel tubing, which increased the weight of the propeller to fifty-five pounds. It was this heavier, but stronger, hub that underwent testing at North Island.[16]

Personnel at Rockwell Field installed the Hart and Eustis propeller on a Curtiss OX-5 engine mounted in a Curtiss JN-4 biplane. Lt. Robert

[12] "Eustis Adjustable Aeroplane Propeller," September 25, 1917.

[13] "New Airplane Propeller Our War Contribution," *Los Angeles Times*, March 28, 1918, section 2, p. 1; "Airship Device Made Here."

[14] S. Hart and R. I. Eustis, "Adjustable Propeller," U.S. Patent No. 1,301,052, April 5, 1919.

[15] G. Ray Boggs to A. J. Farwell, September 25, 1917, Folder "Eustis Adjustable Airplane Propeller, 1917," Box 12, RD 3085, RG 342, NARA.

[16] "Lieutenant Mairesse's' Report on the Propeller with Adjustable-Pitch," n.d. [February 1918], File D52.43/93, WFTD, NASM; McGinty and Lampton, "Characteristics of the Hart Adjustable and Reversible-Pitch Airplane Propeller," 4.

Mairesse, an Aéronautique Militaire officer assigned to the French Aviation Mission to the United States, was the primary pilot for the tests. Mairesse's approximately four hours in the air corroborated Daugherty's test report. The lieutenant believed further experimentation and improvements would correct the inherent design flaws, which centered on the "difficult adjustment of the propeller" and its weight. Drawings and a model of the propeller traveled to Washington, DC, with one of the other French aviation officers around February 1918.[17]

Impressed with Mairesse's test reports, the Army invited Seth Hart and Carl E. McStay to McCook Field in March 1918. They submitted photographs, drawings, and a complete propeller for evaluation. McStay's interview and subsequent negotiations with Colonel Vincent and other Air Service officers revealed the Army's belief that the Hart and Eustis propeller possessed merit and warranted development "from a practical standpoint."[18]

The Army sent another Hart and Eustis propeller to the whirl rig at Westinghouse for testing beginning on March 30, 1918. Representatives of the British, French, and Italian air services were present to observe the tests. Inspection after the ten-hour whirl test revealed that the blades had separated from the ferrules and there was excessive wear in the ball bearing races in the blade roots. Despite those particular design flaws, the Army's specialists recommended that the Engineering Division design and construct a variable-pitch propeller based on the Hart and Eustis design.[19]

Flight tests of what the Army called the "new type adjustable-pitch propeller" began at McCook Field on May 1, 1918. Caldwell and his staff bored out the hub of the propeller flown in San Diego and fabricated new blades for installation on a 150-horsepower Hispano-Suiza engine mounted in a Curtiss JN-4H biplane. The tests compared the performance of the new design with a standard fixed-pitch propeller of the

[17] McGinty and Lampton, "Characteristics of the Hart Adjustable and Reversible-Pitch Airplane Propeller," 4; Henry H. Arnold, *The History of Rockwell Field*, n.p., 1923, HRA; "Lieutenant Mairesse's' Report on the Propeller with Adjustable-Pitch," n.d. [February 1918].

[18] R. H. Young to J. G. Vincent, February 17, 1921; R. H. Young to H. O. Marmon, February 19, 1921, Folder "Propellers–Hart, 1921," Box 77, RD 3102, RG 342, NARA.

[19] "New Airplane Propeller Our War Contribution"; W. A. Hale to Airplane Engineering Department, "Report on Destructive Whirling Test of Hart and Eustis Adjustable-Pitch Propeller," April 5, 1918, File B5-442000-01, PTF, NASM; McGinty and Lampton, "Characteristics of the Hart Adjustable and Reversible-Pitch Airplane Propeller," 5.

same diameter and blade shape and illustrated the performance benefits of variable-pitch propellers during climb and cruise.[20]

The encouraging tests results gained from the whirl and flight tests of the Hart and Eustis propeller led to a new government research and development program. Caldwell and his colleagues began work on a controllable and reversible propeller series designed for the Air Service's primary aeronautical power plants, the Hispano-Suiza 150-, 180-, and 300-horsepower V-8 engines, built by the Wright-Martin Company in New Brunswick, New Jersey, and the 400-horsepower Liberty V-12 engine manufactured by various automobile manufacturers.[21] Air Service aircraft powered by the Wright-Martin engines included the Curtiss JN-4H and JN-6H series and the Vought VE-7. The primary airframe applications for the Liberty were the DH-4, Engineering Division USD-9A, and the Packard Le Peré LUSAC-11.[22] Vincent gauged the importance of the new technology to the Air Service by stating to the propeller unit, "Let's rush the job."[23]

Caldwell believed it was necessary to support development of the new variable-pitch propeller for two primary reasons. First, the propeller required a long, thorough, and costly development period before it could be approved for service use. The only appropriate organization in the United States to supervise the program was the propeller unit at McCook Field. Second, because the program was "entirely a propeller research matter," it was difficult to conduct the required work without government facilities. The only "outside commercial facilities" – meaning the whirl-testing rig at Westinghouse – that Hart could rely upon was already inundated with wartime production work evaluating fixed-pitch propellers, and was not available if he did not have government sponsorship.[24]

[20] F. W. Caldwell to T. H. Bane, "Outline of History of Development of Hart Propeller," January 21, 1921, Folder "Propellers–Hart, 1921," Box 77, RD 3102, RG 342, NARA; McGinty and Lampton, "Characteristics of the Hart Adjustable and Reversible-Pitch Airplane Propeller," 5–6, 8; "New Type Adjustable-Pitch Propeller," *Bulletin of the Experimental Department, Airplane Engineering Division, United States Army* 1 (June 1918): 78–79.

[21] Caldwell to Bane, "Outline of History of Development of Hart Propeller," January 21, 1921.

[22] James C. Fahey, *U.S. Army Aircraft, 1908–1946* (New York: Ships and Aircraft, 1946), 8, 9.

[23] J. G. Vincent to O. H. Skinner, "Variable-Pitch Propeller of Messrs. Hart and Eustis," May 25, 1918, File D52.43/93, WFTD, NASM.

[24] Caldwell to Bane, "Outline of History of Development of Hart Propeller," January 21, 1921.

The new program initiated a close relationship between the propeller unit, Seth Hart, and his partners from the Los Angeles Chamber of Commerce. In Hart, Caldwell wanted "to get the services of a high class man with a great deal [at] stake in the invention." He convinced the Air Service to employ Hart as a civilian researcher in August 1918 with a small salary to defray some of his living costs.[25] Combining knowledge of the Army's interest with "patriotic motives," Hart and his associates incorporated the Adjustable and Reversible Propeller Corporation in Delaware during the fall of 1918 to control the patent rights and manufacture their self-styled "revolutionary invention," which received the abbreviated name of Hart propeller.[26]

By 1920, Caldwell and Hart had propellers ready for evaluation by operational flying units. Hearing that the propellers had "progressed beyond the experimental stage," Col. Henry H. Arnold, the Air Service representative for the Army's western department, recommended that at least one example be sent to March Field, Riverside, California, in March for three reasons. First, he believed field tests under actual operating conditions were crucial to understanding the capabilities of the technology. Second, tests outside of Dayton would promote communication between the experimental and operational components of the Air Service regarding the value of new technologies. Finally, demonstrations of a new innovation invented in southern California would "secure for the Air Service a large number of additional boosters." Arnold's recognition of the propeller's marketing value to the image of the Air Service was intentional.[27]

In response to Arnold's suggestion, Brig. Gen. William "Billy" Mitchell of the Office of the Chief of Air Service ordered the fabrication and distribution of thirty propellers in April. Ten propellers for the 150-horsepower Hispano-Suiza engine in JN-4H and JN-6H training aircraft were to go to McCook Field, March Field, and Carlstrom Field, Florida. The Air Service installed Hart propellers on training aircraft for the sole purpose

[25] Ibid.

[26] T. H. Bane to W. E. Gillmore, December 30, 1920, Folder "Reversible Propellers, 1920," Box 52, RD 3096, RG 342, NARA; C.E. McStay to COAS, "Proposal for Sale, Non-Exclusive Rights, Hart-Eustis Adjustable and Reversible Aircraft Propeller," January 8, 1921, Folder "Propellers–Hart, 1921," Box 77, RD 3102, RG 342, NARA; McGinty and Lampton, "Characteristics of the Hart Adjustable and Reversible-Pitch Airplane Propeller," 7.

[27] H. H. Arnold to the Director of Air Service, "Reversible-Pitch Propeller," March 3, 1920, Folder "Reversible Propellers, 1920," Box 52, RD 3096, RG 342, NARA.

of acquiring information because they provided a stable and safe platform to conduct evaluation flights.[28]

To make sure the test program succeeded, Seth Hart traveled to March Field in June 1920. While there, he took the opportunity to showcase the propeller to the public in late August. The Air Service sent five aircraft from March Field, including two JN-4Hs equipped with Hart propellers, to fly over Los Angeles. The seventy-mile flight was the first cross-country flight made with the propellers. They landed at Cecil B. DeMille's Mercury Aviation Field. There, the Air Service pilots demonstrated the propeller's abilities by stopping seventy-five feet after their wheels touched the ground.[29] The noticeable change in the sound of the engines as the Jennies neared the ground, the unprecedented short landing, and the great cloud of dust created by the propellers encouraged the newsreel cameramen to shoot "a great many feet of film."[30]

Air Service pilots at McCook, March, and Carlstrom fields learned to fly with the Hart propeller during 1920. They obtained the best results in landing by gliding at normal, or forward, pitch until about ten feet from the ground. From there, the pilot gradually reduced pitch to a zero degree angle to slow the airplane down. As the airplane approached an altitude of three feet, the pilot applied full reverse pitch. Even at a high landing speed, the effect of the reversible propeller brought the tail down and facilitated a good three-point landing. The procedure was a precarious and difficult maneuver. If the pilot waited to apply full reverse at two feet above the ground, the time it took the engine and propeller to react would be too late and the airplane would hit the ground hard and bounce into the air. Once on the ground, the pilot had to anticipate the direction the airplane would take in full reverse to prevent it from swerving around into a ground loop. The pilots and maintenance personnel learned that landing with a reversible-pitch propeller was hard on shock absorbers and landing gear as well.[31]

[28] W. Mitchell to Chief, Air Service Supply Group, April 1, 1920, Folder "Reversible Propellers, 1920," Box 52, RD 3096, RG 342, NARA; L. W. McIntosh to COAS, "Reversible-Pitch Propellers," November 15, 1921, Folder "Propellers–Hart, 1921," Box 77, RD 3102, RG 342, NARA.

[29] T. H. Bane to Commanding Officer (hereafter cited as CO), March Field, June 9, 1920, Folder "Reversible Propellers, 1920," Box 52, RD 3096, RG 342, NARA; "Variable-Pitch Prop Is Tested," *Los Angeles Times*, September 5, 1920, section 6, p. 8.

[30] S. Hart to T. H. Bane, September 4, 1920, Folder "Reversible Propellers, 1920," Box 52, RD 3096, RG 342, NARA.

[31] H. Brand, "Report on the Performance of Hart Variable-Pitch Propeller," November 25, 1920, Folder "Reversible Propellers, 1920," Box 52, RD 3096, RG 342, NARA.

Continued flight tests of the Hart propeller made the Air Service question the true practicality of reversible pitch. Flight evaluations by test pilot Lt. John A. Macready at McCook in November 1920 revealed that the thrust of the reversed propeller forced the nose of the airplane down at a dangerous angle, which could turn the airplane end over end or somersault onto its back. Unfortunately, fellow pilot, Lt. Muir S. Fairchild, confirmed that belief when he reversed the propeller at about 100 feet and crashed into the nearby Great Miami River. Fairchild was lucky to escape with a broken arm and ribs.[32]

The test pilots experienced difficulty in operating the propeller controls. In order to operate the gear wheel and change pitch, the pilot had to combine the exertion of fifty pounds of physical force with the same delicate "feel" and concentration needed on the airplane's control stick. The pilots could not turn the wheel fast enough for the landing run, with the result that there were more than a few bad landings. The reverse mechanism needed to operate as easily as the engine spark and throttle controls before the propeller would be of any practical value. As for using the propeller as a controllable-pitch propeller in forward flight, one test pilot noted that it was easy to reduce the pitch in flight, but it could not be increased if the engine was running faster than 500 rpm.[33] The limitations of mechanical actuation were becoming clear.

The installation of the propellers on the JN aircraft required considerable modifications. It was not a simple matter of bolting a Hart reversible propeller to the crankshaft of the Hispano-Suiza engine. The propeller had to be part of an integrated propulsion system. To provide clearance for the control linkage from the cockpit to the propeller hub, technical personnel had to remove a considerable area of the radiator surface. As a result, the aircraft engines suffered from overheating during the late summer of 1920. Hart went so far as to drain the emergency gas tank in

[32] Frank W. Caldwell, "American Progress in Adjustable-Pitch Propellers," in *International Air Congress, London, 1923*, ed., W. Lockwood Marsh (London: Royal Aeronautical Society, 1923), 588, 594; H. R. Harris to CO, McCook Field, "Test of Reversible Propeller," November 13, 1920, Folder "Reversible Propellers, 1920," Box 52, RD 3096, RG 342, NARA; Maj. Gen. Howard C. Davidson, Interview by Brig. Gen. George W. Goddard, August 20, 1966, interview K239.0512-996 C.1, transcript, United States Air Force Oral History Program, United States Air Force Historical Research Agency, 10.

[33] B. K. Yount to COAS, "Report on Reversible-Pitch Propeller," October 28, 1920; F. A. Johnson to COAS, "Hart Reversible Propeller," November 2, 1920; R. Royce to F. A. Johnson, November 4, 1920, Folder "Propellers–Hart, 1921," Box 77, RD 3102, RG 342, NARA. Brand, "Report on the Performance of Hart Variable-Pitch Propeller," November 25, 1920; C. R. Melin to CO, March Field, "Report on Reversible-Pitch Propellers," October 27, 1920, Folder "Propellers–Hart, 1921," Box 77, RD 3102, RG 342, NARA.

the upper wing to use as an additional water tank to alleviate the problem, but it was not until the advent of cooler weather in October that the overheating stopped.[34]

The operation of the Hart propeller was potentially destructive to the operation of the engine. The transition from forward to reverse pitch meant that, for a split second, the propeller was not producing effective thrust, which meant engine speed increased at a rapid and dangerous rate. The jump to maximum revolutions caused engine valves to "blow," or explode when engine rpm became so high that it disrupted the intake and ignition cycles. Intake valves could not adequately deliver the vaporized air and gas mixture to the cylinder and raw gasoline would collect in the area of the valve and valve seat. At the moment of ignition, the raw gasoline combusted and destroyed the valve and valve seat. The transition from forward to reverse pitch also required a more responsive carburetor. Air Service personnel found that the current Zenith model lagged and choked the motor when the propeller was set from forward to reverse. Mechanics at March Field installed a more consistent Stromberg model to alleviate the problem.[35]

The service tests revealed what it would be like to operate and maintain variable-pitch propellers in the field. Air Service mechanics learned that "great care" had to be taken in preventing dirt from getting into the exposed Hart mechanism and causing wear. Used to wiping down wood propellers with cloth rags, mechanics found that they had to wash the new propeller in gasoline and then coat the exposed parts with lightweight oil to prevent rust, which attracted a large amount of dirt. The oil and centrifugal force was supposed to carry the dirt away, but in reality, the practice required additional routine cleaning. The use of variable-pitch propellers meant that mechanics had more maintenance duties to perform in the field.[36]

Despite the problems that arose during the tests, the Air Service pilots were still enthusiastic for variable-pitch propellers. Two JN-6H biplanes had flown for a total of thirty-five hours and made approximately fifty

[34] T. H. Bane to Commandant of March Field, "Hart Reversible Propellers," July 1, 1920; T. H. Bane to Commandant of Carlstrom Field, "Hart Reversible Propellers," July 2, 1920; Hart to Bane, August 15, 1920, Folder "Reversible Propellers, 1920," Box 52, RD 3096, RG 342, NARA; Brand, "Report on the Performance of Hart Variable-Pitch Propeller," November 25, 1920.

[35] Brand, "Report on the Performance of Hart Variable-Pitch Propeller," November 25, 1920.

[36] Brand, "Report on the Performance of Hart Variable-Pitch Propeller," November 25, 1920; "The '1920' Model Hart Reversible Propeller Manual and Parts List," *Air Service Information Circular*, no. 239 (July 15, 1921).

landings at March Field by October. One of the aircraft flew a sixty-mile cross-country flight while landing and taking off in small fields along the way.[37] The personnel at March Field were especially praiseworthy of the Hart propeller. Second Lt. Harold Brand, an instructor and one of the test pilots, exclaimed, "Personally, I am very enthusiastic over the performance of these propellers and look forward to the time of their adoption on service-type ships." Maj. Barton K. Yount, the commanding officer, recognized that "the idea is excellent, and the mechanical construction of the propeller appears to be sound and satisfactory."[38]

The aeronautical community took note of the performance benefits offered by the Hart propeller during the flight test programs of 1919 and 1920. Maj. Rudolph W. "Rudy" Schroeder, the chief test pilot at McCook Field who became "Uncle Sam's highest flier" by flying to a record 32,000 feet, estimated that a high-altitude airplane equipped with the propeller could possibly reach 50,000 feet. The *New York Times* proclaimed that with its introduction, the "dreams of an airplane that could take off and land in a space the size of an ordinary backyard virtually were realized."[39] Second Assistant Postmaster General Otto Praeger was busy building a transcontinental aerial network through the US Air Mail Service. He believed that with the new device, mail planes could operate from the roofs of Post Office buildings or landing fields that were smaller than regular airfields and closer to urban areas. In Praeger's view, McCook Field's work on variable-pitch propellers and supercharged engines were the "greatest recent contributions to commercial aeronautics."[40] *The Ace* described the tests as an "unqualified success" that promised unprecedented performance during takeoff and cruise over the old fixed-pitch, or "stationary blade." Speaking before the Aero Club of New York City, Gen. "Billy" Mitchell stressed that it enabled airplanes to achieve unprecedented altitudes and speeds up to 400 mph when used in conjunction with a supercharged engine. As a result, an aerial crossing of the Atlantic Ocean could be achieved in six to seven hours.[41]

As the flight test programs progressed during the summer of 1920, Col. Thurman H. Bane, the Chief of the Engineering Division, advocated

37 Melin to CO, March Field, "Report on Reversible-Pitch Propellers," October 27, 1920.
38 Brand, "Report on the Performance of Hart Variable-Pitch Propeller," November 25, 1920; Yount to COAS, "Report on Reversible-Pitch Propeller," October 28, 1920.
39 "New Propeller Stops Airplane in 50 Feet," *New York Times*, October 30, 1919, n.p.
40 O. Praeger to T. H. Bane, March 29, 1920, Folder "Reversible Propellers, 1920," Box 52, RD 3096, RG 342, NARA.
41 "Adjustable and Reversible Propeller: Robert Eustis' Invention Nears Perfection," 12, 24.

outright production of the Hart variable-pitch propeller.[42] He related to Hart:

> I don't like to push the issue, but I am very much in favor of putting two or three hundred of these propellers into production at once and supplying them to all fields. I rather think at times that some of us are inclined to be a little too conservative, myself in particular. Of course, if we want to continue the engineering of this propeller, we would eventually get it perfect, but by that time the airplane may be abandoned for some other more rapid means of transportation.

Bane encouraged Hart to contact Maj. Rueben H. Fleet, the soon-to-be Air Service Chief of Procurement, to persuade him to initiate the process of getting variable-pitch propellers into regular service.[43]

The partnership that began in May 1918 between Hart and the propeller unit resulted in a variable-pitch propeller developed under direct sponsorship by the US government. The design featured wood blades and was intended for use on a 150-horsepower Hispano-Suiza engine installed in JN-4H training airplane. The Air Service called it the "1920 Model Hart Reversible Propeller" and issued a manual to each aircraft crew chief assigned to an airplane equipped with the new propeller (Figure 8). The voluminous instruction book included over forty photographs illustrating the parts, installation, methods, and operation of the propeller that the crew chief had to learn in addition to his regular flight line duties. Up to December 1920, the Army had spent over $150,000 in the propeller's development program.[44]

By January 1921, the Army had received approximately fifty Hart propellers.[45] Many of those propellers were in general use at the various Air Service flying fields. The Engineering Division notified the Adjustable and Reversible Propeller Corporation in late 1920 that it was "the intention of the Air Service to acquire the rights for and to put into general use" the Hart propeller. By October 1921, Lt. Col. James E. Fechet, Chief of

[42] T. H. Bane to S. Hart, August 24, 1920, Folder "Reversible Propellers, 1920," Box 52, RD 3096, RG 342, NARA.

[43] T. H. Bane to S. Hart, September 9, 1920, Folder "Reversible Propellers, 1920," Box 52, RD 3096, RG 342, NARA.

[44] "The '1920' Model Hart Reversible Propeller Manual and Parts List," 5; F. W. Caldwell to Seth Hart, June 24, 1918, Box 12, Folder "Propellers–Variable Pitch, 1918–1919," RD 3085, RG 342, NARA; Caldwell, "American Progress in Adjustable-Pitch Propellers," 578–582; T. H. Bane to W. E. Gillmore, December 30, 1920, Folder "Reversible Propellers, 1920," Box 52, RD 3096, RG 342, NARA.

[45] Examples of those propellers are now in the collections of the Smithsonian National Air and Space Museum and the National Museum of the US Air Force.

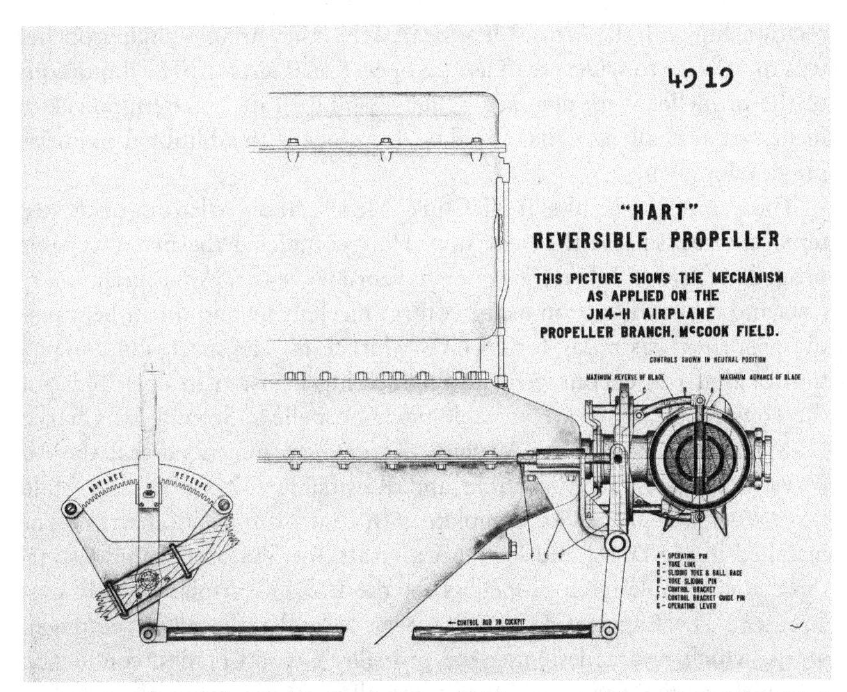

FIGURE 8 The US Army Air Service's 1920 Hart reversible and variable-pitch propeller proved to be a dead end due to its mechanical design. Smithsonian National Air and Space Museum (NASM WF-4919).

the Air Service's training and operations group, urged the Engineering Division to "continue to devote attention to this propeller" due to its value to both supercharged aircraft and the small field operations during wartime conditions.[46]

The corporation had three specific demands before that could happen. First, the government would pay $175,000 for a license that would allow the manufacture, contracting, use, and sale of the Hart propeller by the government. Second, the Adjustable and Reversible Propeller Corporation also desired the privilege of referring authorized representatives of "friendly foreign powers" to McCook Field for detailed information about the propeller. Finally, taking into account the government's support to that point, the corporation wanted to retain its cooperative

[46] J. E. Fechet to Chief, Engineering Division, "Report on Reversible-Pitch Propellers," October 26, 1921, Folder "Propellers–Hart, 1921," Box 77, RD 3102, RG 342, NARA.

relationship with the Army.[47] It appeared that the variable-pitch propeller was on its way to widespread use on operational aircraft. The limitations of the propeller were not seen as detrimental to its long-term development, but as challenges that could be overcome with additional engineering development.

The promising results at McCook, March, and Carlstrom fields fostered enthusiasm for the next step. Hart completed the first reversible propeller for the Liberty engine in February 1921. After preliminary tests and the modification of the control mechanism and thrust bearings, the propeller was ready for its final whirl tests. The successful completion of that test encouraged the Engineering Division to open bids for the construction of Hart variable-pitch propellers. Second Lt. Charles N. Monteith, Chief of the Airplane Section, recommended that the Air Service contract the Adjustable and Reversible Corporation to build three reversible propellers, complete with controls for the Liberty engine installed in the DH-4 and USD-9A aircraft, for $24,000. An additional three controllable-pitch propellers for the USD-9A airplane would cost $12,000. The Engineering Division was to undertake all development work, which meant designing the propeller's wood blades, conducting all whirl and flight tests, and furnishing all necessary materials including shell stock, ball bearing steel, and blade retention ferrules.[48]

The intensive 1920–1921 development program led to a lull during 1922–1923. Unfortunately, the technical flaws in the Hart design could not be overcome. As engines and airplanes increased in size, the system of mechanical actuation proved woefully inadequate. Within the performance regime offered by smaller aircraft with 200-horsepower or less, the Hart propeller proved to be a promising design. When designed, built, and installed for a high-horsepower engine, however, the mechanical difficulties of adapting the basic design to the new performance environment created immense challenges. In the long run, continued difficulties with friction in the hub caused by undersized ball-bearings, the inability to synchronize the propeller with the engine, and an inadequate blade retention system ensured the ultimate demise of the Hart propeller.[49]

[47] C. E. McStay to COAS, "Proposal for Sale, Non-Exclusive Rights, Hart-Eustis Adjustable and Reversible Aircraft Propeller," January 8, 1921, Folder "Propellers–Hart, 1921," Box 77, RD 3102, RG 342, NARA.

[48] C. N. Monteith to T.H. Bane, "Reversible Propeller for Liberty," February 23, 1921; C. N. Monteith to Contract Section, "Reversible and Variable-Pitch Propellers," February 28, 1921, Folder "Propellers–Hart, 1921," Box 77, RD 3102, RG 342, NARA.

[49] Caldwell, "American Progress in Adjustable-Pitch Propellers," 578–582.

Despite the failure of the program, Caldwell believed the original Hart and Eustis design was "a valuable step in the progress of propeller development."[50] He and his colleagues worked with Seth Hart to solve its problems through successive designs without discarding it entirely. Caldwell credited Hart with performing a "good deal of the pioneering work" in variable-pitch propellers.[51]

Collaboration with Industry

The Adjustable and Reversible Propeller Corporation collaborated with the American military in the creation of a variable-pitch propeller. The Engineering Division looked to other private concerns for the simultaneous development of variable-pitch propeller designs that were either original or based on the basic Hart mechanism. The earliest of these projects involved the Army inviting pioneer Spencer Heath of the American Propeller and Manufacturing Company of Baltimore to design a variable-pitch propeller in 1919.[52]

Heath's controllable and reversible propeller, called the Universal and designed for the 150-horsepower Hispano-Suiza engine, appeared in early 1922. According to Heath, his propeller possessed five unique features. First, the hub used small brake drums driving through an elaborate chain of gears to control each blade, which eliminated continuously running gears, collars, or bearings in the pitch control mechanism. Second, the propeller used energy from the engine's rotating crankshaft rather than the physical strength of the pilot to change the propeller's pitch. Third, the mechanism allowed an unrestricted range of blade adjustment between two predetermined pitch positions. The blades could rotate a full 360 degrees, which enabled full reversible operation. Fourth, Heath incorporated a blade angle indicator that could be mounted in the cockpit for the benefit of the pilot. Finally, this design allowed automatic throttling of the engine while the propeller passed through the neutral pitch position, compensating for the sudden decrease in thrust.[53]

50 F. W. Caldwell to R. I. Eustis, May 9, 1925, Folder "Hart Reversible, 1925," Box 224, RD 3143, RG 342, NARA.

51 Caldwell, "American Progress in Adjustable-Pitch Propellers," 580.

52 S. Heath to T. H. Bane, May 21, 1919, Folder "American Propeller and Manufacturing Company, 1919," Box 12, RD 3085, RG 342, NARA.

53 David L. Bacon, "The 'Universal Propeller' Built By Paragon Engineers, Inc., Baltimore, Maryland," NACA Technical Memorandum No. 70 (March 1922), 1-4; Caldwell, "American Progress in Adjustable-Pitch Propellers," 582; D. Adam Dickey and O. R. Cook, "Controllable and Automatic Aircraft Propellers," *SAE Transactions* 30 (March 1932): 107–108; Moore, "The Screw Propeller."

Heath demonstrated the Universal to the NACA and the Army throughout 1922. In March 1922, David L. Bacon, a NACA researcher detailed to specialize in propellers, concluded that the Universal design was promising. The time required to change from full speed forward to full speed reverse was approximately five seconds. Bacon noted that it was "obvious that some of its advantages are gained at the expense of additional complication and the question immediately arises whether they are worth it."[54]

Heath's demonstration for Chief of Air Service Maj. Gen. Mason M. Patrick took place at Bolling Field in Anacostia in October 1922. According to one McCook Field officer, the demonstration was "more for publicity purposes than for any effort to sell it either to the Army or Navy."[55] Heath mounted the propeller and engine on a flatbed wagon and he and his assistants drove it back and forth on the field solely by the thrust and changing pitch of the propeller. Patrick participated in the demonstration as a passenger until a mischievous mechanic manipulated the direction of the truck and he fell on his backside.[56] Nevertheless, Patrick was interested and wanted the Engineering Division's opinion on the practicality and potential usefulness of the propeller.

Caldwell and the propeller unit believed that it was not necessary to use engine power to shift pitch on high-performance aircraft. In the event that future aircraft with large engines that generated more than 400 horsepower required variable-pitch propellers, then a propeller like Heath's would be of potential value. Despite his misgivings, Caldwell recommended that the Engineering Division buy one propeller for ground tests at the whirl rig and on an engine, but he felt that no more funds should be allocated for that project.[57]

[54] Bacon, "The 'Universal Propeller' Built By Paragon Engineers," 5-6.

[55] H.W. Harms to Propeller Branch, Engineering Division, "Universal Propeller by Mr. Heath of the American Propeller Company," October 26, 1922, Folder "Propellers, 1922," Box 111, RD 3112, RG 342, NARA.

[56] David L. Bacon, "The 'Universal' Adjustable and Reversible Propeller Built By Paragon Engineers, Inc., Baltimore, Maryland," NACA Technical Memorandum No. 155 (November 1922), 6; Spencer Heath MacCallum, telephone interview by author, March 6–7, 2001.

[57] T. H. Bane to G. W. Lewis, "Report on Paragon Universal Propeller," March 28, 1922; A. H. Hobley to COAS, "Universal Propeller by Mr. Heath of the American Propeller Company," November 1, 1922, Folder "Propellers, 1922," Box 111, RD 3112, RG 342, NARA; C. W. Howard to R. L. Walsh, "Paragon Engineers Incorporated, Letter Dated December 29, 1924," February 10, 1925; F. W. Caldwell to L. MacDill, "Propellers of the Paragon Engineers," June 15, 1927, Folder "Heath Propellers, 1924–1929," Box 5735, RD 3181, RG 342, NARA.

As Caldwell and the propeller unit deliberated over its further involvement with American Propeller, Heath gained an audience with the Aircraft Committee of the House of Representatives on Capitol Hill in February 1925. Heath and his assistants brought the flatbed truck apparatus and ran the Universal Propeller just outside the House office building at the request of Senator Ovington Weller of Maryland. Weller chaired the Committee on Manufactures, which concerned itself with American industry and technology, and navigated the same social circles as Heath in Baltimore. Part of Heath's sales pitch included his explanation of the gearshift analogy. His propeller gave an airplane "all the properties of an automobile with several forward and reverse speeds" smoothly and efficiently without the lurching and choppy "different steps of gears" that everyday earthbound car transmissions offered. Perhaps due to Heath's appearance before Congress, the Army placed an order for one variable-pitch propeller at a cost of $5,000 in March 1925. The new propeller, like many of the other American experimental designs, would be for installation on the ever-present Liberty engine.[58]

The improved Universal propeller arrived at McCook for whirl testing in August 1925, an unfortunate time for Heath because the whirl-testing rig became inoperative due to equipment failure. Balancing the immediate and operational requirements of the Air Service, Caldwell assured Heath that "we will get to your propeller ASAP."[59] Despite this promise, tests on the propeller were not completed until November 1926, more than a year later. Both blades of the propeller were loose in their mountings after ten hours on the whirl rig at 800-horsepower and ten hours mounted on a Liberty engine. Caldwell directed one of his assistants, Second Lt. Glen T. Lampton, to disassemble the propeller after the whirl test to inspect the mechanism and propeller assembly for design flaws. Lampton found that Heath's use of ball bearing races was the cause for the loose blades. Basically, there was too much "play" (about three-quarters of an inch) in the bearings that retained the blades and allowed pitch adjustment. There was a possibility that the propeller could adopt a dangerous pitch setting that could result in the destruction of the airplane. He suggested

[58] Purchase Order No. 54706, "Propeller, (Heath), Adjustable-Pitch, Suitable for Use on Liberty 12 Engine," March 30, 1925, Folder "Heath Propellers, 1924–1929," Box 5735, RD 3181, RG 342, NARA.

[59] S. Heath to F. W. Caldwell, August 31, 1925; F. W. Caldwell to S. Heath, September 21, 1925; F. W. Caldwell to J. F. Curry, "Test of Propeller for Mr. Spencer Heath," November 17, 1925; J. F. Curry to R. L. Walsh, "Heath Variable-Pitch Propeller," January 11, 1926, Folder "Heath Propellers, 1924–1929," Box 5735, RD 3181, RG 342, NARA.

the use of hardened and tapered roller bearings that would ensure a more secure fit. For safety reasons, Caldwell recommended that the problem be corrected before flight testing commenced.[60]

The propeller unit's attitude toward Heath's propeller highlighted the military's role in fostering innovation. As the government's propeller specialist, Caldwell had to believe a new design had a potential and immediate value to the Army and the development of aeronautics overall in order for testing and further development to proceed. It was clear that the Universal propeller did not fall into that category. In a letter to Heath in July 1927, the chief of the experimental engineering section, Maj. Leslie MacDill, reiterated the propeller unit's criticisms of the design. He admitted that the control could be "quite successful," but there was simply no need for it.[61]

Another pioneer propeller company, the Standard Steel Propeller Company, was eager to receive an Army contract to construct an early experimental controllable and reversible-pitch propeller. Standard Steel was part of a larger American metalworking community and originally entered the propeller business to manufacture steel, fixed-pitch propellers. Located in the world's premier metalworking city of Pittsburgh, Standard Steel committed itself to the production of one item – metal propellers.[62] The cultural influence of steel-making on the naming and direction of Standard Steel reflects a broader trend in which United States was transforming itself into a "nation of steel" characterized by railroads, skyscrapers, armored warships, and the infrastructure that brought them about.[63]

The chief engineer of Standard Steel was Thomas A. Dicks, a sixty-year old English émigré who came to Pittsburgh as a young boy and had previously worked at Westinghouse as a tool designer. At Westinghouse, Dicks and a coworker, James B. Luttrell, worked with Frank Caldwell to build the first propeller whirl test rig. They founded the Dicks-Luttrell Propeller Company in May 1918 to experiment with fixed-pitch propellers with

[60] S. Heath to F. W. Caldwell, March 26, 1926; F. W. Caldwell to L. MacDill, "Propellers of the Paragon Engineers," June 15, 1927; G. T. Lampton, "Universal Paragon Propeller," November 29, 1926; W. E. Gillmore to M. M. Patrick, August 22, 1927, Folder "Heath Propellers, 1924–1929," Box 5735, RD 3181, RG 342, NARA.

[61] L. MacDill to S. Heath, July 18, 1927, Folder "Heath Propellers, 1924–1929," Box 5735, RD 3181, RG 342, NARA.

[62] D. H. Horne to C. N. Monteith, "Request for Loan of Control for Hollow-steel Reversible Propeller," September 11, 1922; Delos C. Emmons to Standard Steel Propeller Company, September 12, 1922, Folder "Propellers, 1922," Box 111, RD 3112, RG 342, NARA; *Polk's Pittsburgh City Directory, 1920* (Pittsburgh, PA: R. L. Polk, 1920), n.p.

[63] Thomas J. Misa, *A Nation of Steel: The Making of Modern America, 1865–1925* (Baltimore, MD: Johns Hopkins University Press, 1995), xx–xxiv.

hollow-steel blades under government contract. The reorganization of the company by Pittsburgh investors resulted in the creation of Standard Steel on January 2, 1919. The company's managerial leadership came from Harry A. Kraeling, president and general manager of the company, who was the entrepreneurial force and aggressive marketer behind the financial success of the company.[64]

Discussions between the propeller unit and the officers and engineers of the company began in October 1919. The mechanical hub design was attributed to Dicks, who received the assistance of Caldwell. The design incorporated what Caldwell called a "very ingenious" synchronizing device for shutting off the throttle when the propeller was in neutral position. However, the propeller suffered from excessive friction and uncontrollable pitch adjustment. There was also a strong tendency for the propeller to go in neutral position in either forward or reverse. Even though the Dicks-Caldwell design appeared satisfactory from the standpoint of mechanical engineering after considerable development, Caldwell noted that its excessive weight made it "prohibitive for airplane use."[65]

As a result, the Engineering Division expressed an interest in offering Standard Steel a contract not for the company's hub, but instead for developing steel blades for the Army's reversible propeller in January 1920.[66] To Caldwell, the use of metal blades offered an "attractive solution to the adjustable-pitch propeller problem" because they attached to the metal hub much more easily.[67] In 1923, Standard Steel abandoned its design of a variable-pitch mechanism.

[64] "Duralumin Airplane Propeller," *The Metal Industry* 24 (1926): 440; "Thomas A. Dicks Dies," *Aviation News* 2 (October 9, 1944): 13–14; William F. Trimble, *High Frontier: A History of Aeronautics in Pennsylvania* (Pittsburgh, PA: University of Pittsburgh Press, 1982), 116; "Brief Histories of the Development of Companies Exhibiting at the International Aeronautical Exhibition," *Aviation* 25 (December 1, 1928): 1759; Eugene E. Wilson, *Slipstream: The Autobiography of an Air Craftsman* (Palm Beach, FL: Literary Investment Guild, 1967), 165; "Hamilton's Thomas A. Dicks in Propeller Field Since 1917," *The Bee-Hive* 18 (June–July–August 1943): 14; Russell Trotman, "Turning Thirty Years," *The Bee-Hive* 24 (January 1949): 21.

[65] H. W. Harms to H. A. Kraeling, "Desirability of Having Mr. Dicks Remain at McCook Field For a Week or Two," October 15, 1919; T. H. Bane to Standard Steel, "Development of Steel Blades for Reversible Propellers," December 24, 1919, Folder "Propellers, Standard Steel Propeller Company, 1919," Box 13, RD 3085, RG 342, NARA; Caldwell, "American Progress in Adjustable-Pitch Propellers," 582, 586; Dickey and Cook, "Controllable and Automatic Aircraft Propellers," 107.

[66] T. H. Bane to Standard Steel Propeller Company, "Development of Steel Blades for Reversible Propeller," January 2, 1920, Folder "Reversible Propellers, 1920," Box 52, RD 3096, RG 342, NARA.

[67] Caldwell, "American Progress in Adjustable-Pitch Propellers," 582, 586.

Competition with Industry

By far, the most successful variable-pitch propeller program to address the design flaws inherent in the Hart reversible propeller concentrated on a mechanism designed by Caldwell and the propeller unit at McCook Field. The propeller unit began the project in 1920 as a simultaneous development to the other propeller programs. The so-called Engineering Division propeller design, intended for Liberty and Hispano-Suiza engines, was different than the Hart, Heath, and Standard Steel mechanisms.

Getting a variable-pitch propeller to "change gears" was the biggest challenge for any would-be designer. Caldwell's first mechanically actuated controllable and reversible propeller design, patented in January 1922, was not a success because overwhelming centrifugal and thrust forces jammed the blade actuation system.[68] He realized that he and his associates had to solve the interrelated problems of countering the forces that prevented blade actuation, decreasing the physical force required by the pilot to change pitch, and retaining the blades in their roots.

Caldwell understood that a thorough knowledge of centrifugal force, thrust, and twisting forces, or moments, that acted upon a whirling propeller was "of the greatest importance" to the design of a practical variable-pitch propeller and its control system. The control of centrifugal force, which acted radially outward on the rotating blade, was the key to a practical mechanically actuated variable-pitch propeller. Caldwell and his assistant for the project, Ernest G. McCauley, placed counterweights where the blade joined the hub to neutralize that force and keep the propeller where it needed to be during operation. In theory, the counterweights, which extended forward of the propeller, diverted centrifugal force to allow the pilot to exert the necessary physical force to change the propeller's pitch. Caldwell and McCauley filed a joint patent application in May 1922.[69]

As a result from the preliminary flight tests of the Hart propeller at Air Service training fields, the propeller unit began to concentrate its efforts on producing a propeller that required less force on the part of the pilot to operate. Caldwell incorporated large, commercially available ball bearings to minimize friction and facilitate further the ease of blade

[68] F. W. Caldwell, "Variable-Pitch or Reversible Propeller," US Patent No. 1,404,269, January 24, 1922.

[69] Caldwell, "American Progress in Adjustable-Pitch Propellers," 571, 575, 578; Dickey and Cook, "Controllable and Automatic Aircraft Propellers," 107; Wilbur C. Nelson, *Airplane Propeller Principles* (London: John Wiley and Sons, 1944), 48–50; E. G. McCauley and F. W. Caldwell, "Reversible Propeller," US Patent No. 1,658,385, February 7, 1928.

movement. The bearings alleviated the effect of whirling centrifugal force and air pressure generated on the blades during the creation of thrust that acted upon the actuation mechanism in flight.[70]

Originally developed for the Hart 1920 propeller, Caldwell and McCauley's screw-actuation mechanism gave the pilot, in the words of later observers, a "favorable mechanical advantage." An "intermediate control unit" locked the cables and cockpit controls in place to offset all physical loads until the exact moment of pitch actuation.[71] Caldwell and McCauley based the control system for the propeller on the lever principle. He believed that it offered almost instantaneous response time for the pilot, which made the propeller effective when the pilot reversed its pitch to act as a brake.[72]

Caldwell and McCauley's design for the pitch-throttle synchronizer offered an improvement over previous designs. When the pilot set the propeller in the forward pitch position and opened the throttle, a block dropped behind a small pin on the control lever, which prevented it from going below the minimum safe pitch for flying. As the pilot shut off the throttle, the block raised to enable the pilot to shift the propeller into neutral or reverse pitch position. While the propeller was in neutral position, the pin lay underneath the block so the throttle could not be opened and over-rev the engine. After reaching the full reverse position, a small hook permitted the throttle to be thrown on or off by a slight motion of the pitch control lever. Caldwell and McCauley filed a patent application for their combined pitch control lever and pitch-throttle synchronizer in October 1922.[73] Caldwell asserted that the combination of the counterweight, ball-bearings, and screw-actuation mechanism created a mechanical arrangement where "the operating friction in this propeller is so low that the pitch may be shifted with two fingers on the lever when the engine is run at full throttle."[74]

Caldwell experimented with new blade materials in an attempt to solve the blade retention problem that plagued the Hart and Heath programs. The Engineering Division propellers featured blades made from wood,

[70] L. W. McIntosh to COAS, "Reversible-Pitch Propellers," November 15, 1921, Folder "Propellers–Hart, 1921," Box 77, RD 3102, RG 342, NARA; Caldwell, "American Progress in Adjustable-Pitch Propellers," 571, 575, 578, 582, 587, 588.

[71] Dickey and Cook, "Controllable and Automatic Aircraft Propellers," 107.

[72] Caldwell, "American Progress in Adjustable-Pitch Propellers," 588.

[73] E. G. McCauley and F. W. Caldwell, "Propeller Control Mechanism for Reversible or Adjustable-Pitch Propellers," U.S. Patent No. 1,693,451, November 27, 1928.

[74] Caldwell, "American Progress in Adjustable-Pitch Propellers," 571, 575, 578, 582, 587, 588.

aluminum alloy, and Bakelite Micarta, a plastic-like phenolic resin that Westinghouse marketed first for use as an electrical insulator. Caldwell found limited success with Micarta blades mounted in a propeller hub designed for the 180-horsepower Hispano-Suiza engine installed in the JN-4H biplane. Another propeller designed for the Liberty had aluminum alloy blades manufactured by Standard Steel with hollow shanks fitted over steel extensions from the hub and mounted on frictionless bearings.[75]

By 1923, Caldwell claimed that the successive refinements incorporated into the Engineering Division's design resulted in the "most completely satisfactory adjustable propeller developed up to the present."[76] The Air Service never flight-tested the first propeller. Six were intended for use on the Liberty engines of the airship *Roma*, but it crashed earlier in February 1922. The propellers that followed and their successive modifications continued to fly at McCook and Wright fields through the 1920s and into the early 1930s with both Engineering Division wood and Micarta blades and new aluminum alloy blades manufactured by Standard Steel. They were an expression of Caldwell's belief that specific "refinements in design" would result in a practical variable-pitch propeller.[77]

Leading members from the aeronautical community believed that the variable-pitch propeller developed at McCook Field was an unqualified success and would be the point of departure for future development. Dr. Sanford A. Moss, the pioneer in the development of superchargers at GE and McCook Field, applauded the prospect of the immediate introduction of a practical variable-pitch propeller in June 1922.[78] According to Maj. Harold S. Martin, the chief engineer of McCook Field, however, the variable-pitch propeller developed by Hart and Caldwell in the propeller unit was not a proven design. They believed that the "first thing to do is get the mechanical part so it works satisfactorily."[79] There was limited success with the Hart propeller designed for the 150–180 horsepower Wright Hispano-Suiza engine series, especially regarding the

[75] Caldwell, "American Progress in Adjustable-Pitch Propellers," 582, 587, 588; Dickey and Cook, "Controllable and Automatic Aircraft Propellers," 107.

[76] Caldwell, "American Progress in Adjustable-Pitch Propellers," 582.

[77] F. W. Caldwell to COAC, July 20, 1927, Folder "Propellers – January through July, 1927," Box 5663, RD 3163, RG 342, NARA.

[78] S. A. Moss to T. H. Bane, June 28, 1922, Folder "Propellers–Reversible (Hart), 1922," Box 112, RD 3112, RG 342, NARA.

[79] H. S. Martin, "Variable-Pitch Propeller," July 7, 1922, Folder "Propellers–Reversible (Hart), 1922," Box 112, RD 3112, RG 342, NARA.

increased climb and overall performance of the JN aircraft. There was little or no success for the propellers designed for the Liberty and the 300-horsepower Wright Hispano-Suiza engines. The Liberty propellers never made it to the service test stage. The problem required more testing and evaluation.[80]

The Air Service's conservative approach toward developing and introducing new technologies shaped the final analysis of the service's variable-pitch propeller programs. They were not a priority program during World War I and the 1920s, because no one saw a real demand for controllable-pitch on biplanes with low wing loadings and speeds below 150 mph. Reversing the pitch of the propeller to act as a brake for the airplane during landing or as a maneuvering tool for an airship was a major impetus behind developing variable-pitch propellers in the early interwar period. Another synergistic development within the technical system of the airplane negated the usefulness of the reversible-pitch propeller: wheel brakes. The aeronautical community learned to integrate brakes into landing gear systems beginning in the mid-1920s. Wheel brakes performed the same function of slowing the airplane down at landing and were simpler, cheaper, and more practical than reversible propellers.[81]

The same people who praised the initial merits of the variable-pitch propeller were also the first to enforce a slow and deliberate development program. In February 1922, Gardner W. Carr, the Engineering Division's civilian assistant chief of production, stated in *U.S. Air Services* that the investigation of variable-pitch propellers encouraged the belief that they enhanced performance. He acknowledged, however, that the variable-pitch propeller was still far from being a practical component of service airplanes.[82] The assistant chief of the Engineering Division, Maj. A. H. Hobley, noted that "comparatively little work was done on the adjustable type" during the year 1922 due to other propeller projects.[83] The conservatism was justified given the state of variable-pitch propeller development

80 T. H. Bane to S. A. Moss, July 21, 1922, Folder "Propellers–Reversible (Hart), 1922," Box 112, RD 3112, RG 342, NARA.

81 Thomas Carroll and Smith J. DeFrance, "The Use of Wheel Brakes on Airplanes," NACA Technical Note No. 311 (July 1929), 1–2, 8–9; Edward P. Warner, "The SAE Propeller and Power Plant Sessions," *Aviation* 27 (August 31, 1929): 470.

82 Gardner W. Carr, "Organization and Activities of Engineering Division of the Army Air Service," *U.S. Air Services* 7 (February 1922), 25.

83 A. H. Hobley, "Achievements of the Engineering Division for the Calendar Year 1922," *The Slipstream* 4 (January 1923): 23.

at the time and it reflected McCook Field's practical approach toward engineering development.

The current operational requirements of the Air Service always dominated the amount of research and development that any group at McCook, including the propeller unit, could conduct. Testing and approving wood propeller designs for immediate service use was always the priority. When it came to developing new designs, Caldwell had little time to focus on variable-pitch propellers when he had to direct simultaneous developmental and experimental work on fixed-pitch wood, metal, and Micarta propellers. Also, other Air Service programs received a higher priority, especially engine supercharging.

At the national level, the Air Service was finding it hard to continue variable-pitch propeller development within the budgetary restrictions created by the newly elected administration of Calvin Coolidge. Bane told Carl McStay that "the change in administration appears to be detrimental, rather than beneficial, to the Air Service, as the new cabinet officers are not familiar with the situation regarding a separate Air Service."[84] With the future of the service at stake, it was difficult to encourage the administration and the War Department to allocate funds and to make the decisions necessary to bolster a military aeronautical engineering organization. The introduction of new and interactive high-performance technologies, which included radial engines, bombsights, instruments, spark plugs, fuels, and variable-pitch propellers, into operational service, depended on it.

Finally, the propeller unit's work on its own design highlighted a final and important reason why variable-pitch propellers were not in operation in the 1920s. Caldwell and individuals within the propeller unit developed their own designs and patented the successive improvements in an effort to advance the technology. This put the propeller unit into a delicate position with private inventors, specifically Spencer Heath, and the chief of army aviation, Major General Patrick. Just as the Air Service became the Air Corps and with it the reorganization of the Engineering Division into the Materiel Division in 1926, the resulting tension over the government's role in fostering new technologies came to a boiling point in 1927. The propeller unit could either assist industry with its designs or it could pioneer new technologies, but politically it could not do both.

[84] T. H. Bane to C. E. McStay, June 1, 1921, Folder "Propellers–Hart, 1921," Box 77, RD 3102, RG 342, NARA.

According to the Materiel Division, the propeller unit's design offered the attractive combination of "simplicity, lightness, and ease of maintenance."[85] Spencer Heath complained to both the Materiel Division and to Patrick that the propeller unit favored its own design over those developed in the private sector, particularly his Universal propeller. To him it was obvious that the Army made it difficult for the "practical testing of outside apparatus," meaning inventions developed outside the military. Heath urged that the test of his propeller should be "placed in the hands of competent operating personnel, whose time is not taken up with the development and testing of mechanisms of their own design."[86] It appeared to Heath that the propeller unit was not allowing competition to take place. In a letter to Patrick, Heath contended that he had been working to develop a practical variable-pitch propeller at the request of the federal government since 1917 and had already spent more than $80,000 on the project. He alleged that Caldwell was spending "many hundreds of thousands of [the government's] dollars on a parallel development in direct competition" with his work. According to Heath, "the whole difficulty lay in the false policy of allowing Army employees not only to compete against outside industry, but of allowing the same employees to exclude from practical demonstration anything that private enterprise and industry might at any cost develop or produce."[87]

Patrick was a major critic of how the Materiel Division's experimental research and development impacted the struggling aviation industry's technological initiative. The Army employed its own aeronautical experts primarily to work with manufacturers to refine their designs. They also developed new designs based on their individual, as well as collective, research and experience. Unfortunately, any new technology, by private firm or by in-house military design sections, had to be approved by those same experts before it could enter production. Patrick asserted that the arrangement put the Army in direct competition with private industry. The ability to approve or disapprove a potential new design – including their own – gave the advantage to the Air Service experts.[88] Patrick

85 L. MacDill to S. Heath, July 18, 1927, Folder "Heath Propellers, 1924–1929," Box 5735, RD 3181, RG 342, NARA.

86 S. Heath to L. MacDill, July 28, 1927, Folder "Heath Propellers, 1924–1929," Box 5735, RD 3181, RG 342, NARA.

87 S. Heath to M. M. Patrick, October 26, 1927, Folder "Heath Propellers, 1924–1929," Box 5735, RD 3181, RG 342, NARA.

88 Mason M. Patrick, *The United States in the Air* (Garden City, NY: Doubleday, Doran and Company, 1928), 100–102; Robert P. White, *Mason Patrick and the Fight for Air Service Independence* (Washington, DC: Smithsonian Institution Press, 2001), 94–101.

related to Brig. Gen. William E. Gillmore, Chief of the Materiel Division, that the treatment Heath received created the impression that the propeller unit was more concerned with improving devices initiated by its own personnel rather than giving outsiders opportunities to demonstrate their inventions. Regardless of the merits of any potential innovation, Patrick ordered the Materiel Division to treat all inventors "fairly and considerately."[89]

In defense against Heath's accusations against the McCook and Wright field research establishments, General Gillmore, prepared an appropriate rebuttal on behalf of MacDill, Caldwell, and the propeller unit. Regarding alleged government competition with private industry, he made it clear that the "Division's propeller was developed from the old Hart propeller – not by Caldwell, but by the propeller unit of which Mr. Caldwell is chief." He went on to say that the new propeller was part of a series of engines and superchargers that the Army had developed immediately after World War I. It was at the service test stage and was "as far along as the engine development with which it is to be used" while the Universal was a "needlessly complicated" and heavy experimental design for "future engines of much larger horsepower."[90]

Caldwell and the propeller unit were forced to justify their mode of conduct and the technical choices they made because of an angry and rejected inventor who also happened to be a successful propeller manufacturer. Given the fact that the government endeavored to support new technologies, Heath's allegations were untrue. Nevertheless, Caldwell and McCauley had patented the successive improvements found in the Engineering Division propeller. The question of whether the government or the individual had the right to innovate had come to a crisis point for the Army, and it would have a direct effect on the future development of the propeller in the United States.

The propeller community could not foresee the ultimate form of the practical and modern variable-pitch propeller in 1917, but Caldwell and the propeller unit's work can be considered the starting point for that development in the United States. Caldwell worked with private industry to improve the basic Hart and Eustis design, modified it, gave it to other manufacturers to see if they could improve it, and then decided it was impractical, which gave him the experience to move on to more successful

[89] M. M. Patrick to W. E. Gillmore, "Correspondence with Paragon Engineers, Inc.," August 3, 1927, Folder "Heath Propellers, 1924–1929," Box 5735, RD 3181, RG 342, NARA.

[90] W. E. Gillmore to M. M. Patrick, November 9, 1927, Folder "Heath Propellers, 1924–1929," Box 5735, RD 3181, RG 342, NARA.

avenues of research and development. The 1917 Hart and Eustis design and its derivatives were everything the modern propeller would not be – mechanically actuated, built with wood blades, reversible – but they were the first physical manifestations of the idea that variable pitch was a viable possibility. It was the first expression of the propeller community's belief that it could improve the overall performance of the airplane through its chosen aeronautical specialty.

The fates of the individual inventors highlight the problematic government–industry relationship during this period. Driven initially by patriotism during World War I, Seth Hart and Robert Eustis left the aviation business and returned to their normal lives after the government moved on to other propellers. Hart remained in the insurance business until his death in February 1938. Eustis faded into obscurity after leaving the Adjustable and Reversible Propeller Corporation at some point in the mid-1930s. Hollywood aviator Paul Mantz joined the corporation to demonstrate the propeller's utility when installed on small sport aircraft, which raised its public profile. Despite the notoriety, there is no record of the Adjustable and Reversible Propeller Corporation actually selling a production propeller to a military or commercial concern before its disappearance on the eve of World War II.[91]

Heath sold his American Propeller and Manufacturing Company to the Bendix Corporation at a huge profit just before the Wall Street Crash of October 1929. He served as an engineering consultant for two years before leaving aeronautics entirely. Dividing his time between his 100-acre estate southwest of Baltimore and an apartment in Greenwich Village in New York City, Heath moved on to a final career in philosophy and social anthropology, which he pursued until his death in 1963.[92] The factory remained in Baltimore on Grindall Street a few years before being relocated to Toledo, Ohio.[93]

As late as 1927, the Army's propeller community continued to pursue its idea of what the modern variable-pitch propeller would be. Between

[91] "New Aviation Aid Disclosed," *Los Angeles Times*, June 25, 1934, p. A10; "Death Calls Veteran Insurance Man," *Los Angeles Times*, 20; Genealogical Data on Seth Hart, March 2002.

[92] MacCallum Telephone Interview; Alvin Lowi, Jr., "The Legacy of Spencer Heath: A Former Student Remembers the Man," *Journal of Socioeconomics*, 1998, www.logan.com/afi/spencer1.html (Accessed April 16, 2003); and "The Legacy of Spencer Heath: Conclusions," *Journal of Socioeconomics*, 1998, www.logan.com/afi/spencer3.html (Accessed April 16, 2003).

[93] Accession Memorandum for Acquisition Lot NAM 135, March 1, 1930, in "Atwood-Wright Propeller," File A19300031000, RO, NASM.

Hart, Heath, Standard Steel, and the Army's own design, all concerted attempts to develop a variable-pitch propeller had focused on perfecting a mechanical system of actuation and an adequate system of blade retention. Caldwell was confident that the design of a successful variable-pitch propeller was a "matter of working out the details in accordance with the best mechanical engineering, rather than the introduction of an entirely new type of design."[94]

The Army's experience with attempting to create a practical variable-pitch propeller in conjunction with private companies during the postwar period was a major component of the US government's role in technological innovation during the 1920s and 1930s. The Propeller Laboratory's efforts to introduce a practical variable-pitch propeller highlighted questions about the appropriate role of a national government developing fundamental aeronautical technologies. Innovative research conducted in government aeronautical laboratories by civilian and military researchers, especially when it resulted in actual technologies rather than abstract data that could be then used by others to create commercial products, raised the ire of the private aviation industry.[95] Some members of the nascent propeller community viewed the government-subsidized propeller development program as a threat to their existence.

Innovation in Germany, Great Britain, and Canada

While early variable-pitch propellers took to the air in the United States, members of the international propeller community experimented with their own designs during World War I and the Aeronautical Revolution.[96] During the war, the Imperial German Army Air Service operated Zeppelin-Staaken R.VI four engine bombers, the largest production aircraft of the time, in its 1917–1918 strategic bombing campaign against England. The service modified one of the lumbering Riesenflugzeug (giant aircraft) with an extra supercharged Mercedes D.II engine in hope of getting better high-altitude performance. Dr. Hans Reissner took his experience working with Hugo Junkers and developed a ground-adjustable propeller for the bomber. As an engineering professor at the Charlottenburg

[94] F. W. Caldwell to R. I. Carr, April 19, 1927, Folder "Propellers – January through July, 1927," Box 5663, RD 3163, RG 342, NARA.

[95] William F. Trimble, *Wings for the Navy: A History of the Naval Aircraft Factory, 1917–1956* (Annapolis, MD: Naval Institute Press, 1990), xiv, 69–71.

[96] Ronald Miller and David Sawers, *The Technical Development of Modern Aviation* (New York: Praeger, 1970), 76.

Technische Hochschule (the modern-day Berlin Institute of Technology), he was well-qualified to face both the theoretical and practical challenges of variable-pitch propeller design.[97]

The limitations imposed on German aircraft development in the aftermath of World War I shaped variable-pitch propeller development in the war-torn and bankrupt country. Bon vivant John Jay Ide, the NACA's official "technical assistant" and the US government's unofficial aeronautical spy in Europe, visited Reissner in Berlin during the spring of 1920. Reissner had a new mechanical variable-pitch design, but he faced two major development obstacles. First, the dire economic situation in Weimar Germany meant the well-connected Reissner could not gather the needed funds to conduct a flight test or extend his basic patent. Finally, even if Reissner found the capital to conduct a test flight, he would not have been permitted to do so due to the severe postwar restrictions on flying aircraft or developing aeronautical technology for any purposes that could be construed as anything remotely military.[98]

Reissner eventually found assistance from the Helix Propeller Company and produced a prototype that received considerable interest from the US Navy. Ide received an example propeller, minus the all-important control linkage, in September 1921 and shipped it to the United States. Caldwell and his team tested the propeller at McCook Field in April 1924, but the absence of the control mechanism led the American military to lose interest.[99] Reissner moved on to more theoretical pursuits in aircraft propeller and structural design before immigrating to the United States in 1938 to join the faculty of the Illinois Institute of Technology. It is unclear whether Reissner's propeller ever took to the air in Europe or the United States.

By far, the country most involved in propeller development was Great Britain. The British government previously sponsored an experimental variable-pitch propeller program through the Royal Aircraft

[97] Eric Reissner, "Hans Reissner: Engineer, Physicist and Engineering Scientist," *The Engineering Science Perspective*, 2 (December 1977): 103; George W. Haddow and Peter M. Grosz, *The German Giants: The Story of the R-Planes, 1914–1919* (London: Putnam, 1969), 94.

[98] John Jay Ide, "Visit to Germany," May 27, 1920, Folder "John Jay Ide–Paris Office," Box "NACA," Row 2, LEK 9, NASAHO, 2, 9, 10.

[99] G. W. Lewis to Cdr. S. M. Kraus, "Reissner Variable Pitch Propeller," September 28, 1921, Folder "F23, Volume 2," Box 1952, RG 72, NARA; A.H. Hobley to the NACA, "Test of Reissner Variable Pitch Propeller," June 11, 1923, Folder "Propellers, 1923," Box 5501, RD 3122, RG 342, NARA; F. W. Caldwell and D. A. Dickey, "Destructive Whirl Test No. 632: Test of Reissner Adjustable Pitch Propeller," April 24–25, 1924, File "D52.43/557 – Propellers-Reissner," WFTD Library Index Card File, NASM.

Establishment (RAE) at Farnborough, the institution responsible for the World War I British aviation research, development, and production program. Much like the Air Service at McCook Field, the RAE employed civilian research engineers and technicians who specialized in different areas of aeronautical development. Their main goal was to design, test, and construct new military aircraft and modify existing military aircraft to increase performance. Previously called the Royal Aircraft Factory, the facility became the RAE in November 1917 with the creation of the Air Ministry and the Royal Air Force (RAF). E. J. H. Lynam led the RAE's airscrew research and development program during World War I and through the 1920s, which focused on enhancing the efficiency and durability of propellers for military aircraft.[100]

RAE research on high-altitude aircraft performance revealed that the variable-pitch propeller was the key to all-around efficiency. Lynam and his team at Farnborough investigated two types: mechanical, or "hand-controlled," and automatic.[101] They first designed a four-blade, mechanically actuated variable-pitch propeller in June of 1915 and evaluated it at the facility's spinning tower with loads up to 330 horsepower at 1,050 rpm for thirty minutes. They then tested it in the air on a B.E.2c airplane with a 100-horsepower Royal Aircraft Factory IA engine for eleven hours from October 1917 to January 1918. Compared to a "standard propeller," the new design almost doubled the airplane's rate of climb between sea level and 10,000 feet to 190 feet per minute. The pilot controlled pitch with a manual hand wheel in the cockpit that gave the wood blades a range of ten degrees. The propeller and control gear weighed eighty-five pounds, fifty more than its wood fixed-pitch counterpart, which the researchers asserted was a significant disadvantage for the relatively light 1,300 pound airplane. Despite the weight penalty, they did believe the variable-pitch mechanism improved the aircraft's performance.[102]

[100] Ronald W. Clark, *Tizard* (Cambridge, MA: Massachusetts Institute of Technology Press, 1965), 26; C.G. Grey, *A History of the Air Ministry* (London: George Allen and Unwin, 1940), 35, 75, 78, 194.

[101] Royal Aircraft Factory, "Resumé of Work at the R.A.E. on Variable-Pitch Propellers," Report No. B.A.237, June 4, 1918, AVIA 6/1283, National Archives of the United Kingdom (hereafter cited as TNA).

[102] Royal Aircraft Factory, "Preliminary Report on the Performance of the B.E.2c with Variable-Pitch Propeller," Report No. B.A.176, November 23, 1917, AVIA 6/1254, TNA; "The Variable-Pitch Propeller – Experiments Conducted at the Royal Aircraft Factory," Aeronautical Research Council *Reports and Memoranda* No. 402 (January 1918): 515, 517.

The RAE conducted further tests in 1918 to determine whether or not the propeller was strong enough to withstand regular operations. A problem quickly emerged: finding a suitable method of attaching the wood blades to the metal hub. In 1920, the RAE experimented with the propeller on an R.E.8 aircraft with a more powerful supercharged RAF IVD engine. That same year, it also constructed a two-blade propeller for an S.E.5a with a Wolseley Viper Hispano-Suiza-type engine.[103]

The RAE also investigated the feasibility of a constant-speed, or automatic, propeller that changed blade pitch according to varying flight conditions while the engine speed remained the same and without the direct input of the pilot. The organization's interest recognized the availability and advantages of supercharged engines that could deliver "constant power at all altitudes." Moreover, the new propeller seemed ideal for aerobatic maneuvering where "hand control would be impossible." The RAE experimented with two types of mechanism: centrifugal and hydraulic. The centrifugal propeller relied upon weights at each blade root directly connected to a central spring in the hub. The hydraulic propeller employed a release valve to control oil pressure acting upon plungers in the blade roots inside the hub. A novel feature of the hydraulic mechanism was that pilot could override the automatic control mechanism while airborne and alter blade angle for increased fuel economy during cruising flight. The pilot could also preset the blade change between two predetermined speed settings. Neither system offered precise control and the pilot had to be mindful not to overrun, or "race," the engine at full throttle while diving the airplane.[104]

The British government's wartime experimental propeller program, like the one conducted by the US Army, clearly indicated that the variable-pitch propeller was superior over its fixed-pitch counterpart. In both nations, the prevailing attitude toward aircraft design, with its ever-present emphasis on minimum weight and maximum simplicity, combined with the very practical limitations of mechanical design, made the program a limited success. The aeronautical engineering community's adherence to those paradigms, however, would shape the progress

[103] "The Construction of the R.A.E. Experimental Variable-Pitch Airscrew," Aeronautical Research Council *Reports and Memoranda No. 471* (April 1918): 518; K. M. Molson, "Some Historical Notes on the Development of the Variable Pitch Propeller," *Canadian Aeronautics and Space Journal* 11 (June 1965), 179; Robert M. Bass, "An Historical Review of Propeller Developments," *Aeronautical Journal* 87 (August/September 1983): 258–259.

[104] Royal Aircraft Factory, "Resumé of Work at the R.A.E. on Variable-Pitch Propellers."

of the variable-pitch propeller in Great Britain in a way different than it occurred in the United States.

The British government continued its interest in variable-pitch propellers in the early 1920s in conjunction with the increasingly widespread development of supercharged engines. The Air Ministry issued specifications in early 1923 for a reversible variable-pitch airscrew designed for Jupiter and Jaguar engines. There were seven different designs including submissions by Blackburn Aeroplane Company, William Beardmore and Company, Gloster Aircraft Company, Hawker Engineering, and the Metal Airscrew Company.[105] Two years later, in 1925, the Air Ministry solicited proposals for an all-metal variable-pitch propeller. In response, Dr. Henry Selby Hele-Shaw and Thomas Edward Beacham, newcomers to the propeller community, responded with a hydraulically actuated, constant-speed, variable-pitch propeller they had patented in December 1924.[106]

The two inventor/engineers were certainly up to the task. Henry Selby Hele-Shaw had a long and illustrious career in British engineering science and technology. Born in 1854, he was a hydraulics specialist – he had designed transmission gear for automobiles and trucks, steering gear for ships, and pumps and motors for marine applications – and he had reaped the commercial benefits of his inventions. He was also instrumental in the founding of the engineering departments at universities in Bristol, Liverpool, and Johannesburg, South Africa. His work on mechanical integrators won him the Watt Medal from the Institute of Civil Engineers in 1885. He demonstrated the nature of streamlined flow, which contributed to the practical application of hydrodynamic theory to work on real fluids and to his election as a Fellow of the Royal Society in 1899. During his tenure as Chairman of the Institution of Mechanical Engineers Education Committee in the early 1920s, Hele-Shaw established the National Certification Scheme that became the standard for accrediting individual engineers in Great Britain.[107]

[105] Royal Aircraft Factory, "Variable-Pitch Airscrew Designs," Report No. B.A.471, May 15, 1923, AVIA 6/1464, TNA.

[106] G. Geoffrey Smith, "Evolution of the Variable-Pitch Airscrew," *Flight* 40 (August 14, 1941): 85; Bruce Stait, *Rotol: The History of An Airscrew Company, 1937–1960* (Stroud, Gloster, England: Alan Sutton Publishing, 1990), 1; H. S. Hele-Shaw and T. E. Beacham, "Feathering Screw Propellers," UK Patent No. 250,292, April 6, 1926.

[107] D. G. Christopherson, "Henry Selby Hele-Shaw," in *The Dictionary of National Biography, 1941–1950*, ed. L. G. Wickham Legg and E. T. Williams (London: Oxford University Press, 1959), 373–374; Jane Carruthers, "Henry Selby Hele-Shaw LLD, DSc, EngD, FRS, WhSch (1854–1941): Engineer, Inventor, and Educationist," *South African Journal of Science*, 106 (February 2010): 34–39.

Hele-Shaw's ability to adapt his extensive knowledge of hydraulics to new technologies that fascinated him more than compensated for his lack of practical experience in fields such as automotive engineering and aeronautics. His hydraulic clutch answered the need for the more efficient transmission of an engine's horsepower to the wheels of a motorcar. Hele-Shaw aimed to do the same thing for the airplane after World War I through the development of a variable-pitch propeller. At a time when the wood, fixed-pitch propeller was the primary choice for designers, he recognized the need for variable-pitch. Hele-Shaw also believed that his specialty, hydraulics, was the answer.

Thomas Edward Beacham was a young mechanical engineer just beginning a brilliant career. He earned his degree from the City and Guilds Engineering College of London in 1913. Beacham also held a commission in the Royal Army and served as a captain on the Western Front during World War I. After returning from the war, he began his collaborative engineering work with Hele-Shaw on hydraulic pumps and controls, which included their variable-pitch propeller.[108]

Hele-Shaw and Beacham believed they had developed the best method of variable-pitch actuation, what they called "power operated gears in which a governor adjusts the pitch." They rejected contemporary manually actuated designs because they believed that the pilot did not have the physical strength to control the propeller's pitch nor the time to devote constant attention to efficient pitch variation while operating the other complicated systems of the airplane. The governor, the key component of the propeller assembly, automatically regulated the hydraulic pressure that varied propeller pitch in response to different altitudes and engine speeds. Overall, they believed that their hydraulic propeller offered light weight, reliability, and "provision against possible failure of parts."[109]

The two inventors recognized that constant-speed operation offered specific performance advantages. They provided efficient operation for either supercharged or non-supercharged engines at both high and low altitudes. It resulted in greater climbing efficiency, reduced fuel consumption at cruising speed, and allowed full power output with supercharged engines at high altitudes. For multiengine aircraft, variable-pitch

[108] "Captain Thomas Edward Beacham," in *The Aeroplane Directory of British Aviation* (London: Temple Press, 1955), 421.

[109] H. S. Hele-Shaw and T. E. Beacham, "The Variable-Pitch Airscrew: With a Description of a New System of Hydraulic Control," *Journal of the Royal Aeronautical Society* 32 (July 1928): 528–536.

propellers provided safer and more efficient flight in the event of engine failure.[110]

The engineers relied upon their fellow specialists from the automotive and aeronautical communities to fabricate their creation. A. Harper, Sons and Bean Ltd. of Dudley in the West Midlands, a motorcar manufacturer and engineering specialty firm, constructed the hub.[111] Metal Propellers Ltd. at Purley Way in Croydon, provided two hollow-steel blades for the propeller. The company had begun operations in early 1917 based on the partnership of two of Great Britain's leading propeller engineers, Henry Leitner and Dr. Henry C. Watts. Their steel propeller consisted of laminated sheet blades pressed to shape and welded at the edges, which were then joined by a two-piece hub that allowed pitch adjustment on the ground. In the 1920s, Leitner-Watts propellers generated great interest within the fledgling military and commercial air services around the globe. The RAF supplied them to its Bristol Fighter squadrons based in Africa and the Middle East. The Imperial Japanese Navy purchased two- and three-blade versions for its Felixstowe F.5 flying boats and Avro 504L floatplanes and for its first generation of ocean-going aircraft, Gloster Sparrowhawk shipboard fighters and Parnall Panther reconnaissance airplanes. The Zeppelin Company delivered the LZ 126, soon to be christened the ZR-3 USS *Los Angeles*, across the Atlantic from Friedrichshafen to Lakehurst, New Jersey, in October 1924 with Leitner-Watts propellers installed.[112]

The first Hele-Shaw Beacham constant-speed propeller won the Air Ministry competition. The Air Ministry required independent evaluation by the RAE at Farnborough. A team led by E. J. H. Lynam ground tested the propeller and installed it on a Rolls-Royce Condor III engine for flight tests.[113]

Hele-Shaw and Beacham's success in acquiring the Air Ministry contract spurred the interest of the Gloster Aircraft Company in sponsoring the further development of the design. Hugh Burroughes of Gloster began negotiations with the two inventors in 1925. By the following year, the company owned the design and manufacturing rights for the Hele-Shaw

[110] Ibid., 526–529.

[111] British Motor Manufacturers, "Bean/Hadfield-Bean," 2010, www.britishmm.co.uk/history (Accessed January 24, 2010).

[112] "Metal Propellers Ltd.," *Flight* 17 (October 1, 1925): 646; Metal Airscrew Company Ltd., "L.W. (Leitner-Watts) Metal Propellers" (n.p.: January 1922): 2, 5, 6, 15; *Metal Propellers, Limited–Manufacturers of Metal Air Propellers*, n.d. [1926], File B5-554000-01, PTF, NASM.

[113] Hele-Shaw and Beacham, "The Variable-Pitch Airscrew," 529; Stait, *Rotol*, 1.

Beacham propeller. Harry Lawley Milner, affectionately known as "Pop" by coworkers, joined Gloster in 1924 to serve as propeller designer, having previously worked with Hele-Shaw and Beacham beginning in 1922. Educated at the Manchester Institute of Technology and Victoria University, Milner apprenticed at Sir W. G. Armstrong Whitworth Ltd. and later worked on aero engine design at the RAE in 1921.[114]

Under the direction of Milner and chief engineer and designer Henry P. Folland, Gloster Aircraft assumed the monumental research and development costs involved in adapting the new propeller for different types of engines. Folland's interest in the Hele-Shaw Beacham variable-pitch propeller ran contrary to his reputation within the British aeronautical community as a conservative designer. Responsible for the famous World War I British S.E.5a pursuit airplane, and various postwar air racers, Folland was known for his staunch resistance to innovations such as monoplane wings and metal construction, which reflected the overall attitude of his previous employer, the RAE, as well as the mainstream British aviation industry.[115]

Besides the engineers, inventors, and entrepreneurs in the United States and England, there was also a lone inventor in Canada who worked on designing a variable-pitch propeller. Wallace Rupert Turnbull of Rothesay, New Brunswick, had a background in engineering and propeller research similar to Frank Caldwell. Turnbull had earned a mechanical engineering degree with an emphasis in electrical engineering from Cornell University in 1893. He continued graduate study at the universities of Heidelberg and Berlin before working as an experimental engineer in the Edison Lamp Works of GE in Harrison, New Jersey, from 1897 to 1901. For medical reasons, Turnbull became an independent consulting engineer operating from his private laboratory in Rothesay. He was independently wealthy, having received a $750,000 inheritance from his father's estate in 1899.

Turnbull became interested in aeronautics at the turn of the century after reading about the experimental activities of Lilienthal and Langley. He built the first Canadian aerodynamics laboratory – complete with a wind tunnel – in 1902, delivered a prize-winning paper on aircraft efficiency before the RAeS in 1908, and conducted a systematic investigation of aerial propellers from 1908 to 1910. During World War I, he joined

[114] "Harry Lawley Milner," in *The Aeroplane Directory of British Aviation*, 503; Stait, *Rotol*, 2.

[115] Thomas Foxworth, *The Speed Seekers* (New York: Doubleday, 1975), 74–75, 323–325.

the aviation department of Frederick Sage and Company Ltd. in England where he served as an aircraft inspector, experimental engineer, and chief propeller designer.

While in England, Turnbull experimented with new technologies to contribute to the war effort, including hydroplanes, torpedo-defense screens for warships, aerial bombsights, and composite coatings for wood propellers. He began work on a variable-pitch propeller in November 1916. By July 1918, he had constructed a working model of a mechanically actuated propeller. His design used the power of the engine transmitted through a gear train and brake wheels to actuate pitch. Turnbull formed a partnership with the general manager of Sage's aviation department, E. C. Gordon England, to develop the new propeller. The RAF placed an order for three propellers, but quickly cancelled the order after the Armistice. The partnership quickly dissolved after Turnbull returned to Rothesay and England was unable to interest any buyers in the design. Despite the initial setback, Turnbull continued to refine his design.

Turnbull's first full-size propeller, called the No. 1, featured mechanical actuation and was designed for use on a Sopwith Snipe fighter equipped with a Bentley BR.2 rotary engine as stipulated in the original RAF contract. Turnbull completed the propeller at Rothesay in December 1922 and approached the Royal Canadian Air Force (RCAF) to ascertain its interest in flight testing. Turnbull took the No. 1 to the RCAF's Camp Borden, Ontario, and modified it for installation on an antiquated World War I-era Avro 504K trainer with a 130-horsepower Clerget rotary engine. The test pilots experienced difficulty with the mechanical brake control mechanism. Nevertheless, the design showed promise and Turnbull received a silver medal for his design at the February 1923 Inventions Show in New York. A hangar fire at Camp Borden destroyed the No. 1 in October 1923. Unfortunately, by 1924 the extreme financial undertaking of private research and development had depleted Turnbull's inheritance and put him $90,000 in debt.[116]

McCook Field engineers were familiar with Turnbull's work as early as May 1919. Spencer Heath of the American Propeller and Manufacturing Company reported to the Engineering Division of his encounter with

[116] J. H. Parkin, "Wallace Rupert Turnbull, 1870–1954: Canadian Pioneer of Scientific Aviation," *Canadian Aeronautical Journal* 2 (January–February 1956): 4, 8, 10, 39–40; Mac Trueman, "On a Wing and a Prayer: Wallace Rupert Turnbull," *The New Brunswick Reader* 4 (December 7, 1996): 9, 10, 12.

Turnbull in Atlantic City, New Jersey, in May 1919. He described Turnbull's earlier mechanically actuated design in detail and assured the Air Service that it would be very interested in the mechanism, especially since he believed it was very similar to the Hart propeller under development in Dayton.[117]

The difficulties with mechanical actuation led Turnbull to focus on developing an electrically actuated propeller starting in June 1923. Through the auspices of the Associate Air Research Committee of the National Research Council (NRC) of Canada, he received a $1,500 grant from the RCAF to construct and test a new variable-pitch propeller design with an electric motor actuation drive in May 1924. Turnbull completed the propeller, called the No. 2, in February 1925 for another RCAF Avro 504K with a Clerget engine (Figure 9). Unlike other variable-pitch designs from the period, the Turnbull propeller outwardly resembled a conventional wood fixed-pitch propeller. He mounted the electric motor at the front of the hub under a metal dome. A nickel-chrome steel spindle connected the wood blades to the wood hub, and to change pitch the blades rotated on a ball thrust washer. Cockpit controls and instrumentation included a pitch lever, a power switch, and a blade pitch indicator. Vickers Ltd. of Montreal, the Canadian subsidiary of the British arms maker, constructed the propeller while the lone inventor fabricated the control and angle indicator.[118]

Turnbull returned to Camp Borden for tests of the No. 2 from June 29 to July 9, 1927.[119] The primary goal was to determine the strength and practicality of the design. Flight Lt. G. E. Brooks inaugurated the first of thirty-two flights that totaled approximately twelve hours with the No. 2 and the Avro 504K. The seven RCAF test pilots experienced the same performance benefits that so many other pilots described when they flew aircraft equipped with early variable-pitch propellers: quick takeoff and climb and lower fuel consumption at cruise. Furthermore, there were no mechanical defects reported. *Aviation* reported that the Camp Borden tests "clearly indicate what a revolution will occur, for long-distance flying, when variable-pitch propellers come into use."[120]

[117] S. Heath to T. H. Bane, May 21, 1919, Folder "American Propeller and Manufacturing Company, 1919," Box 12, RD 3085, RG 342, NARA.

[118] "The Turnbull Variable-Pitch Propeller," *Aviation* 24 (February 20, 1928): 446–447; Parkin, "Wallace Rupert Turnbull," 40, 41.

[119] "A Promising Variable-Pitch Airscrew," *The Aeroplane* 33 (November 9, 1927): 646.

[120] "The Turnbull Variable-Pitch Propeller," 446–447.

FIGURE 9 Wallace R. Turnbull with his No. 2 electric propeller installed on an RCAF Avro 504K biplane at Camp Borden in 1927.
Courtesy of Canada Aviation and Space Museum/CASM-2432.

Through the support of the RCAF and the NRC, Turnbull's No. 2 design became the first variable-pitch propeller designed, built, and flown in Canada.[121] He returned to his laboratory at Rothesay to investigate the further potential of his creation.[122]

[121] The Turnbull No. 2 survives today in the collection of the Canada Aviation and Space Museum in Ottawa.

[122] Parkin, "Wallace Rupert Turnbull," 40–41.

Success or Failure?

The scramble to reach and maintain aeronautical parity in World War I stimulated the development of the airplane. The embryonic variable-pitch propeller and the community that created it had depended on direct government support for its survival. Their motivation was in response to government specifications and the challenge of developing a new technology. Researchers realized that mechanical actuation, blade retention, structural design, and designing a propeller with performance that negated any appreciable weight increase were the principal obstacles. The absence of a practical design hampered the performance of military aircraft.

Government support, however, could generate unintended consequences. In the United States, the Hart, Heath, Standard Steel, and Engineering Division series of mechanically actuated variable-pitch propellers were joint developments between private industry and the government. The failure of those programs raised the question of the government's role in the innovation and development of new technologies. Within the propeller unit, the same government personnel that evaluated the designs of private parties through whirl testing and engineering support simultaneously took those ideas and attempted to improve upon them or submitted their own designs in competition all in a supposedly unbiased environment. The differing levels of unregulated support and creative initiative blurred the lines between government and industry. The freewheeling style of the Engineering Division and the propeller unit that guided directly the US Army Air Service's first steps toward developing a practical variable-pitch propeller in the 1920s stymied innovation.

In England and Canada, inventors and engineers worked in the private sector to produce potentially viable hydraulic and electric designs. At crucial moments, they received assistance from their respective governments and in the case of Hele-Shaw and Beacham, an aircraft manufacturer, but not at any levels comparable to what occurred in the United States. Hele-Shaw and Beacham in England and Turnbull in Canada continued with their work while Caldwell and his colleagues at McCook reeled from their false start and pursued their goal from another more promising direction.

5

"The Propeller That Took Lindbergh Across"

On May 20, 1927, Charles A. Lindbergh and the *Spirit of St. Louis* took off from New York bound for Paris for the first successful solo nonstop flight from the United States to Europe. The Ryan-built airplane incorporated some of the latest aeronautical developments: a radial air-cooled engine; monoplane configuration; earth-inductor compass; and a metal, ground-adjustable-pitch propeller. Even so, the heavily laden plane barely cleared the telephone wires at the end of Roosevelt Field by twenty feet. The *Spirit*'s spectacular takeoff highlighted a critical limitation of its propeller. Built by the Standard Steel Propeller Company of Pittsburgh, the device's blades could be adjusted on the ground for efficiency for different operating conditions, primarily takeoff or cruise, but once the blades were set they could not be changed in flight. Searching for every bit of economy for the 3,610-mile flight, Lindbergh had two choices. He could have an easy takeoff with a possible fatal landing in the Atlantic short of Europe or attempt a risky one with a much more desirable landing on French soil. The metal ground-adjustable propeller was the best available technology using a promising new design that pointed the way toward more advanced propellers for high performance aircraft of the 1930s (Figure 10).

The development of the metal ground-adjustable propeller was a major technological milestone of the Aeronautical Revolution of the late 1920s and 1930s. It represented almost a decade of work to improve the overall performance of the airplane by propeller specialists working for the federal government and private industry. The two partners in this pioneering development were the propeller unit at McCook Field and the Standard

FIGURE 10 The Standard Steel ground-adjustable propeller selected by Charles Lindbergh for the *Spirit of St. Louis* reflected the state-of-the-art in 1920s propeller design. Smithsonian National Air and Space Museum (NASM A-4819-A).

Steel Propeller Company. Caldwell and his colleagues conducted research and issued specifications and contracts for the new propellers to increase overall aircraft performance. Their civilian partner, Standard Steel, developed new designs, attempted to make them practical, and produced them for profit. Tracing the development of the metal ground-adjustable propeller, known as the "Propeller That Took Lindbergh Across," reveals important themes for the history of aviation: the key role a specific device played in an event that captivated the world, the importance of government–business partnerships in solving technical challenges, and how communities perceived new innovations, especially in terms of the fundamental choices they made to push technology forward.

The Indeterminacy of Materials

Caldwell and the propeller unit at McCook Field started experimentation with metal as a material alongside their work on variable-pitch mechanisms and the establishment of the whirl-testing facilities before the end of World War I. They conducted this simultaneous work against the backdrop of addressing the pressing needs of the wartime production program centered on offering immediate solutions to problems related to materials, design, and production. The style in which they approached that challenge reflected a characteristically conservative and practical attitude toward new designs and materials. Wood was the dominant material, but it had its strengths, including light weight and low cost, and limitations, primarily the lack of long-term durability and the availability of quality stands of high-grade lumber. As a result, they kept other options in mind as they used the technology, knowledge, and creativity at their disposal to contribute to the war effort.

Initially, Caldwell believed plastic would be the ideal propeller construction material. He identified Westinghouse's Micarta electrical insulation product as an alternative to both wood and metal.[1] Micarta consisted of the phenolic resin Bakelite, one of the earliest synthetic plastics, impregnated into layers of cotton duck cloth and molded into shape under high pressure.[2] Besides propellers, Westinghouse manufactured Micarta in the form of aircraft control pulleys, casings for consumer radios, table fans, and cameras, pistol grips, and knife handles. For many in American industry, the futuristic quality of plastics, and their increased use, was an indicator of a new and rapidly approaching "Plastic Age" for humankind.[3]

Caldwell designed a Micarta propeller for the Air Service's workhorse Liberty V-12 engine in April 1918. Comparative flight tests with wood propellers revealed that Micarta propellers with their thinner blades generated higher speeds for aircraft at the same rpms.[4] McCook Field test

[1] See F. W. Caldwell and N. S. Clay, "Micarta Propellers I: Materials," NACA Technical Note No. 198 (August 1924); "Micarta Propellers II: Method of Construction," NACA Technical Note No. 199 (August 1924); "Micarta Propellers III: General Description of the Design," NACA Technical Note No. 200 (August 1924); and "Micarta Propellers IV: Materials," NACA Technical Note No. 201 (September 1924).

[2] J. A. Wordingham and P. Reboul, *Dictionary of Plastics* (New York: Philosophical Library, 1964), 15, 126.

[3] Jeffrey L. Meikle, *American Plastic: A Cultural History* (New Brunswick, NJ: Rutgers University Press, 1995), 2, 64, 68–69, 73, 89–90.

[4] F. W. Caldwell to Major Marmon, "Recommendation that Micarta Propeller Work Continue," April 13, 1918; J. G. Vincent to W. C. Potter, "Report on Development of Micarta Propeller," April 16, 1918, Folder "Propellers – January through April, 1918," Box 12, RD 3085, RG 342, NARA; Propeller Department, "Bakelite Propellers Under Test," *Bulletin of the Experimental Department, Airplane Engineering Division, United States Army* 1 (July 1918): 105–113.

pilot Maj. Rudy Schroeder flew a Le Peré LUSAC-11 biplane equipped with special gasoline, a supercharger, and a Micarta propeller to a new world altitude record of 33,143 feet on February 27, 1920.[5] Air Service DH-4s flew with one-piece, fixed-pitch Micarta propellers, including General Mitchell's personal aircraft, *Osprey*, into the mid-1920s.

Westinghouse began an aggressive marketing campaign in the aviation marketplace. Perhaps the most famous aircraft to fly with Micarta propellers were Fokker trimotors used for pioneering long-distance flights. For instance, Army Air Corps lieutenants Lester J. Maitland and Alfred Hegenberger flew the C-2 *Bird of Paradise* 2,400 miles nonstop across the Pacific from Oakland, California to Wheeler Field, Hawaii, on June 28–29, 1927. One year later, Australian Charles Kingsford Smith flew his F.VII *Southern Cross* from San Francisco to Brisbane, Australia, from May 31 to June 3, 1928.[6]

Despite the promise of the new material, Micarta propellers were not an ideal replacement for wood. A Micarta propeller for the Liberty V-12 engine weighed about eighty-three pounds, about three pounds more than its wood counterpart. Despite the similar weight, a Micarta propeller cost approximately five and a half times more than a $100 wood propeller with metal tipping. The high price cost Westinghouse a major contract with the US Post Office's Air Mail Service and other cost-conscious aerial organizations during the austere budget climate of the 1920s.[7]

The main problem with Micarta concerned its durability, which was supposedly its strongest selling point. A brand new Micarta propeller with its hard, slick, and reddish-brown surface could withstand gravel and wet grass kicked up during taxi and takeoff, as well as rain and changes in humidity or temperature, much better than a wood propeller. Small nicks, gouges, and cracks which developed in the blades over time, however, exposed the duck cloth within the Bakelite to moisture. For instance, if the airplane flew through rain, the canvas absorbed water, expanded, and cracked the Bakelite, resulting in the deterioration of the propeller. Leroy Manning of the Airplane Division of the Ford Motor Company evaluated three Micarta propellers on a Tri-Motor aircraft in January 1929 and found them "decidedly inferior from the standpoint

5 T. H. Bane to W. S. Roberts, "Bakelite Micarta Propeller," March 19, 1920, Folder "Propellers–Bakelite Micarta, 1920," Box 51, RD 3096, RG 342, NARA.

6 "Westinghouse Active at Los Angeles," *U.S. Air Services* 13 (September 1928): 40.

7 T. H. Bane to A. K. Atkins, "Drawings and Specifications for Bakelite Propellers," January 21, 1920; and O. Praeger to A. M. Huyler, April 3, 1920, Folder "Propellers–Bakelite Micarta, 1920," Box 51, RD 3096, RG 342, NARA.

of durability" after flying through "an ordinary rainstorm." He reported that was also the consensus in his conversations with airline and manufacturer representatives at the Chicago Aircraft Show. Manning's boss, Harold A. Hicks, curtly informed Westinghouse that Ford was no longer interested in Micarta propellers.[8] The Air Corps discontinued using Micarta propellers for the same reason in 1930, and Westinghouse ceased production shortly thereafter.[9] In the end, the American aeronautical community did not embrace plastic as a primary material for the modern airplanes of the 1930s and 1940s.

The continued practicality of wood and the perceived promise of Micarta confirmed the propeller unit's reservations about metal during those early days at McCook Field. Caldwell had stated in December 1917 that steel, the material thought to be the best for the purpose, was "not a suitable material for propeller construction," whereas "wood propellers may be designed for any number of rpm as far as stress is concerned."[10]

Nevertheless, the Air Service kept its options open when it contracted with John W. Smith of the Smith Fixed Radial Engine Company of Philadelphia, to construct a tubular steel propeller in March 1918. Smith had an extensive background in early aeronautics. He built his first all-metal monoplane in 1911 and submitted a ten-cylinder radial engine to the British Admiralty in 1915. Engineer William B. Stout of the Army's Technical Advisory Board, who served as McCook Field's liaison to inventors during the war, recommended Smith and his propeller to Caldwell.[11]

Smith believed he could develop a high-speed metal propeller that would eliminate the need for reduction gearing on aircraft engines, specifically the Liberty V-12. Overall, Smith submitted eight ground-adjustable, multipiece propeller designs to the Army. The basic layout of the Smith steel propeller was a forged steel hub connected to separate blades shaped from tubular steel flattened to form an airfoil and welded at the

[8] Leroy Manning to H. A. Hicks, "Westinghouse Micarta Propellers," January 17, 1929; and H. A. Hicks to L. Manning, "Micarta Propellers," January 22, 1929, Box 128, Series I, Accession 18, Aeronautics Collection, Henry Ford Museum and Greenfield Village Research Center (hereafter cited as HFM).

[9] C. W. Howard to Chief, Field Service Section, "Micarta Propellers, Design 073137," May 12, 1930, Folder "Propellers – January through July, 1930," Box 5780, RD 3193, RG 342, NARA; C. W. Howard to G. R. Miller, January 20, 1933, Folder "Propellers – General, 1933," Box 5919, RD 3230, RG 342, NARA.

[10] Frank W. Caldwell, "Destructive Propeller Tests," Propeller Memorandum No. 1, December 21, 1917, File 216.21052-1, HRA, 2.

[11] William B. Stout, *So Away I Went!* (New York: Bobbs-Merrill Company, 1951), 123.

tip. The retention threads in the blade roots and hub allowed the adjustment of pitch from fourteen to eighteen degrees. Smith claimed that his yet-to-be manufactured steel propellers were more efficient than wood propellers at high speeds in the area of 2,000 rpm, lasted longer in service, and cost less to produce as well as operate.[12] Envisioning the use of his invention on all American combat aircraft, Smith believed his propellers would be "thoroughly suited for large production and ... reliable."[13]

The Army's evaluation of the Smith propeller series during the spring and summer of 1918 proved the inventor's assertions wrong. During the ten-hour whirl test at 600 horsepower and 2,000 rpm in May, the tubular blade collapsed from excessive vibration and separated from the hub. Caldwell and the propeller unit oversaw comparative flight tests between a Smith propeller and the Engineering Division's 8–45 wood design on an Air Service DH-4 biplane powered by a Liberty V-12 engine during July and August 1918. The wood propeller proved to be more efficient and capable of rotating at higher rpms.[14]

Smith's conclusions about the tests revealed technological trends that would be persistent throughout the development history of the metal propeller. He believed the tests were inconclusive because the two propellers were not set for the same pitch, which could have easily been done due to the adjustable feature of the steel propeller. Smith contended that his propeller would have outperformed the wood propeller at takeoff and cruise if the pitch setting were correct. Perhaps the most enlightening comment Smith would make about the superiority of metal propellers to their wood counterparts was his conclusion that the thin airfoils he was using on his propeller would demonstrate a higher efficiency than the thick airfoils used on wood propellers.[15]

[12] J. W. Smith to H. E. Blood, December 10, 1918; J. G. Vincent to J. W. Smith, March 18, 1918; W. B. Stout to J. W. Smith, "Propeller Construction," April 18, 1918; J. W. Smith to W. B. Stout, May 23, 1918, Folder "Propellers, Smith, J.W., 1918–1919," Box 5370, RD 3085, RG 342, NARA; Frank W. Caldwell, "Development of the Drop-Forged Metal Propeller," *Aeronautical Engineering: Contributions of the ASME Aeronautics Division* (April–June 1930), 73.

[13] J. W. Smith to W. B. Stout, June 10, 1918, Folder "Propellers, Smith, J.W., 1918–1919," Box 5370, RD 3085, RG 342, NARA.

[14] J. W. Smith to W. B. Stout, "Test of #2 Propeller – Threaded Construction," May 2, 1918, Folder "Propellers, Smith, J.W., 1918–1919," Box 5370, RD 3085, RG 342, NARA; F. W. Caldwell, "Comparison of Thrust of Smith Steel Propeller No. 6 with Propeller 8–45," Propeller-Report No. 17, July 24–25, 1918, File 216.21053–17, BAP, HRA.

[15] J. W. Smith to F. W. Caldwell, August 12, 1918, Folder "Propellers, Smith, J.W., 1918–1919," Box 5370, RD 3085, RG 342, NARA; "Propellers – Smith Steel," File D52.43/55, WFTD, NASM.

Beginning in May 1918, inventors also submitted built-up and plate-type propellers. The builtup type was the result of welding, joining, and riveting various metal components. The Engineering Division evaluated built-up hollow-steel propellers from Frederick H. Luense of Chicago and the Reinforced Propeller and Insulating Company of New York City. It also tested a chrome vanadium steel propeller submitted by James Ingells of Muskegon, Michigan. A plate-type propeller consisted of a series of flat metal sheets stacked on top of each other and bolted together to form the shape of a propeller.[16] Hermann Faehrmann of Brooklyn, New York, submitted an eight-foot six-inch diameter plate-type propeller designed for the OX-5 engine in June 1918. McCook tests indicated that the Faehrmann propeller was inefficient and weighed three times more than a wood propeller designed for the same engine. Considering the immediate demands of meeting the American wartime aviation production program, the propeller unit could not waste time trying to correct a flawed design.[17]

The Army began to reevaluate its preliminary work on steel propellers in September 1918. Colonel Vincent suggested that Caldwell investigate the viability of experimenting with all-metal propellers, especially the Smith design. He ordered Caldwell to prepare a report that incorporated the available data on the subject as well as the opinions of other technical personnel. Caldwell was to make a definite recommendation on whether the Army should proceed with metal propeller development. The chief engineer stressed that "if this steel propeller really has definite possibilities, I think we should go forward with experimental work in a modest way, but on the other hand, I do not want to spend any money foolishly. Unless we can determine that this propeller has real possibilities, I believe the expense should be stopped."[18] While open to experimentation with new materials, the Army's engineering leadership

[16] F. H. Luense to F. W. Caldwell, May 13, 1921, Folder "Propellers – Inventions, 1921," Box 78, RD 3102, RG 342, NARA; T. H. Bane to Reinforced Propeller and Insulating Company, June 25, 1919, Folder "Propellers, Reinforced and Insulating Company–1919," Box 13, RD 3085, RG 342, NARA; T. H. Bane to J. Ingells, "Construction of Four Propeller Blades," June 14, 1919, Folder "Propellers, Ingells, Steel–1919," Box 12, RD 3085, RG 342, NARA; Caldwell, "Drop-Forged Metal Propeller," 73.

[17] H. Faehrmann to F. W. Caldwell, December 31, 1918; H. Faehrmann to Secretary of War, December 7, 1927; T. H. Bane to H. Faehrmann, "Steel Propeller," January 10, 1918; F. W. Caldwell to H. S. Martin, "Faehrmann Propeller," January 13, 1919, Folder "Faehrmann Propeller, 1918–1919, 1927–1928," Box 5698, RD 3173, RG 342, NARA.

[18] J. G. Vincent, "Smith Forged Steel Propeller," September 13, 1918, Folder "Propellers, Smith, J.W., 1918–1919," Box 5370, RD 3085, RG 342, NARA.

was clearly not convinced that metal was an ideal material for propeller construction.

Caldwell asked Col. Thurman Bane, the technical director for the Army's Division of Military Aeronautics during the war, to offer his opinion of the importance of further work on steel propellers. The possible advantages of safety, efficiency, and weight savings that metal could offer were well known, but they were not convincing enough to suggest that the nascent American propeller industry convert to metal construction. Bane's response encouraged further experimentation, but at the same time highlighted the indeterminate place that metal held in a propeller community that favored wood construction. He stated that "outside of advantages in production and gain in efficiency, which we believe are possible, we know of no advantages over wood propellers. Briefly, we consider the steel propeller worth development, but do not believe it a very vital matter."[19] Consequently, Caldwell and the propeller unit continued their experimentation with metal propellers alongside their production-oriented work on wood propellers.

Perhaps the largest stumbling block to the development of the metal propeller in the United States during World War I was that no designer could conceptualize the ideal design. The "give-and-take" of technological negotiation between the inventor and the government on the design details of the Smith propeller produced mixed results. For example, the engineers and technical officers at McCook did not approve of Smith's threaded blade roots and encouraged him to use a bolt and flange fastening system. Stout also reminded Smith that Colonel Vincent favored the bolt and flange system. As result, Smith was to receive the blueprints of the modified bolt and flange system for incorporation into his design. Stout equivocated that the adherence to Vincent's favored system "does not mean we want to dictate the design but merely supplement your ideas with the experience of our Design Department."[20] The Army was essentially telling Smith how to design a crucial component of his propeller. Ultimately, Smith became frustrated with what he interpreted as meddling on the part of Caldwell and the propeller unit and withdrew his participation entirely.[21]

[19] F. W. Caldwell to T. H. Bane, "Steel Propellers," September 20, 1918; T. H. Bane to F. W. Caldwell, October 3, 1918, Folder "Propellers – July through September, 1918," Box 5468, RD 3085, RG 342, NARA.

[20] W. B. Stout to J. W. Smith, "Propeller Construction," April 18, 1918, Folder "Propellers, Smith, J.W., 1918–1919," Box 5370, RD 3085, RG 342, NARA.

[21] Stout, *So Away I Went!*, 123.

The Smith steel propeller was, at best, the start of an ambitious military research and development program that underscored the indeterminacy of technological development. A similar steel design from across the Atlantic, the Leitner-Watts propeller from Metal Propellers Ltd., was another indicator, but whirl testing at McCook Field in 1920 revealed its long-term structural weakness.[22] In the end, the low resistance to torsional, or twisting, shear stresses, plus the expensive handwork involved with fabrication, made the tubular, built-up, and plate designs unfeasible.[23] The failure of these designs, which represented the American propeller community's first experience with metal propellers, would inform Caldwell and his colleagues' continued experimentation.

The absence of clear-cut results proving the superiority of the new metal propeller technology led to the Army's reevaluation of the program. Caldwell soon recognized that it was not the inferiority of metal as a material, but the type of construction that stymied metal propeller development. The propeller unit began work on a drop-forged propeller in 1918. Caldwell believed drop-forging was the best method for fabricating a metal propeller because it could produce a strong metal part economically. Drop-forging involved machine hammering heated metal between two cast impressions, or dies, to a predetermined shape that aligned the metal's grain patterns for maximum strength. Drop-forged products were stronger, lighter, and economically and mechanically more practical for volume production than machined or cast metals.[24]

With the construction process decided upon, Caldwell first attempted to duplicate a wood propeller design in metal. Because wood propellers gain their strength from being thick, a metal example of what was originally a wood design weighed considerably more. For a metal propeller to be competitive, its weight had to correspond with the established technology, the wood propeller. Caldwell reduced the weight of the metal propeller by decreasing its thickness, width, and overall mass. Unfortunately, the reduction of blade width decreased the overall blade area, which in turn affected efficiency. Juggling the already existing design parameters, Caldwell next increased the blade diameter – the length of the propeller

[22] L. W. McIntosh to COAS, "Leitner-Watts Metal Propellers," November 26, 1920, Folder "Propellers–General, 1920," Box 51, RD 3096, RG 342, NARA.

[23] Caldwell, "Drop-Forged Metal Propeller," 73.

[24] John Lord Bacon and Edward R. Markham, *Forge-Practice and Heat Treatment of Steel*, 3rd ed. (New York: John Wiley & Sons, 1919), 154; Samuel E. Rusinoff, *Forging and Forming Metals* (Chicago: American Technical Society, 1952), 65–67.

from tip to tip – in order to provide the necessary area to absorb the power of the engine and offset the loss in performance.[25]

Caldwell identified a thin forged-steel propeller with high aspect ratio and large diameter with low camber ratios at the tip and larger camber ratios at the root as an ideal design. This meant a long, narrow propeller that was thinner at the tips and thicker at the roots. Complementing the well-known "advantages of durability and dependability" that metal had to offer were specific performance advantages. The large diameter enabled the propeller to move a greater volume of air without losing much energy. The Wrights had realized this first, and the early flight community quickly recognized the importance of this configuration long before the momentum theory formally explained the phenomenon.[26]

The use of thinner blades, or airfoil sections, toward the tip increased the propeller's efficiency. By 1920, propeller engineers well knew that thin blade sections were ideal for high-speed applications because they did not suffer from compressibility burble, or a sharp increase in drag at high speeds. Much of that knowledge was the result of Caldwell's collaborative work with Elisha N. Fales. Published as a NACA technical report in 1920, this research represented one of the earliest fundamental investigations of airfoil efficiency at high speeds. In their quest for airfoil data at speeds that propellers encountered in flight, Caldwell and Fales created the McCook Field high-speed wind tunnel. With a test section of fourteen inches, the tunnel allowed them to test airfoils at simulated airspeeds in the 400–500 mph range. With the tunnel, they discovered that wood propellers and their thick airfoils for the purposes of structural integrity suffered severe drag limitations at high speeds. To turn faster, propellers had to be thinner, which required the strength and durability of metal.[27]

With the fundamental parameters identified, Caldwell designed an experimental four-blade, forged chrome vanadium steel propeller for the Liberty engine that was eleven feet, six inches in diameter, weighed 110 pounds, and could absorb 400 horsepower. An early metal identified for use in propellers, chrome vanadium steel was a high-strength lightweight alloy. Calculations revealed that the new design was 4 percent more

[25] Caldwell, "Drop-Forged Metal Propeller," 73.

[26] Grover C. Loening, "Comparison of Successful Types of Aeroplanes," *Aeronautics* 6 (February 1910): 44.

[27] Frank W. Caldwell and Elisha N. Fales, "Wind-Tunnel Studies of Aerodynamic Phenomena at High Speed," NACA Technical Report No. 83 (1920), 5–6, 8–10; Caldwell, "Drop-Forged Metal Propeller," 73; George Rosen, *Thrusting Forward: A History of the Propeller* (Windsor Locks, CT: United Technologies Corporation, 1984), 24, 37.

efficient than existing wood designs. The Army contracted with Standard Steel in 1920 to construct the propeller, which initiated a five-year relationship between McCook Field and the company.[28]

The McCook Field propeller unit conducted destructive whirl tests on four different forged-steel designs. The first two suffered from flutter, or torsional vibrations. In an effort to solve the flutter problem, Caldwell looked to existing knowledge on the stiffness of wood propellers and devised a method of calculating the deflection of a propeller under load. He also attempted to make the tips flexible while stiffening the area near the blade root.[29] Both proved to be major breakthroughs in the design of metal propellers.

The third solid-steel design reflected these two improvements and survived the destructive whirl test on the electric motor-driven testing stand. Caldwell and his assistants then connected the propeller to a Liberty engine, which revealed that the design was structurally weak where the blades joined the hub. A threaded connection and clamps attached the blades to the hub. Even a modified hub design that doubled the area of the threaded section failed at the base of the threads. Experimentation with solid-steel forged blades continued until 1923. Caldwell and Dicks of Standard Steel found the design of both hollow and forged steel propeller blades to be initially impractical.[30] Clearly, the destructive whirl tests showed that the "steel" in Standard Steel was impractical from the standpoint of structural integrity.

Choosing Duralumin

After 1922, Caldwell and Dicks placed a new emphasis on drop-forged propellers made from duralumin, an aluminum alloy generally known as "dural." Composed of aluminum, copper, manganese, and lesser amounts of iron, magnesium, and silicon, duralumin became a critical material during the Aeronautical Revolution. Its increased use by the aeronautical community over the course of the 1920s, particularly for airframes, reflected the overall importance of aluminum in terms of industrial growth in the United States and Europe. There was also an inherent

[28] Caldwell, "Drop-Forged Metal Propeller," 73; Material Section, Engineering Division, "Investigation of Steel Propeller X-14849," *Air Service Information Circular* 2, no. 145 (December 30, 1920): 2; *Pittsburgh Post-Gazette*, September 23, 1929.

[29] Caldwell, "Drop-Forged Metal Propeller," 73, 74.

[30] Ibid., 74.

cultural symbolism that choosing aluminum was the ultimate expression of modernity for technologists.[31]

For Caldwell, dural permitted a closer following of wood-propeller design, a practice he and other propeller specialists knew well. Identifying the "very good mechanical properties" of the alloy, he recognized that dural's lower density allowed a stiffer blade with a greater area than corresponding steel ones. Through a comparison of the blade deflections of wood and steel propellers under load, he attempted to construct a dural propeller with flexibility between the wood and steel and with weight slightly more than wood propellers. Caldwell concluded that he could emulate wood propeller design practice. The lesser density of aluminum alloy allowed the diameter of the blade root, or hub shank, to be much greater than had been possible with a steel propeller, which meant the threaded connection was much stronger. Dicks developed a cylindrical wedge that went between the blades and the hub to divert stress from the threaded blade root and prevent it from bending as had happened with the steel propellers.[32]

The use of steel versus duralumin as a propeller material was still a point of debate in 1922. A published Engineering Division progress report noted that fixed-pitch propellers made from both materials showed "considerable promise."[33] Despite the appearance of parity, the Army began to focus on propeller blades made from duralumin. Standard Steel constructed a new drop-forged solid dural propeller for the Air Service in early 1923 using Caldwell's blade design and Dicks's hub design. The destructive whirl tests of the new propeller produced dramatic results. Designed for the 400-horsepower Liberty V-12 engine, the propeller exceeded the limit of 1,400 horsepower available during a cumulative fifty-hour destructive whirl test without any damage. On a Liberty engine, the propeller ran more than 100 hours without failure. In October 1923, Caldwell and his assistant in the propeller unit, Ernest G. McCauley, filed a patent for the detachable-blade, ground-adjustable design that featured threaded blade ends, a cylindrical fastening wedge, and a solid hub.[34]

[31] Mimi Sheller, *Aluminum Dreams: The Making of Light Modernity* (Cambridge, MA: MIT Press, 2014), 10–11, 66–67, 68.

[32] Caldwell, "Drop-Forged Metal Propeller," 74; *Pittsburgh Post-Gazette*, September 23, 1929.

[33] A. H. Hobley, "Achievements of the Engineering Division for the Calendar Year 1922," *The Slipstream* 4 (January 1923): 23.

[34] Frank W. Caldwell and David A. Dickey, "Fifty Hour Endurance Test of Standard Steel Propeller Company Aluminum Alloy Propeller for Liberty 12 Engine," Whirling Test No. 545, April 23, 1923–February 1, 1924, Folder "Standard 1923–1924," Propeller Subject Files, United States Air Force Museum Research Division, Dayton, Ohio; Caldwell,

Flight tests at McCook in 1923 showed a 7 percent increase in efficiency over wood propellers. Additional tests by the Army, Navy, and commercial operators, however, revealed problems with the threaded connection of the blades to the hub. The blades proved to be noninterchangeable, difficult to keep in balance and to hold at a correct pitch, and difficult to remove if foreign matter such as dirt or grime were present. The "neat fit" of the threaded joint was the problem.[35]

Efforts to alleviate the problem of the serviceability of the threaded hub and blade root led to the complete elimination of the joint. Dicks and the engineers at Standard Steel developed a split, or two-piece, steel hub held together by a clamping ring. The assembly compensated for small errors in the vertical balance of the blades. Instead of screwing the blade into a solid hub, Dicks used retaining shoulders at the blade root, which allowed both interchangeability and pitch changes. The shoulders also absorbed the centripetal force of the engine that was transmitted through the crankshaft and hub. The earliest split-hub propeller used a single shoulder and passed its destructive whirl test, but failed during a mechanical stress analysis test. Caldwell was searching for an equal balance between the shearing strength of the blade shoulder and the tensile strength of the blade root, which it did not exhibit. A second split hub design, one using a double shoulder at the root, exhibited equalized shear, crushing, and tensile strengths.[36] Caldwell and McCauley filed their patent application for a crude detachable-blade ground-adjustable design featuring blade-retaining shoulders, an index scale for pitch adjustment, a clamping ring, and a two-piece hub in October 1923, the same month that they filed their patent for the threaded hub described earlier.[37]

Caldwell compared the results of tests conducted by the Materials Section of the Engineering Division at McCook Field on Standard Steel's threaded, split single-shoulder, and split double-shoulder type propellers. The threaded and single-shoulder types failed to exceed a load of 250,000 pounds under combined shear and tension loads. The double-shoulder did not fail until reaching a load of 320,000 pounds.[38]

"Drop-Forged Metal Propeller," 74–75; F. W. Caldwell and E. G. McCauley, "Propeller for Aircraft," US Patent No. 1,608,754, November 30, 1926.

[35] Caldwell, "Drop-Forged Metal Propeller," 75.

[36] Shear occurs when the structural layers of a substance shift laterally under pressure. A crushing force exerts compression on the structural integrity of a substance. Tensile strength refers to the ability of a substance to resist structural forces acting in opposite directions, or stretching.

[37] Caldwell, "Drop-Forged Metal Propeller," 75; F. W. Caldwell and E. G. McCauley, "Propeller for Aircraft," US Patent No. 1,608,755, November 30, 1926.

[38] Caldwell, "Drop-Forged Metal Propeller," 75.

According to Caldwell, the double-shoulder split-hub was "adopted as standard" because it was "obviously stronger in addition to its other advantages." He declared that "in spite of severe abuse," none of the approximately 20,000 blades produced between 1923 and 1930 "ever pulled out of the hub," a testament to the superiority of the design. The double-shoulder design was the "solution of the technical difficulties in producing an efficient design of moderate weight and satisfactory factor of safety."[39] Through the use of the split-hub, the retaining shoulders, and the clamping ring, the Army and Standard Steel achieved a major milestone in propeller design and construction: the introduction of a standardized hub (Figure 11).

The Army-Standard Steel ground-adjustable hub was strong and simple. The two-piece hub clamped tightly around the blade shoulders, which produced a large amount of friction between the blade shank and the hub to prevent unwanted changes in blade angle. The blade retaining shoulders carried the centrifugal load. The key to the design was separating the centrifugal load concentrated in the blade roots from the clamps that kept the entire assembly together. After continued design refinement, Caldwell filed his patent for the standardized split-hub with dual retaining shoulders and clamping rings, which he assigned to Standard Steel, on March 28, 1925 (Figure 12).[40]

The next step was to standardize the dural blade to make the complete propeller a "shelf item" generally available for the aviation market.[41] The myriad of aircraft and engines on the market in the early 1920s required propellers that varied in diameter from six to sixteen feet and capable of handling between 60 and 800 horsepower. Caldwell and Dicks established the basic design parameters so customers could provide their aircraft and engine design data. Then all Standard Steel had to do was select the best propeller blades for the application from a predetermined table.

Caldwell believed the development of a series of blade ends characterized by differing widths, diameters, and pitch would make the new propeller a valuable addition to the aviation market. In an article published in 1930, he recalled a major breakthrough:

> Fortunately, it was found from model tests in the [McCook Field] wind tunnel that the efficiency of the propeller was not appreciably reduced when the pitch

[39] Caldwell, "Drop-Forged Metal Propeller," 75; *Standard Steel Propellers* (Pittsburgh, PA: Standard Steel Propeller Company, 1929), 10.

[40] F. W. Caldwell, "Propeller Hub," US Patent No. 1,653,943, December 27, 1927.

[41] R. K. Smith, "Better: The Quest for Excellence," in *Milestones of Aviation*, ed. John T. Greenwood (New York: Hugh Lauter Levin Associates, 1995), 240.

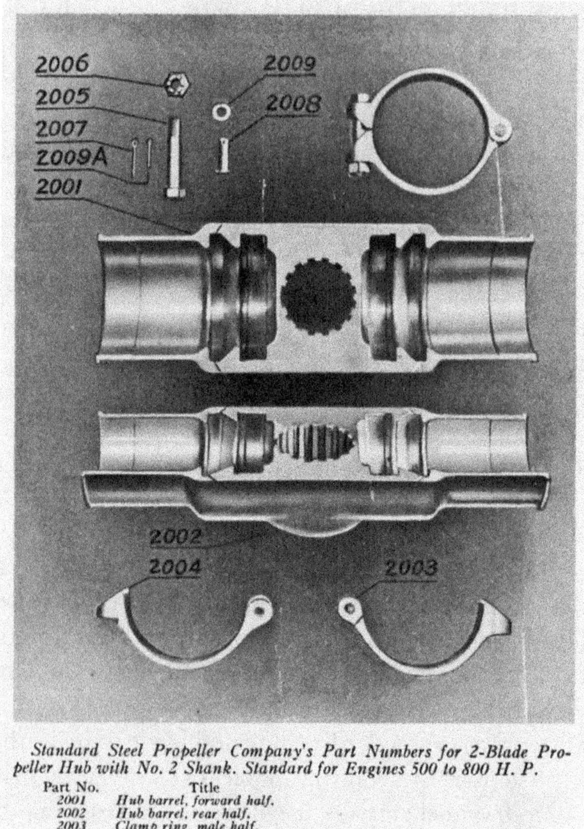

Standard Steel Propeller Company's Part Numbers for 2-Blade Propeller Hub with No. 2 Shank. Standard for Engines 500 to 800 H. P.

Part No.	Title
2001	*Hub barrel, forward half.*
2002	*Hub barrel, rear half.*
2003	*Clamp ring, male half.*
2004	*Clamp ring, female half.*
2005	*Clamp ring bolt.*
2006	*Clamp ring bolt nut.*
2007	*Cotter pin for clamp ring bolt.*
2008	*Clevis pin for clamp ring.*
2009	*Washer for clamp ring clevis pin.*
2009A	*Cotter pin for clamp ring clevis pin.*

FIGURE 11 The split hub and clamping ring design of the Standard Steel ground-adjustable propeller. United Technologies Corporation Archive via Smithsonian National Air and Space Museum (NASM 9A07404).

was altered from the designed value by rotating the blades in the hub sockets. Consequently, it was possible to produce a wide range of designs without an excessive number of dies so that the cost of individual designs has been held within the practical range of requirements.[42]

The ability to change blade pitch enabled construction of propellers for designated power ranges. Standard Steel did not have to design

[42] Caldwell, "Drop-Forged Metal Propeller," 75.

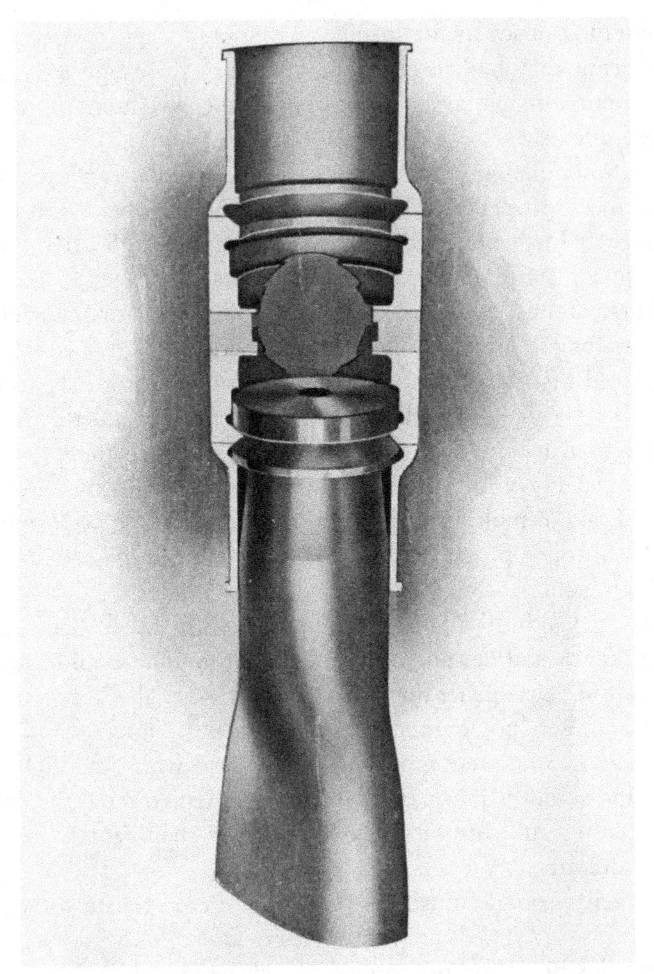

FIGURE 12 The double shoulder retention system found in the hub and blade root of the Standard Steel ground-adjustable propeller. United Technologies Corporation Archive via Smithsonian National Air and Space Museum (NASM 9A07405).

propeller blades for individual engines, only for power ranges, which cut down on the number of different blade designs. Most important, the company could "standardize" a two-blade propeller in three different lengths for the 500–800 horsepower range and suggest the appropriate pitch setting for a particular engine, airplane, and operating condition. From the business standpoint, these advantages facilitated reduced production costs and increased profits. Wood propeller manufacturers had to design their products for specific engines and single operating ranges, which meant they had to either manufacture according to demand or else maintain a much larger inventory in expensive humidity-controlled storage facilities.

Aircraft competing in the October 1923 National Air Races featured the new metal propellers. The Engineering Division fielded three DH-4s in the Liberty Engine Builder's Race, each equipped with a two-blade, ten-foot-diameter Standard Steel ground-adjustable propeller. The aircraft placed first, second, and third. The Army lent the US Air Mail Service the propellers for the Air Mail Pilots Race, where the DH-4s equipped with them placed first, second, and fourth. An adaptation of the basic two-blade Liberty engine propeller was the eight-foot-diameter three-blade propeller used by the Navy-Wright F2W-1 racers in the 1923 Pulitzer Trophy Race. Navy Lt. Steven W. Calloway placed fourth with an average speed of 230 mph. In addition to their remarkable performance, the ground-adjustable propellers used by the American military exhibited a "very high factor of safety" overall.[43]

By 1925, Caldwell and Dicks, working with the propeller unit and Standard Steel, had developed a practical ground-adjustable, aluminum alloy, multipiece propeller for the aviation market. The standard hub and blade for different power ranges reflected the best-known wood propeller design practice, but were more durable, able to withstand higher power ranges, and could be more easily produced. Metal offered the durability, strength, and a measure of safety required by the new commercial and military aircraft.

The metal ground-adjustable propeller offered several other advantages over wood fixed-pitch propellers. Overall, the adjustable hub provided flexibility for pilots flying in different operating conditions. Standard Steel guaranteed that a user of their propeller could adjust the pitch to "save on his gasoline bill" for cruising or alter the pitch for improved takeoff performance when flying from a small airfield. With the appropriate tools, a pilot could make the adjustments in the field. The propeller resisted the effects of rain, hail, and other climatic conditions to "give the same excellent service whether in frigid Alaska, in dry, hot, Arizona, or in the humid climate of Panama." Finally, during a landing

[43] Engineering Division to Information Officer, Tenth School Group, Kelly Field, Texas, "Data Re: Duralumin Propeller," June 26, 1924, Folder "Heath [Reed], 1922–1925," Box 5582, RD 3143, RG 342, NARA; J. Parker Van Zandt, "The 1923 Pulitzer Air Classic," *U.S. Air Services* 8 (November 1923): 15, 21.

accident metal propellers simply bent back rather than disintegrating, and could be reformed by the manufacturer and used again.[44]

Once in service, the new propellers required a new method of inspection called etching to reveal manufacturing defects and operational fatigue. Both the military services and the commercial operators used this process, which was neither a safe or environmentally friendly operation. Individuals tasked with performing the operation had to be very careful with the dangerous chemicals it required. Despite the hazards, etching effectively revealed structural cracks that could not be seen by the naked eye. Inspectors used cloth wrapped on the end of a stick to coat the propeller with a 10 to 20 percent sodium hydroxide solution (caustic soda or lye). The solution left a heavy black coating on the surface of the blade as it oxidized a thin layer of aluminum. Removing the sodium hydroxide solution from the propeller's surface with nitric acid left visible black traces in any structural cracks or deformities in the metal. If the process revealed cracks, inspectors were instructed to discard the propellers. The exposure of dynamic stresses to the blades required that operators perform regular etching inspections every 100 hours because there was no extensive knowledge of just how long a drop-forged dural blade could remain in service.[45]

Despite the achievement of introducing the metal multipiece propeller into operational service, vibration caused by rotation, the impulses of the engine, irregular airfoil sections, and variations in aerodynamic load remained a persistent problem. Caldwell's and Dicks's knowledge of blade aerodynamics, mechanical loads, material strengths, and better blade retention techniques solved most of the vibration problems. Still, an unbalanced and unaligned propeller with unequal pitch settings posed a real danger to the airplane and its passengers. Another continued cause of vibration was that the blades resonated in frequency with the engine.[46] Caldwell stated that the use of a different propeller often offset

[44] *Standard Steel Propellers*, 5–6.

[45] "Instructions for the Inspection of Curtiss-Reed Metal Propellers (Twisted Type), n.d. [1926], Folder "Reed Propellers, 1926–1929, 1932," Box 5867, RD 3216, RG 342, NARA; Caldwell, "Drop-Forged Metal Propeller," 75; *Instruction Manual for the Ford Tri-Motor* (Detroit, MI: Ford Motor Company, 1929; reprint, Appleton, WI: Aviation Publications, 1973), 101.

[46] Fred E. Weick, Interview by Harvey H. Lippincott, March 4, 1981, Folder "Interview: United Aircraft and Hamilton, 1929-1930," Box 3, FWP, NASM, 7, 26, 65; United States Department of Commerce, *Trends In Airplane Design*, 15; Thomas Foxworth, *The Speed Seekers* (New York: Doubleday, 1975), 74; *Standard Steel Propellers*, 9; Rosen, *Thrusting Forward*, 40.

the problem, but it was commonly "a matter that has to be taken care of by the airplane designer" rather than the responsibility of the propeller manufacturer. If a wood propeller exhibited those traits, the manufacturer had to scrap it. Caldwell believed careful and periodic inspections of the metal multipiece propeller alleviated those problems. In his view, propeller design problems now became propeller maintenance problems, easily solved because all maintenance personnel had to do was replace a defective blade or hub component.[47] Maintenance meant careful adherence to inspection schedules and sophisticated inspection techniques.

Metal was quickly becoming the most attractive alternative to wood in the immediate postwar period, but it was not universally adopted by the aeronautical community. A 1925 Engineering Division internal report on future trends in aeronautical development identified the material's superior durability and resistance to aerodynamic and climatic conditions. It also noted that metal propellers were "less subject to injury from long grass and similar obstructions when taxiing on the ground." But, the report made clear, the future of metal propellers was not yet assured. While the introduction of metal propellers paralleled the increased introduction of metal into overall aircraft construction, the report asserted, "metal construction has not as yet obsoleted [*sic*] wood construction."[48]

The Standard Steel Propeller and the Rise of American Aviation

The appearance of the new propeller in the mid-1920s came at just the right time for military and commercial aircraft. During a routine takeoff exercise at the Norfolk Navy Yard, one of the Martin T3M torpedo-bomber scouts slated for use on the US Navy's new aircraft carriers *Saratoga* and *Lexington* lost its wood propeller on takeoff. The accident investigation revealed that the propeller weakened under the power of the 575-horsepower Wright T-3B Typhoon V-12 engine. In reaction, the head of Navy's Bureau of Aeronautics (BuAer) engine section, Lt. Cdr. Eugene E. Wilson, authorized procurement of 100 duralumin propellers for use on an improved variant of the aircraft, called the T4M, which featured equally powerful, but lighter, radial engines. This purchase initiated

[47] Caldwell, "Drop-Forged Metal Propeller," 76.

[48] Herbert L. Wilson, "Trend of Aeronautical Development," 1925, Air Service Technical Data Section, McCook Field, Dayton, Ohio, Box, "Units: McCook/Wright Field – General," United States Air Force Flight Test Center History Office, Edwards Air Force Base, California.

the service use of the Standard Steel ground-adjustable propeller by the US government.[49] Standard Steel advertising argued that its new propeller was standard equipment for the T4M because the improved aircraft relied on "elements having performance for the keynote."[50]

At the same time the US government and commercial operators were increasingly accepting metal propellers in the mid-1920s, Charles A. Lindbergh, a relatively obscure airmail pilot flying the St. Louis-to-Chicago route, began planning a solo transatlantic flight from New York to Paris. He wanted to win the much-publicized $25,000 Orteig Prize for the first nonstop flight between the two world cities, but he also saw it as an opportunity to showcase the capability of aviation. Possessing a visionary yet practical understanding of the limits of 1920s aeronautical technology, Lindbergh believed that a flight of over 3,000 miles was possible.[51] He recognized the synergy of aircraft design where the overall technology – the airplane – was actually a system of technologies intermeshed to best achieve its designer's and operator's airminded purpose.

Lindbergh contracted Ryan Airlines of San Diego, California, in February 1927 to build a single-seat, long-range monoplane. It was to be powered by a single Wright Whirlwind J-5C radial engine and had to be capable of carrying more than 400 gallons of fuel. Much has been written about the importance of the Whirlwind radial engine and its role in the success of the *Spirit of St. Louis*. Celebrated as one of the cornerstone technologies of the Aeronautical Revolution, the nine-cylinder, 225-horsepower engine was capable of performing up to 1,500 hours with regular overhauls.[52] But any engine, no matter how good, is only half of an aircraft's propulsion system. No airplane could get off the ground without the propeller to absorb and transmit the engine's energy into forward thrust.

Founded in 1922, Ryan Airlines employed thirty-five men and women in an abandoned fish cannery on the San Diego waterfront, and had developed a good reputation from building single-engine monoplanes for West Coast mail services. The small company's chief engineer, Donald Hall, had only joined the staff the same month that

[49] Eugene E. Wilson, *Slipstream: The Autobiography of an Air Craftsman* (Palm Beach, FL: Literary Investment Guild, 1967), 77–78, 155–156; Rosen, *Thrusting Forward*, 38, 40.

[50] "Standard Steel Propellers Are Noted for Their Performance," *U.S. Air Services* 12 (June 1927): n.p.

[51] Richard P. Hallion, "Charles A. Lindbergh and Aviation Technology," in *Charles A. Lindbergh: An American Life*, ed. Tom D. Crouch (Washington, DC: National Air and Space Museum, 1977), 39.

[52] "Mechanical Features of the Trans-Atlantic Plane," *American Machinist* (June 2, 1927): 929.

Lindbergh first contacted the company. Hall and Lindbergh's design logic focused on long-range and economical performance that maintained a balance between a myriad of design considerations. Of the latter, safety was a primary concern for Lindbergh. The new airplane had enough power to lift its over 5,000 pound weight off the ground. Lindbergh wanted the cockpit located behind, rather than in front of, all gas and oil tanks so that the pilot would not be crushed in a crash. In addition, the airplane contained the most advanced long-range navigation instrumentation available, including a state-of-the-art earth inductor compass. The Ryan Airlines-Lindbergh collaboration resulted in a highly-advanced long distance airplane known as the NYP (New York-to-Paris), and more widely known by its name: the *Spirit of St. Louis*.[53]

By the time Lindbergh started planning his legendary solo transatlantic flight in the mid-1920s, the ground-adjustable propeller was widely available in both the commercial and military markets. Important governmental legislation in the United States had created a market for new aircraft and equipment. The Air Mail Act of 1925, sponsored by Congressman Clyde Kelly of Pennsylvania, turned over government operated air mail routes to private carriers through postal contracts. This move, also known as the Kelly Act, proved to be the beginning of the American commercial aviation industry. One of the first companies to receive an airmail contract, Western Air Express, operated Douglas M-2 mail planes equipped with Standard Steel propellers on the Los Angeles-to-Salt Lake City route beginning in April 1926.[54] The Army Air Corps Act of 1926, which modified the organization of the old Air Service to broaden its representation within the War Department, authorized a five-year plan for expanding personnel and aircraft procurement. The Army's new Keystone LB-5 biplane bombers and Douglas O-2 observation planes, which formed the nucleus of the Air Corps'

[53] Charles A. Lindbergh, *The Spirit of St. Louis* (New York: Scribner, 1953; reprint, St. Paul, MN: Minnesota Historical Society Press, 1993), 82–83, 85–86; Ryan Aeronautical Company, "The Story Behind the *Spirit of St. Louis*," 1957, Ryan NYP *Spirit of St. Louis* Notebooks, Vol. 4, San Diego Aerospace Museum Library and Archives; Donald A. Hall, "Technical Preparation of the Airplane 'Spirit of St. Louis,'" in *Charles A. Lindbergh: An American Life*, 84, 86; Roger E. Bilstein, *Flight Patterns: Trends in Aeronautical Development in the United States, 1918–1929* (Athens: University of Georgia Press, 1983), 3; Hallion, "Lindbergh and Aviation Technology," 40.

[54] Smithsonian National Air and Space Museum Collections Database, "Douglas M-2," 2005, http://collections.nasm.si.edu (Accessed December 29, 2009); Donald Dale Jackson, *Flying the Mail* (Alexandria, VA: Time-Life Books, 1982), 99, 155.

embryonic strategic and tactical visions of the future, featured Standard Steel propellers, as well.[55]

Two of Lindbergh's transatlantic competitors, René Fonck in the Sikorsky S-35 *New York/Paris* and Noel Davis in the Keystone Pathfinder *American Legion*, also had Standard Steel propellers installed on their ill-fated trimotor behemoths that crashed in September 1926 and April 1927, respectively.[56] The tragedy of those crashes notwithstanding, Standard Steel was a known quantity in the aeronautical marketplace. When it came time to decide on what he wanted for his transatlantic airplane, Lindbergh simply stated, "I want a metal propeller."[57]

By the time Lindbergh had contacted Standard Steel in the spring of 1927, the company had relocated from the city to a new plant and general offices in nearby West Homestead, which was the former home of a cap-pistol manufacturer, the Federal Buster Corporation.[58] Southeast of Pittsburgh on the Monongahela River, West Homestead was part of the massive complex of steel mills and other factories that spurred on American industrialization during the late nineteenth and early twentieth century.

Standard Steel sent four nine-foot-diameter dural ground-adjustable propellers built according to their specification No. 1519 to Ryan Airlines. The 1519 incorporated a scaled-down version of the pioneering 070717 blade designed for use on Liberty V-12 engine-powered aircraft such as the DH-4 and Douglas M-2 mail plane and featuring the widely popular Clark Y airfoil section. Inspected by Standard Steel's assembly foreman and inspector, Alexander F. Mannella, the propellers bound for Ryan consisted of twelve pieces designed specifically for use with the Wright Whirlwind J-5C radial engine.[59]

Like all aspects of the design of the *Spirit of St. Louis*, Lindbergh carefully calculated the role of the propeller through his systems approach.

[55] Maurer Maurer, *Aviation in the U.S. Army, 1919–1939* (Washington, DC: Office of Air Force History, 1987), 191.

[56] David Nevin, *The Pathfinders* (Alexandria, VA: Time-Life Books, 1980), 53–54, 66, 72.

[57] Lindbergh, *Spirit*, 81.

[58] *Polk's Pittsburgh City Directory, 1926* (Pittsburgh, PA: R. L. Polk, 1926), n.p.; *Polk's Homestead City Directory, 1927* (Pittsburgh, PA: R. L. Polk, 1927), n.p.; *Pittsburgh Post-Gazette*, September 23, 1929; "Standard Steel Propeller Company," *Aviation* 25 (December 1, 1928): 1759; *Homestead Messenger* (Golden Jubilee Edition), October 9, 1930.

[59] G. T. Lampton to Chief, Experimental Engineering Section, "Bomber Propellers," June 29, 1927, Folder "Propellers – January through July, 1927," Box 5663, RD 3163, RG 342, NARA; Hamilton Standard Company Press Release, "Seventy Years Ago A Hamilton Standard Propeller Powered the *Spirit of St. Louis* to Le Bourget," (n.p., n.d. [1997]).

Lindbergh and Hall asked Standard Steel to recommend a specific blade pitch that would help them achieve the range needed to reach Paris. They received the following response:

> 15.5 degrees setting probably necessary on your monoplane to get takeoff with heavy load. Fuel economy will be improved on higher pitch setting. If takeoff is satisfactory with 15.5 setting suggest try 16.5 as this will improve fuel economy.[60]

Standard Steel simply calculated the best blade angle – the propeller itself was essentially "off the shelf." Acting on Standard Steel's recommendation and influenced by time constraints, Lindbergh accepted the compromise setting that favored cruise, which was a gamble because it provided minimum efficiency at takeoff. The metal ground-adjustable propeller offered him the flexibility he needed when making the technological decisions to ensure the success of the flight. Wood and metal fixed-pitch propellers could not. Hall and Lindbergh were able to set the blade angle at 16.25 degrees to increase cruise efficiency. The Wright J-5C produced 190 horsepower at 1,545 revolutions per minute, which generated a static thrust of 700 pounds. The maximum cruise efficiency of the *Spirit*'s propeller was 74 percent.[61]

Ryan Airlines completed the *Spirit of St. Louis* on April 28, 1927, and Lindbergh quickly put it through its flight trials. He performed the tests to confirm the theoretical performance curves calculated by Hall and to determine takeoff distances with fuel loads up to 300 gallons. Anxious to beat his competitors, Lindbergh left for New York by way of St. Louis on May 10, 1927, and, on the way, succeeded in breaking a new transcontinental record. Delayed by rainy weather in New York, Lindbergh was not able to leave Roosevelt Field until the morning of May 20, 1927.[62]

At over a mile long, Roosevelt Field was the only suitable place for heavily laden airplanes to take off in the New York area. Unlike the hardscrabble Camp Kearney airfield near San Diego, which Lindbergh had used for his heavy takeoff tests, obstacles surrounded Roosevelt Field – telephone wires and low, tree-covered hills – and he was quite aware of the dangers involved. On the positive side, the Long Island field was near sea level and smoother. The *Spirit* had to use almost every inch of the

[60] Lindbergh, *Spirit*, 116.

[61] "The Ryan Transatlantic Monoplane," *Jane's All the World's Aircraft, 1927* (London: Sampson Low, Marston and Company, 1927), 269; Ev Cassagneres, *Spirit of Ryan* (Blue Ridge Summit, PA: TAB Books, 1982), 53, 54; Hamilton Standard, "Seventy Years Ago A Hamilton Standard Propeller Powered the *Spirit of St. Louis*."

[62] Hall, "Technical Preparation," 4–5; Walter S. Ross, *The Last Hero: Charles A. Lindbergh* (New York: Harper and Row, 1968), 93–98.

field to lift its 5,000 pound load into the air. Not only did running at full throttle for so long place an enormous strain on the engine but the weight could have easily damaged the landing gear, affecting chances of making a safe landing 3,600 miles away in Paris. Lindbergh recalled his approach to the problem years later, "I decided to 'feel' the plane off the ground just as I had often done in underpowered planes while barnstorming. If I felt I was not going to get off the ground, my plan was to simply cut the throttle."[63] This understatement represents an extreme element of risk in an otherwise carefully planned flight. As it happened, the straining *Spirit of St. Louis* lifted off the ground within 1,000 yards of the telephone wires at the end of the field.[64] Perhaps the greatest drama of Lindbergh's journey was his takeoff, because, as he wrote in *The Spirit of St. Louis*, his 1953 Pulitzer Prize-winning memoirs of the flight: "My propeller is set for cruising, not for takeoff."[65]

This account of the takeoff differs greatly from Lindbergh's previous account of the flight, *WE*, written in the immediate wake of his arrival in Paris. At the time, the visionary flyer did not want to expose the vulnerabilities of existing aeronautical technology in a book that he knew would be widely read. While still declaring that he cleared a tractor by fifteen feet and the power lines by twenty "with a fair reserve of flying speed," Lindbergh recounted that it was clear from early in the takeoff roll that he was going to get off the ground.[66]

From the moment he returned to the United States from France, Lindbergh regarded his transatlantic flight as the culmination of a twenty-five-year collaborative effort conducted by a complex and intertwining aeronautical community. Lindbergh and Ryan Airlines used the products of more than one hundred manufacturers in the construction of the *Spirit of St. Louis*.[67] He believed that the evolution of those integrated technologies, among them the Standard Steel metal ground-adjustable propeller, were the key to the future success of aviation.

For Standard Steel, the excitement over Lindbergh's flight generated marvelous publicity while creating a willing market for its products.

[63] Cassagneres, *Spirit of Ryan*, 54.

[64] Robert R. Osborn, "On The Atlantic Flight Preparations," *Aviation* (May 23, 1927): 1084; Cassagneres, *Spirit of Ryan*, 53; Lindbergh, *Spirit*, 112–113, 125, 128, 149, 155, 187; Rosen, *Thrusting Forward*, 40, 42.

[65] Lindbergh, *Spirit*, 183.

[66] Charles A. Lindbergh, *WE* (New York: Putnam, 1927), 216.

[67] John W. Ward, "The Meaning of Lindbergh's Flight," in *Studies in American Culture: Dominant Ideas and Images*, ed. Joseph J. Kwiat and Mary C. Turpie (Minneapolis: University of Minnesota Press, 1960), 33, 37.

And the company did not lose the opportunity to connect itself with the solo transatlantic flight. All correspondence leaving the front office at Homestead bore an imprint from a large rubber stamp proclaiming, "The Propeller That Took Lindbergh Across." It also listed other "famous flights made with Standard Steel Propellers," including long-distance, exploratory, and endurance flights completed by the Army, Navy, and famous aviators of the era such as Amelia Earhart, Ruth Elder and George Haldeman, Arthur C. "Art" Goebel, Navy Lt. William V. Davis, and William S. Brock and Edward F. Schlee (Table 1).[68] Advertisements in *Aero Digest, Western Flying*, and other periodicals urged consumers to "Follow the choice of experience!" and to choose the same propeller Lindbergh had used on the *Spirit of St. Louis*.[69] Standard Steel's connection with Lindbergh and other well-known pilots, including Frank Hawks, Wiley Post, and Clarence Chamberlin, continued to enhance the company's prestige.

A survey of the leading newspapers, aviation periodicals, and surviving photographs of the late 1920s also reveals the conspicuous presence of Standard Steel propellers, highly polished and displaying the easily recognizable red, silver, and gold oval company decal, in both commercial and military aviation. Standard Steel propeller sales doubled in 1927, and by the end of 1928 the company had sold over a million dollars' worth of propellers. *Aviation* reported in October 1928 on the purchase of 140 Standard Steel two- and three-blade propellers by the Fokker Aircraft Corporation alone for use on its trimotor transports.[70]

The increased manufacturing demands influenced two major changes. First, Standard Steel built a new three-story plant beside the older factory in West Homestead. Second, the company reorganized as a corporation, went public, and listed stock on Wall Street beginning in January 1929.[71]

The success of Lindbergh's flight contributed to advancing aviation in the United States and exhibited the growing superiority of American aviation technology. By the early 1930s, ground-adjustable propellers were an essential part of the expansion of aviation around the world.

[68] Standard Steel Propeller Company to R. S. Carter, August 10, 1929, Author's collection.

[69] "Standard Steel Propellers: Follow the Choice of Experience!" *Aero Digest* 12 (May 1928): 787; "The *Spirit of St. Louis* Is Equipped with a Standard Steel Propeller," *Western Flying* 3 (June 1927): 35; "Propeller on Spirit of St. Louis: It Pulls To Victory," *Greater Pittsburgh* (June 25, 1927): 6.

[70] "Orders 140 Propellers," *Aviation* 25 (October 13, 1928): 1194.

[71] "Standard Steel Propeller Company," *Aviation*, 1759; *Pittsburgh Post-Gazette*, September 23, 1929; *Homestead Messenger* (Golden Jubilee Edition), October 9, 1930; Russell Trotman, "Turning Thirty Years," *The Bee-Hive* 24 (January 1949): 22.

TABLE 1 *Famous flights made with Standard Steel ground-adjustable propellers*

Charles A. Lindbergh	Ryan NYP *Spirit of St. Louis*	First solo transatlantic crossing, May 20–21, 1927
US Army Air Service	Fokker C-2A ? (Question Mark)	Demonstration of in-flight refueling and 150 hour endurance record, January 1929
Amelia Earhart	Fokker F.VII	First woman to cross the Atlantic Ocean by air, June 1928
Ruth Elder and George Haldeman	Stinson Detroiter *The American Girl*	Transatlantic flight (unsuccessful), October 1927
US Army Air Service	Loening OA-1A Amphibian	Pan-American goodwill flight, 1926
Lt. C. C. Champion	Wright XF3W Apache	World altitude record of 38,418 feet, July 1927
Arthur C. "Art" Goebel	Travel Air 5000 *Woolaroc*	Winner, Dole Race to the Hawaiian Islands, August 1927
William S. Brock and Edward F. Schlee	Stinson Detroiter *Pride of Detroit*	Around the world flight (unsuccessful), August-September 1927
John Henry Mears and Capt. Charles B. D. Collyer	Fairchild FC-2W *City of New York*	Around the world flight (partial), June-July 1928
Arthur C. "Art" Goebel	Lockheed Vega 5 *Yankee Doodle*	Transcontinental Los Angeles to New York flight, August 1928
Capt. George H. Wilkins	Lockheed Vega 1	Transarctic flight, April 1928
Frank Hawks	Lockheed Air Express	Transcontinental record, February 1929
Cmdr. Richard E. Byrd	Ford Tri-Motor	Antarctic expeditions, 1928–1929
Dale Jackson and Forest O'Brine	Curtiss Robin C-1 *St. Louis Robin*	420 hour endurance flight, July 1929

Note: Standard Steel Propeller Company to R. S. Carter.

They were also proving to be a durable and reliable technology. As Pan American Airways expanded from the Caribbean southward into South America in the late 1920s, it operated a small fleet of Sikorsky S-38 amphibian powered by two 400-horsepower Pratt & Whitney Wasp radials. Pan American's Chief Engineer Andre A. Priester retired a Standard Steel propeller from one of those aircraft in late 1933 and donated it to the Smithsonian Institution to prove a point. That single propeller, like many in the Pan American inventory, had accumulated 2,500 hours in the air over 275,000 miles in Mexico and Central America. Air-cooled radial engines, considered to be the epitome of reliability and design, had a maximum operational life of 2,000 hours.[72] The combination of the two technologies into the propulsion system offered the promise of the airplane as a global technology.

The Army acknowledged its role in fostering a fundamental technology that facilitated the growth of aviation. The Chief of the Airplane Branch at Wright Field recognized metal propeller development, specifically the ground-adjustable propeller, as the most outstanding contribution to the airplane for the period 1926 to 1931. In comparison, the second and third outstanding technologies were the all-metal monocoque fuselage and wing, or airframe, which reduced the need for heavy internal bracing and frame members, and landing wheel brakes, which allowed larger airplanes to land on significantly shorter runways. Moreover, these Army aeronautical engineers viewed metal as a superior material for quantity production because of the type of machinery and processes used, along with their view that there was "a practically unlimited source of supply" of metal, as opposed to the limited sources of aviation-grade wood suitable for propellers.[73] The Airplane Branch based its views on the need for a standardized aeronautical technology that would contribute to the success of future aerial military campaigns.

The Airplane Branch's assessment of the ground-adjustable propeller as an outstanding innovation offers a contrast to the development of the second-ranked metal airframe, which was another crucial element of the

[72] Pan American Airways to P. E. Garber, December 26, 1933; W. H. Hunton to P. E. Garber, August 14, 1934; and D. Wendt, "Memorandum to Accession Record," November 8, 1983, in "Standard Steel Propeller," File A19350003000, RO, NASM.

[73] F. O. Carroll to Chief, Engineering Section, "The Three Outstanding Developments of the Last Five Years Relative to the Airplane Structure," April 27, 1931, Folder 400.112, "Experimental Projects and Developments, 1931," Box 5812, RD 3202, RG 342, NARA.

modern airplane that had its beginnings at McCook Field.[74] It has been argued that a culturally charged "progress ideology of metal" shaped the aeronautical community's transition from wood to metal aircraft structures. To its believers, wood represented outdated preindustrial craft traditions and metal symbolized the glorious technological future of aviation. Advocates for metal, which included the Army's influential engineering and procurement personnel at McCook and Wright fields, embraced its symbolism, used progress-oriented rhetoric to argue its introduction was inevitable, and influenced the development of metal aircraft. In the process, they let further development of wood aircraft structures lapse while possessing little evidence of metal's superiority in terms of data and practical experience.[75]

While the ground-adjustable propeller ended up being made from metal, Frank Caldwell and the propeller unit's simultaneous consideration of wood, plastic, and metal designs belie the existence of that type of ideology driving technological innovation. For them, the debate centered on impracticality versus practicality, not wood versus metal or the past versus the future for that matter. A dural propeller was simply a better piece of equipment than a wood propeller. Metal propeller blades could be designed to be thinner, which made them more efficient than their wood counterparts. Whirl testing proved by generating concrete results that metal offered the durability, strength, and safety required by the new commercial and military aircraft with large-horsepower engines. The existence of this indeterminate and pragmatic context also reveals that these historical actors at McCook and Wright fields did not exhibit a belief in the inevitability of one material over another as they followed disparate technical paths to make their contribution toward making airplanes fly higher, faster, and farther. Once they found a promising path, however, they moved forward from that point. Both technical arguments and practical experience convincingly demonstrated the superiority of metal for use in aircraft propellers.

[74] Aeronautical Research Department, "Research Department Memo No. 314: Report on the Recent Developments of Steel and Duraluminum in Airplane Construction," September 6, 1918, File 216.21031–314, USAFHRA; and "Development of Metal Wings and Control Surfaces," *Bulletin of the Experimental Department, Airplane Engineering Division* 2 (January 1919): 72.

[75] Eric M. Schatzberg, *Wings of Wood, Wings of Metal: Culture and Technical Choice in American Airplane Materials, 1914–1945* (Princeton, NJ: Princeton University Press, 1998), 4, 9–10, 21.

Points of Departure

Those most responsible for the development of the ground-adjustable propeller continued their relatively unknown work. After twelve years of civil service, Caldwell went through what would become known as the "revolving door" between government and industry when he became Standard Steel's chief engineer in June 1929.[76] McCauley followed Caldwell and left the propeller unit to work as a consulting aeronautical engineer in Dayton. He started his own company, the McCauley Aviation Corporation, in Dayton in the late 1930s to market fixed- and ground-adjustable propellers with solid-steel blades for small aircraft.[77] McCauley estimated that he and Caldwell received approximately $50,000 for the commercial rights to their ground-adjustable propeller patents.[78] Dicks left Standard Steel in 1928 to form the Dicks Aeronautical Corporation and develop his own variable-pitch propeller. His new endeavor involved a successful return to the hollow-steel blades that he intended to develop all along.[79]

The story of the development of the metal ground-adjustable propeller broadens the understanding of the process of technological innovation during the Aeronautical Revolution. Frank Caldwell and the propeller unit at McCook Field worked directly with a private company, Standard Steel, led by Thomas Dicks, to experiment and encourage metal ground-adjustable propeller development in the United States. The apparent "favored company" status for Standard Steel, which ensured its economic survival during the 1920s, highlights the existence of a propeller community and the environment in which it emerged. This group used its own unique experiences and terminology to address the problems of continual propeller development.

Development of the metal propeller began in the United States during World War I, at the same time engineers and inventors were trying to create a variable-pitch propeller with wood blades. Before the American propeller community could fully address the perfection of the variable-pitch

[76] "Frank Walker Caldwell," *Collier's Encyclopedia*, 4 (New York: Collier, 1957), 322–323.

[77] "McCauley Solid Steel Airplane Propellers," March 20, 1937, File B5-573000-01, PTF, NASM; McCauley-Textron, Inc., "McCauley History," 2002, www.mccauley.textron.com (Accessed November 3, 2002).

[78] Weick Oral Interview, 69.

[79] Thomas A. Dicks, "Propeller and Method of Making Same," US Patent No. 1,713,500, May 14, 1929; "Work Started Year Ago on Curtiss-Wright Plant Here," *The* [Beaver and Rochester] *Daily Times*, April 22, 1942, p. 6; "Thomas A. Dicks Dies," *Aviation News* 2 (October 9, 1944): 13–14; William F. Trimble, *High Frontier: A History of Aeronautics in Pennsylvania* (Pittsburgh: University of Pittsburgh Press, 1982), 116; Rosen, *Thrusting Forward*, 44.

mechanism, it had to address a required intermediate step, the search for new materials and construction methods to facilitate changing blade angle. The failure of the Army's variable-pitch propeller program was a stark contrast to its simultaneous but separate successful effort to create a ground-adjustable propeller. The ground-adjustable propeller program that focused solely on getting metal propellers into operational service solved the problems of blade retention, material strength, and structural design that plagued the Hart and Eustis, Heath, Standard Steel, and Engineering Division designs. The propeller unit at McCook Field and the Standard Steel Propeller Company were critical to the development of this new technology.

The innovators' work was only half-finished as Lindbergh's dramatic takeoff clearly illustrated the limits of aeronautical technology. The ground-adjustable propeller had limitations since it was still ultimately a fixed-pitch propeller. The inability to change pitch in flight meant that heavily laden aircraft such as Lindbergh's required long takeoff runs to get off the ground. Variable-pitch offered short takeoffs, fast climb, and economical cruising speeds, characteristics Lindbergh would have gladly accepted even if the propeller weighed more. The fact of the matter was that there was no variable-pitch propeller available in 1927.

The lessons learned with the ground-adjustable propeller provided valuable design knowledge. The ground-adjustable propeller's multipiece construction offered valuable experience in working with metal, a new and important material for aircraft manufacturing, and in joining two complex structures, the hub and blades, together. The stage was set for the next technological point of departure for the American propeller community: the introduction of a practical variable-pitch mechanism in the early 1930s. Despite the logical impetus for that development, there was another alternate and concurrent pathway that challenged its success, which was not rooted in technical superiority, but in the spectacular symbolism of progress.

6

"The Ultimate Solution of Our Propeller Problem"

On September 28, 1923, two Curtiss air racers flown by American naval aviators took first and second place in the Schneider Trophy seaplane contest at Cowes, the Isle of Wight, in southwestern England. Lt. David Rittenhouse won the competition with an average speed of 177.38 mph, almost three miles per minute, and a good 20 mph over the slower third place flying boat design from Great Britain. His diminutive gray CR-3 biplane equipped with floats represented the latest concepts in high-speed design, primarily aerodynamic streamlining and drag reduction (Figure 13). Its modified Curtiss D-12 engine was one of the most powerful in the world at 500 horsepower. The newest component of the Curtiss racing system was a duralumin propeller developed by Dr. S. Albert Reed, which to Stanley Spooner, the founder and editor of *Flight* weekly, "undoubtedly contributed considerably" to the Navy's victory.[1] The dramatic success of American and European air racers equipped with Dr. Reed's propeller led the international aeronautical community to embrace it as an airminded symbol of "Progress" through technological innovation.

Reed's approach to propeller innovation offered an attractive alternative to the ground-adjustable-pitch, or detachable-blade, propeller created by the Engineering Division and Standard Steel. If a fixed-pitch propeller made from a single piece of metal could turn faster and propel the airplane at ever-higher speeds as well as withstand the elements, then it would be a significant step forward in the development of propulsion technology. The Reed propeller did not offer the flexibility of being able

[1] Stanley Spooner, "Editorial Comment," *Flight* 15 (October 4, 1923): 589–590.

FIGURE 13 1923 Schneider competition winner Lt. David Rittenhouse stands on the float of his Curtiss CR-3 Racer at Cowes, the Isle of Wight. The thin Reed duralumin propeller was a key component of the Curtiss racing system. Smithsonian National Air and Space Museum (NASM 81–8206).

to change its blade pitch on the ground. Despite that limitation, the use of Reed propellers in the aerial spectacles of the 1920s, primarily air racing, electrified the aeronautical community with their promise of increased speed and durability. Reed the inventor and his corporate backers, the international aeronautical community, and the specialists at McCook Field reacted to that enthusiasm in different ways. The resultant tension revealed the important place of communal perceptions of what constituted ultimate success or failure in the evolution of technology.

A Lone Inventor and America's Largest Aviation Manufacturer

S. Albert Reed had a long, varied, and lucrative life and career as an engineer and inventor before entering aeronautics. Reed's English ancestors traveled to North America during the second half of the seventeenth century, some of them on the *Mayflower*, and settled New England and New York. They participated in the wars and battles that shaped colonial and revolutionary America. That involvement earned the Reed family membership in leading social groups that represented the founding families of the United States.[2] By the 1850s, the Reed family was established, successful, and part of upper-class Northeastern society.

The Reverend Sylvanus and Caroline Reed settled in Albany, New York, where Sylvanus Albert was born on April 8, 1854. Reed graduated from Columbia College (now Columbia University) with an A.B. degree in 1874 and from the college's School of Mines with a master's degree in engineering three years later. After a period of study at the universities of Berlin and Wurzburg, Reed received his doctorate in physics from Columbia in 1880. Reed worked as a professional engineer, insurance entrepreneur, and inventor in New York City until his retirement in 1912. His most lucrative invention was an electrical signaling system that railway companies used as safety devices.[3] Reed came to national prominence as an "insurance engineer" for his analysis of the fire that ravaged San Francisco after the epochal 1906 earthquake. The insurance and building industries embraced many of his suggestions for new fire protection methods and technologies.[4]

Reed started a second career of conducting research in areas he found of interest during the period 1915–1917. One of the potential projects that intrigued him was acoustics. His observation of a foghorn operated in the vicinity of his summer home at Woods Hole on the southern coast of Massachusetts stimulated that interest. Reed believed he could develop a more practical and economical way to generate acoustic noise than the monstrous and highly inefficient steam boiler. He experimented with electric motors to whirl duralumin vanes that generated high-decibel noise. According to one biographer, the eighteen-inch vanes "resembled anything from dinner knives to hamburger spatulas."[5] The loud, moaning noise emanating from the attic laboratory of his apartment on exclusive Riverside Drive in Manhattan ensured that the local police were frequent visitors. In deference to his neighbors and requiring larger facilities, Reed searched for an alternate workshop in 1920.[6]

[2] "Mrs. S.G. Reed Dies at 93," *New York Times*, November 18, 1914, n.p.

[3] Roger Ward, "The Propeller Pioneer," *Flying* 63 (December 1958): 30–31, 74, 76, 78, 80; Ritchie Thomas, "Sylvanus Albert Reed: Inventor," *American Aviation Historical Society Journal* 37 (Summer 1992): 103.

[4] S. Albert Reed, *The San Francisco Conflagration of April, 1906: Special Report to the National Board of Fire Underwriters Committee of Twenty* (New York: The Committee, 1906); "S. Albert Reed," *Insurance Engineering* 12 (1906): 162.

[5] Ward, "The Propeller Pioneer," 74.

[6] C. G. Grey, "Sylvanus Reed," *The Aeroplane* 49 (October 16, 1935): 461; Thomas, "Sylvanus Albert Reed," 103; W. E. Valk, Jr., "The Reed Propeller," in "Reed D-1 Propeller," File A19430003000, RO, NASM; Ward, "The Propeller Pioneer," 30–31.

Reed resumed his experiments with the electric motors and curious looking metal vanes in a rented space on the grounds of the Curtiss Aeroplane and Motor Company factory on Nassau Boulevard in Long Island's Garden City. Reed often visited the Curtiss engineers during their lunchtime breaks to ask persistent and inquisitive questions about aeronautics. During one of his visits in March 1920, he mentioned that his equipment was not producing the desired effect. Power plant project engineer, Charles Hathorn, arranged to sell Reed a war-surplus OX-5 engine and then assisted him in constructing a test stand. The ninety-horsepower engine produced more than enough power to whirl four-foot-diameter and ⅜-inch thick duralumin vanes at supersonic tip speeds.[7] Factory personnel became frequent visitors to Reed's shed to investigate the much louder noise. On one occasion, Reed asked William Waite, the company's self-taught designer of air racers and fighters, his opinion of a thin and straight duralumin vane attached to the OX-5's propeller hub. Waite wryly remarked, "As a foghorn that thing would make a good propeller."[8]

Intrigued by Waite's comment, Reed had a new research challenge: the design of a light, strong, durable, and efficient aeronautical propeller made from duralumin. He began a dialogue with the Curtiss propeller engineers. They informed Reed that their attempts at the design and construction of metal propellers proved unsuccessful due to excessive weight and fatigue failure. Reed countered with a new suggestion based on his experience with his duralumin vanes. Whereas traditional wood propellers were rigid in construction, his idea for a propeller featured flexible thin blades angled forward that could spin at high speeds. The propeller's ability to withstand higher centrifugal forces kept the thin blades from deflecting under aerodynamic load, which made them more efficient and the airplane faster.[9]

Following the model of the late nineteenth and early twentieth century inventor, Reed filed two patent applications for an "Aeronautical Propeller" in 1920 and 1922 that covered these initial findings. Issued July 31, 1923, the first described a one-piece aerial propeller featuring flexible blades "thinner than previously customary" and "dependent upon the stiffening effect of centrifugal force to make the propeller practically operative." The second patent issued on December 9, 1924, illustrated

[7] S. Albert Reed, "Technical Development of the Reed Metal Propeller," *Transactions of the ASME* 48 (1928): 57; Valk, "The Reed Propeller"; G. W. Rich to A. Wetmore, December 2, 1935, "Aeromarine 1914 Propeller," File A19330044000, RO, NASM.

[8] Ward, "The Propeller Pioneer," 31, 74, 76, 78.

[9] Valk, "The Reed Propeller"; Ward, "The Propeller Pioneer," 74, 76, 78.

the blades featured in the earlier patent and their construction from "certain metals or alloys," meaning duralumin. Reed patented his propeller in Great Britain, France, and Germany as well.[10]

Reed's idea was not entirely new. Orthodox propeller design reflected the reasoning that propeller blades had to be stronger, or thicker, as the rotational tip speeds of the propeller increased. Unfortunately, wood propellers with their thick airfoil sections proved unsuitable for high rotational speeds. Many French propeller manufacturers were known for angling, or raking, forward thin and flexible wood blades.[11]

Reed initiated the construction of a duralumin aeronautical propeller in May 1921. Under Reed's guidance, George Hilliker, the Curtiss factory's wood propeller fabricator, took two ⅞-inch slabs of rolled duralumin obtained from the Bausch Machine Tool Company of Springfield, Massachusetts, annealed them, twisted them into a rough aerodynamic planform, and then finished them with a milling machine, coarse file, and a sanding disc. Mounting the thin propeller to the front of an airplane engine required one more design element. Reed and Hilliker added circular wood spacer blocks to the center to create a hub that allowed the use of a wood propeller's front and rear retaining plates and eight bolts. They were able to both fasten the propeller together as well as the completed assembly to an engine crankshaft. Hilliker completed the first duralumin propeller prototype on August 12, 1921. Designated the D-1, it was nine-foot-diameter with a hub diameter of eleven inches and a blade chord of six inches.[12] Reed characterized the crude-looking D-1 as "my curious contraption of twisted metal."[13]

The next step was to flight test the propeller. Despite the enthusiasm displayed by Reed, the Curtiss engineers were skeptical of the potential performance of the propeller because of its unorthodox design and construction. Hathorn supervised the installation of the D-1 on a Curtiss-Standard biplane equipped with a 160-horsepower Curtiss K-6 engine.

[10] S. A. Reed, "Aeronautical Propeller," US Patent No. 1,463,556, July 31, 1923, and US Patent No. 1,518,410, December 9, 1924; Report of the Patent Commissioner, March 18, 1940, *Reed Propeller Company v. United States*, USCC No. 42,133, Folder 168.7329-3, DAD, HRA, 10; W. Fowler to W. E. Valk, Jr., June 20, 1924, Folder 5, Box 14, Curtiss-Wright Corporation Records (hereafter cited as CWC), NASM.

[11] "Description of the Curtiss Reed Metal Propeller," *Aviation* 15 (November 19, 1923): 630; "The Reed Duralumin Airscrew," *The Aeroplane* 25 (November 21, 1923): 510.

[12] Reed, "Technical Development of the Reed Metal Propeller," 59; Valk, "The Reed Propeller"; Ward, "The Propeller Pioneer," 78.

[13] "Dr. S. Albert Reed Accepts the Collier Trophy from NAA," *U.S. Air Services* 11 (May 1926): 27.

FIGURE 14 Albert Reed (second from left), Casey Jones (in cockpit), and in all probability Charles Hathorn and George Hilliker commemorate the flight of the D-1 propeller in August 1921. Smithsonian National Air and Space Museum (NASM 2001–4229).

Casey Jones, the manufacturer's famous test pilot, performed the first flight test of the D-1 propeller at nearby Curtiss Field on August 30, 1921 (Figure 14). Even though the propeller had only undergone a dubious ground test, Jones encountered no difficulty during the flight. Post-flight data revealed that the D-1 propeller was 10 percent more efficient than the wood propeller designed for the K-6.[14]

Eager to sell his new invention to the military services, Reed actively sought contacts with the appropriate procurement officers. At the elite University Club of New York in 1920, Reed had a chance encounter with former chief signal officer, Maj. Gen. George Owen Squier, who referred him to the Engineering Division at McCook Field. Reed wrote to Col. Thurman Bane in October 1920 informing him of his early work

[14] S. A. Reed to C. M. Keyes, April 8, 1925, Folder 5, Box 11, Clement M. Keys Papers (hereafter cited as CMK), NASM; Valk, "The Reed Propeller"; Thomas, "Sylvanus Albert Reed," 103.

on a "propeller adapted for super tip speeds" that was "of metal very thin and structurally non-rigid and depending upon centrifugal force for a virtual rigidity against tangential and axial deflecting stresses." He invited the Engineering Division to evaluate his new invention.[15] The Army responded that it welcomed any information on Dr. Reed's "experimental high-speed propellers."[16] On March 11, 1922, Reed informed the Air Service that he had completed the experimental stage of his propeller development and that he was ready to manufacture propellers of his design under contract to the Army.[17]

The Navy considered the replacement of wood propellers with metal propellers in early 1922. As the first metal propeller on the market, the fixed-pitch duralumin Reed propeller was the only viable candidate. Reed contacted BuAer chief, Rear Adm. William A. Moffett, to inform him of the new propeller. In February 1922, he sent a sample propeller to BuAer for its engineers to evaluate. Ensign Frank Miller, head of BuAer's Propeller Design Section, and Lt. Cdr. Bruce G. Leighton, head of the Power Plant Section, issued the first contract order for six Reed propellers during the spring of 1922.[18] The purchase initiated the use of metal propellers on operational American military aircraft.

BuAer believed that the Reed propeller was an important technological component of modern naval aircraft and it wanted it in production as soon as possible. From January to July 1923, the Navy assisted Reed, his patent lawyer, Willis Fowler, and Curtiss patent attorney, William E. Valk, Jr., through the issuance process for his first patent. Both Curtiss and Reed were anxious for the patents to be issued as well. Valk convinced the Navy to request the patent office to issue a classification of a "privileged case" on behalf of Reed to quicken the process. The Navy argued that future orders of Reed propellers were dependent on the granting of the patent so that an unnamed "manufacturer," presumably the Curtiss Aeroplane and Motor Company, could begin production. The

[15] S. A. Reed to T. H. Bane, October 26, 1920, Folder "Reed Propellers, 1921," Box 77, RD 3102, RG 342, NARA.

[16] L. W. McIntosh to S. A. Reed, "Propeller Adapted for Super Tip Speeds," November 30, 1920, Folder "Reed Propellers, 1921," Box 77, RD 3102, RG 342, NARA.

[17] S. A. Reed to the War Department, March 11, 1922, Folder "Heath [Reed], 1922–1925," Box 5582, RD 3143, RG 342, NARA.

[18] S. A. Reed to W. A. Moffett, February 21, 1922, Folder "F23, Volume 3," Box 1952, RG 72, NARA; "Dr. S. Albert Reed Accepts the Collier Trophy," 27; "Bruce Gardner Leighton," *Who's Who in American Aeronautics* (New York: Gardner Publishing Company, 1925), 73–74.

Navy regarded the quick granting of the Reed patents to be "in the interest of the naval air service."[19]

Reed knew that he could not produce the propellers himself. He granted Curtiss the exclusive license to manufacture and sell propellers of his design in the United States on April 26, 1923. Curtiss acknowledged the "practicality" of the propeller through its ability to increase performance efficiency and the fact that the Navy had ordered six propellers for service aircraft. As part of the agreement, Curtiss had to recognize that Reed created the propeller with his "sole engineering and scientific skill and at his sole expense." The agreement included the right to manufacture and use the profile cutting and blade twisting machinery used to make his propellers as well.[20]

Reed and Curtiss created the Reed Propeller Company, Inc., as a subsidiary of the Curtiss Aeroplane and Motor Company in January 1924 and assigned control of his patents to that organization. If Curtiss acquired the patents outright, they would be bound by the industry-wide Manufacturer's Aircraft Association (MAA) cross-licensing agreement to share proprietary data with other companies. The agreement was a holdover from the infamous Wright and Curtiss patent battles and their detrimental effect on the American aviation production program during World War I. The Wright and Curtiss patents covered the basic definition of what constituted an airplane, which included the wood, fixed-pitch propeller. Other aircraft makers faced the threat of paying substantial royalties that could prove detrimental to the economic health of the fledgling aviation industry. Membership in the MAA alleviated the problem, but meant that any new innovations put forth by its members had to be shared with all. Curtiss formed the separate company holding title to the Reed patents to offset that possibility.[21]

Curtiss, which was America's largest aviation manufacturer, purchased the primary stock for the company from Reed and his investors for $475,000. All operating expenses would come from Curtiss, which was also responsible for the actual manufacturing and distribution. Reed's

[19] W. Fowler to W. E. Valk, Jr., February 17, 1923; W. E. Valk, Jr. to F. H. Russell, April 6, 1923; Anonymous to Honorable Commissioner of Patents, April 12, 1923, Folder 5, Box 14, CWC, NASM.

[20] Extract from License Contract Between S. A. Reed and Curtiss Aeroplane and Motor Company, Inc., 1923, Folder 5, Box 11, CMK, NASM; Report of the Patent Commissioner, *Reed Propeller Company v. United States*, 11.

[21] H. G. Hotchkiss to F. H. Russell, November 21, 1922, Folder 5, Box 14, CWC, NASM; Jacob Vander Meulen, *The Politics of Aircraft: Building an American Military Industry* (Lawrence: University Press of Kansas, 1991), 26–28.

initial investment of $6,000 for use of the Curtiss factory at Garden City for the experimentation, development, and manufacturing of the early twisted-slab propellers had reaped a substantial dividend.[22]

The aeronautical community took notice of the Reed propeller and started placing orders. By late summer 1924, Curtiss had sold over 100 D-type propellers after taking over production and sales activities from Reed. Aircraft that used the propeller included: Army and Navy air racers such as the Curtiss R2C and the Verville-Sperry R-3; service aircraft such as the Army's Curtiss PW-8 pursuit and NBS-4 bomber and the Navy's Vought VE-7 trainer and Curtiss F4C-1 fighter; the US Air Mail Service's DH-4B aircraft; general aviation aircraft such as the Farman Sport and the Curtiss Oriole; and working aircraft such as the Curtiss Seagull and the Night Mail.[23] Curtiss had delivered 410 Reed propellers to the Army, Navy, and the Post Office by December 1925. By February 1926, aircraft designers and operators had used the Reed propeller in more than eighty different engine and airframe combinations.[24] The popular name for the D-type was the "whistling pig" because of the original forging's resemblance to an iron pig and the high-pitched noise the angle of the blade near the hub emitted during operation.[25]

The Spectacle of "Progress"

The Reed propeller filled a significant niche as the choice of the Army and Navy racing teams in the 1920s for the spectacular Pulitzer and Schneider Trophy contests. All of the major racing engines were direct-drive, meaning the propeller was directly connected to the engine crankshaft. The ear-splitting roar of the high output engines with their short exhaust manifolds combined with the high-pitched "banging" and "clanging" sound of the Reed propeller tips reaching supersonic speeds accentuated the excitement and spectacle of government-sponsored air racing in the 1920s.[26]

Contemporary observers recognized that the aeronautical community conducted the major races to cultivate an environment of competition

[22] Report of the Patent Commissioner, *Reed Propeller Company v. United States*, 10–11.

[23] "Curtiss-Reed Metal Propeller Orders," August 13, 1924, Folder 5, Box 14, CWC, NASM.

[24] "The Reed Propeller – 1925," Folder 5, Box 11, CMK, NASM; "Reed Propeller Wins Collier Trophy," *Aviation* 20 (February 22, 1926): 256.

[25] "The Third Part of a Plane: Curtiss Electric Propellers," n.d. [1942], File B5-260140-01, PTF, NASM.

[26] Reed, "Technical Development of the Reed Metal Propeller," 55; Thomas Foxworth, *The Speed Seekers* (New York: Doubleday, 1975), 74.

that led to the further development of the airplane in a linear and progressive evolutionary process. In the case of both the Pulitzer and the Schneider contests, military pursuit and fighter aircraft were seen as the beneficiary. From the standpoint of Curtiss, Reed propeller technology could be transferred directly to pursuit aircraft. The PW-8 pursuit biplane, which reflected the design influence of the Curtiss racers, used a duralumin Reed propeller.[27] This form of technology transfer from racing to operational use reflected the intention of the air races in the 1920s and a reason why the US government sponsored Army and Navy participation.

America's premier closed-course high-speed air race was the Pulitzer Trophy Race, which was the marquee event during the annual National Air Races. Publishing magnates Ralph, Joseph, Jr., and Herbert Pulitzer of the *New York World* and the *St. Louis Dispatch* initiated a series of races in 1919 to promote aviation. While open to international competitors, the Pulitzer Race witnessed annual contests between American Army and Navy fliers as they attempted to push the boundaries of aeronautical technology in the name of interservice rivalry.[28]

The Reed propeller made its racing debut during the 1923 National Air Races at Lambert Field near St. Louis where it quickly made a favorable impression with the aeronautical community and the race-attending public. The Navy's Curtiss R2C-1s equipped with Reed propellers dominated the Pulitzer Race on October 6. In front of 85,000 excited spectators, Lt. Alford J. "Al" Williams won at an average speed of 244 mph, an unbelievable four miles per minute pace, over the 124.27 mile course. His fellow naval aviator, Lt. Harold J. Brow placed second at 242 mph. The fifth-place contestant, Army First Lt. Walter Miller, flew his Curtiss R-6 racer at an average speed of 219 mph.[29]

The R2C-1s incorporated the latest aeronautical developments aimed toward high-speed flight. Curtiss equipped them with special D-12

[27] Russell Shaw, "The Races and Their Purpose," *U.S. Air Services* 9 (September 1924): 24, 27; E. W. Dichman to P. M. Bates, "Curtiss Racing Propeller R3C," June 8, 1926, Folder "Reed Propellers, 1926–1929, 1932," Box 5867, RD 3216, RG 342, NARA; H. S. Martin to CAMC, "Reed Metal Propellers," March 1, 1923, Folder "Heath [Reed], 1922–1925," Box 5582, RD 3143, RG 342, NARA.

[28] Foxworth, *Speed Seekers*, 29–31.

[29] S. A. Reed to L. MacDill, "Reed Duralumin Propeller D-27," September 25, 1923, Folder "Heath [Reed], 1922–1925," Box 5582, RD 3143, RG 342, NARA; J. Parker Van Zandt, "The 1923 Pulitzer Air Classic," *U.S. Air Services* 8 (November 1923): 15, 21; "Curtiss Stands for Speed with Safety: Navy Curtiss Racer," *U.S. Air Services* 8 (November 1923): 5; Terry Gwynn-Jones, *Farther and Faster: Aviation's Adventuring Years, 1909–1939* (Washington, DC: Smithsonian Institution Press, 1991), 295; Foxworth, *Speed Seekers*, 456.

503-horsepower water-cooled engines, flush wing radiators, and a sleek fuselage and airfoil shapes that led aviation observer Dr. J. Parker Van Zandt to state that "streamlining has been carried out to what appears to be the last minute detail." Curtiss claimed that the incorporation of the Reed propeller increased the R2C-1's top speed by 10 mph. Van Zandt considered the Reed twisted-slab propeller to be the outstanding technical advance to have appeared during the races. That was no small assertion considering the other important aeronautical advances being used by both the Army and the Navy. Almost a month after the Pulitzer victory, Williams and his R2C-1 shattered the standing world speed record with an average of 267 mph at Mitchel Field on Long Island in November 1923.[30]

With the Army and Navy committed to achieving speed records, the Reed propeller adopted its singular technical and historical importance as a component technology. The aeronautical community firmly believed that use of the Reed propeller expanded the boundaries of speed. The ability to increase tip speeds without affecting overall efficiency influenced both propeller and engine design, especially for high-speed aircraft. The implications for the future development of the airplane were exciting. Optimistically, *The Aeroplane* declared in November 1923 that the influence of the Reed propeller on high-speed aircraft "may prove to be of very much greater importance in the near future."[31]

The Curtiss product line of airframes, engines, and propellers was impressive. First Lt. Russell J. Maughan of the Army Air Service flew from New York to San Francisco in a much-publicized one-day "dawn-to-dusk" flight in June 1924. Curtiss designed Maughan's propeller for optimum cruising efficiency. Curtiss took the opportunity to assert that the Reed propeller was the "safest and most efficient propeller ever tested" and that, overall, the company's PW-8 pursuit airplane, the D-12 engine, and the Reed propeller had "set new standards for plane, motor, and propeller."[32] Curtiss sought to sell its full range of aeronautical products in combination with each other, but the advertising rhetoric also epitomized popular views of the airplane as a spectacular technical system.

[30] Van Zandt, "The 1923 Pulitzer Air Classic," 15, 21; Foxworth, *Speed Seekers*, 222.

[31] "The Reed Duralumin Airscrew," 510.

[32] "Curtiss: Speed with Safety," *U.S. Air Services* 9 (September 1924): 10; G. C. Kenny to CAMC, "PW-8–Coast to Coast," June 7, 1924, Folder "Propellers – January through August, 1924," Box 5541, RD 3132, RG 342, NARA; C. R Roseberry, *The Challenging Skies: The Colorful Story of Aviation's Most Exciting Years, 1919–1939* (Garden City, NY: Doubleday, 1966), 49–50; Foxworth, *Speed Seekers*, 209.

The Reed propeller also debuted at the Schneider Trophy races in 1923. French industrialist and early aviator, Jacques P. Schneider, encouraged the development of commercial seaplanes by creating the Coupe d' Aviation Maritime Jacques Schneider in 1912. Schneider's interest in high-speed boats and his rationalization that 70 percent of the earth's surface was water influenced his efforts. The first race in 1913 featured only two aircraft, but the Schneider Trophy contest by 1920 became what historian Terry Gwynn-Jones called, "the world's most fabled air race."[33]

Having cultivated a hypercompetitiveness in the Pulitzer races and seeing the public relations potential of winning the Schneider, the US Navy brought their Curtiss CR-3 racers, equipped with floats, to Cowes, the Isle of Wight, England, in September 1923. Lt. Rittenhouse flew his racer 34 mph faster than the winner of the previous year's competition.[34] According to C. G. Grey, the outspoken longtime editor of *The Aeroplane*, the Curtiss racers represented a "revolution in design" that marked the beginning of completely new ideas of streamlining, fuselage design, and propulsion technology for the British.[35]

The British reacted to the American victory at Cowes in two ways. First, they created a high-speed aircraft program that was a joint partnership between industry, government research, and the military. The RAF provided the test pilots of the High Speed Flight; overall leadership came from the Air Ministry. Second, the Fairey Aviation Company Ltd. of Middlesex, gobbled up the rights to the Reed propeller and D-12 engine.[36] It was a clear example of the transfer of American high speed aircraft propulsion technology to Great Britain.

Unable or unwilling to attempt to break the Navy's records of the previous year, the Air Service executive decreed that the four aircraft that made up the competition field for the 1924 Pulitzer in Dayton would simply give the public an exciting race with evenly matched aircraft. Part of the performance matching included the installation of Engineering Division-designed wood propellers because they were cheaper than Reed propellers. During the exciting diving start of the race on October 4, 1924, Capt. Bert E. "Buck" Skeel nosed his silver Curtiss R-6 racer over from above 4,000 feet into a sixty-degree angle. As the aircraft broke through the clouds into the sunlight, it burst apart and disintegrated,

[33] Gwynn-Jones, *Farther and Faster*, 52–53.

[34] Spooner, "Editorial Comment," 590.

[35] C. G. Grey, "On the Schneider Trophy Contest," *The Aeroplane* 41 (September 10, 1931): 616.

[36] "Curtiss-Reed Metal Airscrews in Europe," *Flight* 16 (April 24, 1924): 236.

killing Skeel. The fuselage dug ten feet into a creek bank as the wings and other debris floated harmlessly to earth. An Air Service investigation ruled that the Engineering Division birch propeller disintegrated under the high power of the D-12 engine. The resultant imbalance from the propeller failure led to catastrophic engine damage, which ripped the racer apart. The Army accident board banned the diving start and the use of wood propellers on its air racing and other high-speed aircraft. C. G. Grey declared that it was "singularly imbecile to fit a wooden airscrew to an engine of such power designed for such speed." Arthur Nutt, designer of the D-12 engine, agreed.[37] The aeronautical community turned to the Reed propeller as standard air racer and pursuit equipment, which validated its perceived importance as a milestone aeronautical innovation.

Reed began work on his second major propeller design, the R-type, in January 1925, in preparation for the continued use of his propeller for air racing and aircraft with engines of 200 horsepower or more. The R-type was a technological step above the twisted-slab D-type. Rather than having workers hammer and twist the propeller into shape, Reed used two heated dies to stamp the hub and the root area of the blades into shape from a solid duralumin forging. The outer areas of the blades remained unfinished until workers trimmed and machined them into shape with metal-cutting tools. In its initial shape, the propeller was annealed, twisted to its aerodynamic pitch, and then heat-treated.[38] As a result, it was stronger, more efficient, and resembled a wood fixed-pitch propeller even more than the original D-type.

Curtiss designers chose the new propeller, called the "racing type propeller" by the aeronautical community, for use on the 1925 air racers. Between June and September, a frantic development program took place centered on the Garden City manufacturing plant and the whirl rig at McCook Field. The first propeller, the R-1, flew on a Curtiss Oriole on September 8, 1925. Reed and Curtiss slated the next two R-types for the 1925 Pulitzer Trophy and Schneider Cup entrants for the United States.[39]

The record of the Reed propeller, and the metal propeller in general, was steadily increasing. Of the eighty-five aircraft participating in the October 1925 National Air Races at Mitchel Field, fifty used R-type

[37] C. G. Grey, "The Dayton Flying Meeting – II," *The Aeroplane* 27 (October 29, 1924): 409–410; Grey, quoted in Foxworth, *Speed Seekers*, 214–216, 458.

[38] "The Schneider Cup Race and Curtiss Reed Propellers," *U.S. Air Services* 11 (December 1926): n.p.; "'Curtiss-Reed' Means Long Life," *U.S. Air Services* 13 (January 1928): 2; Reed, "Technical Development of the Reed Metal Propeller," 58, 61.

[39] S. A. Reed to C. G. Grey, November 24, 1925, Folder 5, Box 11, CMK, NASM.

propellers. Twelve featured Standard Steel ground-adjustable-pitch propellers and twenty-three used conventional wood propellers. Every airframe in the two major high-speed races, the Army Pursuit Race and the Pulitzer Race, featured Reed propellers. Among the nine winners of the major races, six used the Reed propeller.[40] One of the more memorable souvenirs for that year was a nine inch-diameter D-type propeller made from duralumin and embossed with the date and the company's logo.[41]

At the 1925 Schneider Trophy Race on October 26 in Baltimore, every entry except for the Italian team used Reed propellers manufactured by Curtiss or by licensed companies. The Army's winning entrant, a Curtiss R3C-2 floatplane racer flown by First Lt. James H. "Jimmy" Doolittle, featured a seven foot-eight inch-diameter R-type design EX-32995 designed for use on the Curtiss V-1400 engine. A day later, Doolittle broke the world seaplane record at 246 mph. The same airplane and propeller, in the R3C-1 configuration with wheeled landing gear, won the 1925 Pulitzer at an average speed of 249 mph with First Lt. Cyrus Bettis at the controls. Overall, at the end of 1925, the aircraft that held the world speed records for straightaway and closed course for both land- and seaplanes featured Reed propellers as part of the propulsion system. Efficiency tests of a three-foot model of the R3C-1's R-type propeller conducted at the Stanford University wind tunnel for the NACA determined that its efficiency was an unprecedented 87.3 percent.[42]

The US government canceled the Army and Navy racing programs after the 1926 season. Air racing had been a good way to garner attention and public interest in aviation, but it proved to be too expensive. A panel organized by the Chief of the Air Service Mason M. Patrick, under the leadership of Dexter S. Kimball of Cornell University, recommended that the Army stop competing in air races in March 1925.[43] Patrick did acknowledge that air racing had a technical value, much like automobile

[40] "The Reed Propeller – 1925"; "Speed with Safety: Curtiss-Reed Metal Propellers at the Pulitzer Races," *U.S. Air Services* 10 (December 1925): n.p.

[41] See NASM artifact A19820514000, "Propeller Souvenir, 1925 National Air Races, Curtiss-Reed."

[42] S. A. Reed to C. G. Grey, November 24, 1925; "The Reed Propeller – 1925"; W. L. Gilmore to Chief, Engineering Division, "Whirling Test of 1925 Racer Propeller," July 31, 1925, Folder "Heath [Reed], 1922–1925," Box 5582, RD 3143, RG 342, NARA; "Speed with Safety: Curtiss-Reed Metal Propellers at the Pulitzer Races," n.p.; Reed, "Technical Development of the Reed Metal Propeller," 61; Foxworth, *Speed Seekers*, 476–477.

[43] Maurer Maurer, *Aviation in the U.S. Army, 1919–1939* (Washington, DC: Office of Air Force History, 1987), 172–173.

racing stimulated passenger car development, but that time had passed.[44] Racing was just too expensive and Congress was not forthcoming with funding in a cash-strapped political landscape.[45]

Nevertheless, the Reed propeller and its licensed versions overseas persisted as a major component of European air racers and speed record attempts. Italian pilot Mario de Bernardi achieved 318 mph on a 1.864 mile course at Venice, Italy, in March 1928 in a Macchi M.52 equipped with a modified Reed propeller. The two aircraft that wrenched the record away from the Italians were British Supermarine air racers. Augustus H. Orlebar reached 358 mph in an S.6 in September 1929 and George H. Stainforth reached 408 mph in an S.6B in September 1931. Both aircraft used Reed propellers license-built by Fairey.[46] The world racing community fully recognized the importance and advantages of using Reed-designed propellers on their aircraft.

Reed always contended that his propeller was more than a mere accessory to be added to specialized racers. He envisioned his propeller as having a more important role in increasing the efficiency and durability of propellers for operational aircraft. According to Reed, his propeller simply replaced wood propellers in the same performance regime with a slight increase in efficiency of 2 to 5 percent, but with a dramatic increase in durability. A ten-foot-diameter Reed propeller mounted on a DH-4B aircraft with a Liberty V-12 engine generated the same tip speeds as a wood propeller. Realistically, the Reed propeller was taking the place of wood propellers for the reason of durability, not speed.[47]

The first nonmilitary organization to use the Reed propeller was the US Air Mail Service operated by the Post Office Department. From 1918 to 1926, the service pioneered regularly scheduled long-distance commercial flight in North America through its daily transcontinental airmail route from New York to San Francisco and its night service from New York to Chicago. The service adopted Reed propellers as standard equipment for its DH-4Bs in 1924. A veteran airmail pilot stated that he

[44] Mason M. Patrick, *The United States in the Air* (Garden City, NY: Doubleday, Doran and Company, 1928), 114.

[45] "Navy Funds Are Limited," *The Washington Post*, December 26, 1926, p. F11.

[46] S. A. Reed to C. M. Keys, May 2, 1928, Folder 5, Box 11, CMK, NASM; Reed, "Technical Development of the Reed Metal Propeller," 56–57; Gwynn-Jones, *Farther and Faster*, 285.

[47] S. A. Reed, "The Reed Propeller," February 5, 1926, Folder "Reed Propellers, 1926–1929, 1932," Box 5867, RD 3216, RG 342, NARA; Reed, "Technical Development of the Reed Metal Propeller," 56; "Reed Propeller Wins Collier Trophy," 256.

and his colleagues "no longer consider them as metal propellers–they are just propellers."[48]

The performance of the Reed propeller in air racing, commercial operations, and exploratory flights led to the adoption of progress-oriented hyperbole in advertising. The Curtiss Aeroplane and Motor Company firmly believed that the Reed propeller was the pinnacle of design. Curtiss advertisements in the aeronautical trade press extolled the virtues of the new propeller and its importance to the overall progress of technology when they declared, "Metal propellers have been talked of for years. They are now available."[49]

At the root of this rhetoric was the propeller's importance to the safety of the airplane while retaining high operating efficiency. Overall, the Reed propeller negated the "uncertainty and frailness" of wood propellers by its resistance to climatic and environmental conditions such as rain and ocean spray, hail, grass, dust, dirt, and sand as well as rough handling and storage. Curtiss advertising attested that Reed propellers had been in service for two years with no durability problems. The Reed propeller was the most efficient at speeds unobtainable with a wood propeller. As a result, Reed-equipped aircraft dominated races and captured world speed records. Those achievements, in the opinion of the Curtiss advertising staff, proved that the Reed propeller was the safest and most efficient propeller ever produced. The Reed propeller was "the choice of pilots who demand high efficiency and absolute dependability in the face of severe operating conditions."[50]

The Curtiss Aeroplane and Motor Company, the nation's largest and most powerful aviation combine, assertively went about selling "Progress" in the guise of the Reed propeller. A 1925 advertisement entitled, "The Luxuries of One Generation Become the Necessities of the Next," stated:

> In this age of rapid scientific development, the line between luxuries and necessities is so finely drawn, that anything which tends towards progress and advancement is soon adopted as necessary ... In the humble field of aeroplane propeller construction, the above holds true, and the metal propeller in the remarkably short period of four years has become a real necessity. Certainly no manufacturer can afford to turn out any commercial machines without careful consideration

[48] "The Reed Propeller – 1925"; "Speed with Safety: Curtiss-Reed Metal Propellers at the Pulitzer Races," n.p.; S. A. Reed, "The Reed Propeller," February 5, 1926, Folder "Reed Propellers, 1926–1929, 1932," Box 5867, RD 3216, RG 342, NARA.

[49] "Curtiss Stands for Speed with Safety: Curtiss Reed One-Piece Duralumin Propeller," *U.S. Air Services* 9 (March 1924): 6.

[50] "Curtiss-Reed Propellers Withstand Spray, Rain, Hail, Heat, Cold at the North Pole," *U.S. Air Services* 11 (June 1926): n.p.

of this wonderful new invention, and even commercial operators should give it serious thought because of added safety, increased performance and payload.[51]

The Reed metal propeller had arrived, and with it a new aerial age rooted in technological progress.

The aeronautical community recognized Reed for the achievement embodied in his propeller. The National Aeronautic Association (NAA) awarded Reed the Collier Trophy for 1925 during a ceremony at Bolling Field on March 19, 1926. The solid duralumin propeller entered into the pantheon of other aeronautical technologies such as the flying boat (Glenn Curtiss, 1912), the automatic stabilizer (Orville Wright, 1913), and the gyroscopic control and the drift indicator (Elmer A. Sperry, 1914, 1916). The award committee consisted of its chairman Orville Wright; George Lewis, the NACA's director of research; Earl N. Findlay, editor of the journal *U.S. Air Services*; Godfrey L. Cabot, President of the Board of Governors, NAA; and Porter H. Adams, a founding member of NAA. The dramatic increase in the use of Reed's propeller since 1921, especially its "general adoption" in American high-speed racing, influenced the committee's selection. In his acceptance address, Reed mentioned that he was fortunate enough to "add to the progress of aviation" by designing a propeller to increase efficiency and safety for the benefit of the aeronautical community.[52]

The award of the Collier Trophy to Reed firmly established the momentum for the propeller within the aeronautical community. The Reed propeller was a standard item to incorporate into advanced aircraft design. The aviation trade press had welcomed Reed and his propeller since its introduction. To *The Aeroplane*, the Reed propeller represented "a very striking and distinctly courageous departure from standard methods of airscrew construction."[53] After the 1923 National Air Races, Curtiss vice president Frank H. Russell told the Army that the adoption of the Reed propeller was "the ultimate solution of our propeller problem in aeroplane operation."[54] The communal recognition and praise of Reed's

[51] "The Luxuries of One Generation Become the Necessities of the Next," *U.S. Air Services* 10 (July 1925): 8.

[52] "Reed Propeller Wins Collier Trophy," 256; "Dr. S. Albert Reed Awarded Collier Trophy," *U.S. Air Services* 11 (March 1926): 38; "Dr. S. Albert Reed Accepts the Collier Trophy," 27; "Porter Hartwell Adams," *Who's Who in American Aeronautics* (New York: Gardner Publishing Company, 1925), 15.

[53] "The Reed Duralumin Airscrew," 510.

[54] F. H. Russell to L. MacDill, October 17, 1923, Folder "Heath [Reed], 1922–1925," Box 5582, RD 3143, RG 342, NARA.

achievement only fortified the belief that the propeller was the definitive expression of American progress through technological achievement.

A Flawed Design

The advertising and technical arguments for the Reed propeller were persuasive. As the aeronautical community learned, however, the Reed propeller could not support the expansion of the airplane's role into a transcontinental and transoceanic environment for two primary reasons. First, the high-rpm engine directly connected to a small-diameter, fixed-pitch propeller was only suitable for purpose-built racing applications during a short, exciting, and dramatic period. It was inappropriate when an aircraft equipped in that manner had to carry passengers or cargo for long distances and in and out of airfields of varying sizes and altitudes.

Second, the Reed propeller was neither strong nor durable enough for safe and continued use in a rigorous military and commercial environment. The aeronautical community had placed its faith in the wrong propeller. While the spectacle of air racing and excitement over the use of metal encouraged the perception that it was a symbol of continued aeronautical innovation, it was the daily use of the propeller by the military and commercial operators that truly proved that the Reed propeller was a false start. The propeller was structurally weak in construction due to the basic flaw in its design – it was too flexible. The fact that the Reed propeller was completely flexible to offset the centrifugal and thrust loads was the technical reason for its downfall. The continual flexing ensured that the blades weakened over time. Duralumin was not elastic enough to withstand the stress. The premature introduction of both the D- and R-type propellers damaged Curtiss and Reed's relationships and credibility with the military services and civilian operators.

From the start, the Engineering Division ignored the public spectacle of the Reed propeller and never fully supported the design. Caldwell encouraged all-metal propeller research, but he subtly suggested that Reed investigate ground-adjustable propellers. While the performance of the original D-1 was impressive, the D-8 propeller for the Liberty V-12 engine represented more accurately their opinion of the design. In a spring 1922 test, the D-8 turned too quickly for the 400-horsepower engine to produce effective thrust, and it failed at the thirty-two-inch station. Reed also did not provide the required documentary drawings for the Army. Full-scale drawings facilitated the inspection of propellers so that Army inspectors could gauge the uniformity of the manufactured product to the original design.[55] The inability to design a practical propeller for each engine application and to provide the documentary drawings indicated to Caldwell that Reed's design methodology was not sophisticated or realistic, especially because he believed that the original test of the D-1 indicated the disadvantages of his design.

Caldwell and the Engineering Division became increasingly skeptical about the basic design of the Reed propeller. They warned Curtiss in March 1923 that the development of the Reed propeller had to be "handled very carefully if we are to minimize the possibilities of accident in using these propellers." Caldwell identified the presence of too many variables in the process of forming and heat-treating the propellers, which required continuous checks of material strengths and test specimens from each individual blade. Moreover, the usual inspection of form, balance, and track, standards that determined the operational suitability of an individual propeller had to be accomplished in an "unusually careful manner."[56]

Nevertheless, Caldwell, as chief of the propeller unit, made recommendations that improved the overall design of the Reed propeller. The Army wanted a second source of supply for metal propellers, an issue ever-present in the mind of procurement officers, and for the sake of encouraging innovation. The first source was Standard Steel with its ground-adjustable propeller. The Army bought D-type propellers for the Vought VE-7 and the Verville-Sperry M-1 Messenger for evaluation purposes beginning in April 1923 and on through early 1924. Caldwell suggested that narrowing the blade chord and increasing the thickness near the hub could significantly strengthen the propeller.[57] The gesture indicated the Engineering Division's support of the development of the Reed propeller with more than just whirl testing and evaluation.

[55] F. W. Caldwell to T. C. McMahon, January 1, 1922; T. H. Bane to S. A. Reed, "Propeller Tests," February 15, 1922; F. W. Caldwell to CAMC, "Reed Duralumin Propellers," March 22, 1923; T. H. Bane to S. A. Reed, "Duralumin Propellers D-1 and D-8," April 26, 1922; T. H. Bane to S. A. Reed, "Duralumin Propeller," May 10, 1922, Folder "Heath [Reed], 1922–1925," Box 5582, RD 3143, RG 342, NARA.

[56] Caldwell to CAMC, "Reed Duralumin Propellers," March 22, 1923.

[57] F. W. Caldwell to L. MacDill, "Method for Strengthening the Reed Propeller," February 15, 1924; CAMC to Chief, Engineering Division, "Reed Duralumin Propellers," April 28, 1923; F. W. Caldwell to C. N. Monteith, "Reed Dural Propellers," May 14, 1923; F. J. Koerner to CAMC, "Purchase Order 48601," October 15, 1923, Folder "Heath [Reed], 1922–1925," Box 5582, RD 3143, RG 342, NARA; F. W. Caldwell to W. E. Donnelly, "Sperry Messenger Duralumin Propeller," February 19, 1924; F. W. Caldwell to C. W. Howard, "Purchase of Metal Propeller for Sperry Messenger," March 4, 1924, Folder "Propellers – January through August, 1924," Box 5541, RD 3132, RG 342, NARA.

Reed was not a cooperative member of the propeller community. Despite the appearance of professional courtesy, Reed was impatient in his dealings with the Army. He was a prolific correspondent to the Engineering Division and was well-known for his persistent inquiries about data he requested, which included data from whirling tests of his own propellers or those from other manufacturers. Reed would also refute the conclusions made by Caldwell and his engineers and would demand that he or one of his representatives be present when his propellers were tested.[58] It was ironic that Reed needed the Army to confirm his results and to ultimately buy his propellers, but he never went beyond treating them as his inferiors.

More importantly, Reed believed that destructive whirl testing was excessive and unnecessary for the evaluation of his invention. He contended that a brief ten-hour whirl test at 50 percent overload was a sufficient evaluation of his propeller's factor of safety. The causes of failure in previous metal propeller designs were due to design mistakes such as the use of composite construction, which he characterized as incorporating a multitude of "junctions, connections, welds, rivets, etc." Since his solid duralumin propeller was the "first and only lightweight single piece propeller that has ever been flown … made or tried," he boasted that "no precedent can be applied to my propeller without qualification."[59]

Caldwell's response was an enlightening synopsis of Reed, his propeller, and the involvement of the Curtiss Aeroplane and Motor Company in the development process. He asserted that the introduction of new materials like metal dictated extreme destructive whirl testing to evaluate structural integrity and design. Before being "thoroughly satisfied" with the Reed propeller, Caldwell wanted a test of twenty-five hours at 1,000 horsepower. Reed's zealous desire to flight test the propellers without thorough whirl testing and the "failure of the Curtiss Company to establish a satisfactory standard of workmanship" were the principal reasons why he had no confidence in their product. Unlike Standard Steel and Westinghouse, Reed and Curtiss were not following a rational development plan aimed at solving the "real problem" in metal propeller development, which were "satisfactory mechanical features, such as the factor of safety, strength, and endurance." A failure in the air only decreased

[58] S. A. Reed to Technical Data Section, McCook Field, July 23, 1921; S. A. Reed to T. H. Bane, February 13, 1922, Folder "Heath [Reed], 1922–1925," Box 5582, RD 3143, RG 342, NARA.

[59] S. A. Reed to H. S. Martin, April 2, 1923, Folder "Heath [Reed], 1922–1925," Box 5582, RD 3143, RG 342, NARA.

their chances of developing a successful propeller. Caldwell was exasperated for having to remind them of that. Overall, he believed performance and reliability were one and the same. In his opinion, "thinner sections will produce a greater efficiency if the propellers are properly designed and carefully made."[60] The gauge of a strong and safe design was whirl testing, which the Reed propeller was barely able to withstand.

The tension over the Reed propeller reached a climax in March 1924 when the Chief of the Engineering Division, Maj. Lawrence W. McIntosh, requested background information on the Army's relationship with Curtiss. The company expressed to him the opinion that the Army was not supporting the development of the Reed propeller and called the Engineering Division's testing procedures into question. Caldwell stressed that he and the propeller unit had accommodated the many requests of Reed and Curtiss since early 1921 where they provided technical data and whirling and engine tests at government expense. Whereas Curtiss was concerned with meeting a 100 percent overload during a test, Engineering Division-designed propellers made by Standard Steel were exceeding 250 percent overload during test.[61] The problem was not Caldwell and the propeller unit; it was Reed and the Curtiss Aeroplane and Motor Company.

The Reed propellers used on military air racers were not safe or practical from Caldwell's standpoint of strength, safety, and endurance. During the 1923 Pulitzer Race, blade flutter was chronic. Lieutenant Brow's second-place racer had a propeller that the Engineering Division rejected as unsafe from excessive flutter. Curtiss rationalized that the Army's testing procedures were too extreme and since the R2C-1 flew, there was no reason why it should be rejected. Engineering Division personnel learned from Brow after the race that the excessive flutter of the propellers made it impossible for him to make a right-hand turn in flight. The propeller installed on Lieutenant Miller's fifth-place R-6 fluttered badly while the aircraft revved up on the ground. Ever mindful of the big picture in propeller development, Caldwell commented that,

[60] F. W. Caldwell to H. S. Martin, "S.A. Reed's Propellers," April 10, 1923, Folder "Heath [Reed], 1922–1925," Box 5582, RD 3143, RG 342, NARA. The army conveyed the message to Reed in H. S. Martin to S. A. Reed, "Duralumin Propellers," May 3, 1923, Folder "Heath [Reed], 1922–1925," Box 5582, RD 3143, RG 342, NARA.

[61] F. W. Caldwell to H. A. Sullivan, "Reed Propeller," February 28, 1924; R. E. Ellis to L. W. McIntosh, "Mr. S.A. Reed and the Curtiss-Reed Propeller Development at Engineering Division," March 17, 1924, Folder "Heath [Reed], 1922–1925," Box 5582, RD 3143, RG 342, NARA.

"it is questionable ... whether the propeller has a satisfactory factor of safety."[62] The propeller for the Army's Verville-Sperry R-3 Racer cost an astonishing $1,500. Lt. Alex Pearson had to leave the race during the first lap because it was so unbalanced that it damaged the engine from extreme vibration.[63]

The use of the Reed propeller in air racing, its primary claim to fame, was an interesting episode regarding choice and indeterminacy of new technologies during the Aeronautical Revolution. It was an integral part of the spectacle of air racing during what historians have called the "golden age of aviation," which for many was an era that was full of flash but with no long-term substance. The Reed propeller was a temperamental and dangerous specialty item. Nevertheless, no air racer equipped with a wood propeller ever beat an airplane equipped with a Reed propeller.

The need for a second source of supply for metal propellers and the momentum created by the air racing success of the Reed propeller influenced the Army's decision to continue using the unsafe propeller on its pursuit aircraft. During diving tests conducted at Selfridge Field in September 1924 flown by Lieutenant Doolittle in a PW-8A, the aircraft's Reed duralumin propeller failed at the hub.[64] The Engineering Division requested a halt in production of D-type propellers for the PW-8 series "pending further investigation as to the strength of these propellers" in May 1924.[65] The major weakness was where the blades flexed as they emerged from the hub blocks. That combined with cracks forming around the sharp-cornered bolt holes and wear between the hub and the propeller itself resulted in blade failure near the root. Maintenance personnel for the Army's improved PW-9 aircraft powered by D-12 engines

[62] F. W. Caldwell to Technical Data Section, "Propellers," June 26, 1924, Folder "Heath [Reed], 1922–1925," Box 5582, RD 3143, RG 342, NARA.

[63] R. E. Ellis to L. W. McIntosh, "Mr. S.A. Reed and the Curtiss-Reed Propeller Development at Engineering Division," March 17, 1924, Folder "Heath [Reed], 1922–1925," Box 5582, RD 3143, RG 342, NARA; K. de Fastenau to A. A. Verville, September 14, 1923, Folder "Propellers, 1923," Box 5501, RD 3122, RG 342, NARA; Foxworth, *Speed Seekers*, 456.

[64] J. F. Curry to CO, Selfridge Field, "Propeller," September 24, 1924; W. E. Donnelly to Chief, Flying Section and Power Plant Section, "Accident-PW-8A (Lt. Doolittle)," September 25, 1924, Folder "Propellers–August through December, 1924," Box 5541, RD 3132, RG 342, NARA; T. E. Tillinghast to COAS, June 2, 1924, Folder "Propellers–January through August, 1924," Box 5541, RD 3132, RG 342, NARA.

[65] T. E. Tillinghast to COAS, May 29, 1924, Folder "Propellers – January through August, 1924," Box 5541, RD 3132, RG 342, NARA.

observed that the D-type blades stretched lengthwise under centrifugal force at high rpms in a dive.[66]

Although more efficient, the R-type was not an improvement regarding structural strength. The Navy grounded fourteen aircraft, primarily Curtiss F6C-3 Hawks, at Naval Air Station Hampton Roads in August 1927 due to propeller "surface defects and cracks." More important, the chief of BuAer's design section, Cdr. Eugene E. Wilson, crashed in an aircraft equipped with a Reed propeller. BuAer was fearful the cause was corrosion generated by intercrystalline embrittlement in the saltwater environment. The problem plagued the thin duralumin sheets used to cover the first generation of American naval aircraft and airships. The Bureau of Standards investigation led by Henry S. Rawdon revealed the thicker duralumin forgings used for propellers yielded "no important embrittlement problem." Rawdon concluded that Wilson's Reed propeller failed from fatigue, which continued to be the primary concern in duralumin propeller design.[67]

Specifically, the stamping of the patent date near the hub had weakened the propeller and caused complete structural failure. The surface marking on the blade, which was a characteristic feature of Reed propellers, only worsened the fact that the propeller was required to operate at stresses rapidly approaching its endurance limit. The factor of safety for the propeller was small, and the absolute values of the stresses were not known. There was also an alarming inconsistency in the quality of the duralumin forgings used to make the propellers. The Navy found that newer R-types made from Aluminum Company of America 25S forgings were better than the Bausch forgings.[68]

The other major user of the Reed propeller, the US Air Mail Service, also experienced considerable difficulty with the Reed propeller. Its DH-4Bs experienced propeller failures in flight due to structural failure throughout a two-year period beginning in July 1925. The failures caused

[66] L. MacDill to D. B. Colyer, "Curtiss Reed Aluminum Alloy Propellers, Drawing No. EX-32915," June 23, 1926; G. H. Brett to Chief, Procurement Section, "Unsatisfactory Performance Reports," March 24, 1927, Folder "Reed Propellers, 1926–1929, 1932," Box 5867, RD 3216, RG 342, NARA.

[67] Henry S. Rawdon, "Corrosion Embrittlement of Duralumin," NACA Technical Note No. 282 (April 1928), 10.

[68] H. A. Backus, "Report of Investigation of Curtiss-Reed R-Type Propellers at Hampton Roads, Virginia," August 12, 1927, Folder "Reed Propellers, 1926–1929, 1932," Box 5867, RD 3216, RG 342, NARA; Bureau of Standards, "Report on Two Duralumin Propellers," September 30, 1927, Folder "F23, Volume 12," Box 1956, RG 72, NARA; George K. Burgess, "Bureau of Standards Report on Two Duralumin Propellers Submitted by BuAer," September 30, 1927, File B5-260215-01, PTF, NASM.

concern throughout the aeronautical community. For the military, it was apparent that the same problems could develop on their aircraft. BuAer chief Moffett was eager to learn from Carl F. Egge, the Air Mail's General Superintendent at Omaha, Nebraska, the nature and character of the problems.[69]

Rudy Schroeder, who had joined the Airport Division of the Ford Motor Company in Dearborn, Michigan, after leaving the Army Air Service, stated in July 1926 that his organization viewed the airmail failures with "a great deal of alarm." They were eager to learn the maximum hours that they should fly their Reed propellers and what type of inspection should be conducted to ascertain their suitability for flight. The Engineering Division's policy on the use of Reed propellers, according to Maj. Harold S. Martin, the chief of production engineering at McCook, was to limit the length of service on the Liberty V-12 engine to 500 hours after which units had to ship them to a maintenance depot for thorough inspection.[70] The ruling cast serious doubt on the long-term durability and strength of the Reed propeller.

From 1926 to 1928, the military services and commercial operators instituted the procedures for periodically inspecting metal propeller fatigue. Reed believed that the inspection process in addition to what he and Curtiss recommended was unnecessary because his metal propellers were overbuilt. Reed refused to recognize that the basic design of the D-type was defective. To the inventor, once "a single error in the design" was fixed, a Reed propeller possessed "an unlimited life."[71]

While it was apparent that the Reed propeller was structurally unsound, its inability to vary its pitch was the major reason why it did not survive the decade of the 1920s and contribute to the final outcome of the Aeronautical Revolution. The path the Engineering Division wanted to take regarding propellers was clear by the fall of 1924. In a letter to

69 H. W. Huking to J. B. Johnson, July 31, 1925, Folder "Heath [Reed], 1922–1925," Box 5582, RD 3143, RG 342, NARA; D. B. Colyer to Chief, Engineering Division, June 15, 1926; L. MacDill to D. B. Colyer, "Curtiss Reed Aluminum Alloy Propellers, Drawing No. EX-32915," June 23, 1926, Folder "Reed Propellers, 1926–1929, 1932," Box 5867, RD 3216, RG 342, NARA; W. W. Webster to Carl F. Egge, "Curtiss-Reed Duralumin Propellers," September 1, 1925, Folder "F23, Volume 9," Box 1954, RG 72, NARA.

70 R. W. Schroeder to Engineering Division, July 21, 1926; H. S. Martin to Chief, Field Service Section, Fairfield Air Intermediate Depot, "Data for Technical Order," July 12, 1926; G. H. Brett to Chief, Engineering Division, July 17, 1926, Folder "Reed Propellers, 1926–1929, 1932," Box 5867, RD 3216, RG 342, NARA.

71 Reed, "Technical Development of the Reed Metal Propeller," 59; S. A. Reed to L. MacDill, December 13, 1926, Folder "Reed Propellers, 1926–1929, 1932," Box 5867, RD 3216, RG 342, NARA.

Frank Russell, the chief of the Engineering Division, John F. Curry, stated that, "We believe in metal propellers, especially the adjustable-pitch type."[72] Caldwell authored that correspondence for Curry's signature. Even though the word "adjustable" to that point indicated a variable-pitch propeller, it is clear in this correspondence that Caldwell is referring to a detachable-blade ground-adjustable-pitch propeller.

The value of a propeller that enabled pitch variation on the ground or in the air was not obvious. Through the mid-1920s, two major propeller manufacturers, Curtiss and Westinghouse, believed that fixed-pitch propellers were the most efficient even though they were suitable for only one operating regime. The flexibility and relative efficiency over a number of operating regimes, primarily takeoff and cruise, offered by variable-pitch propellers were not yet apparent. The airplanes and engines the propeller manufacturers designed their products for were basically one to two-seat, single-engine aircraft. That paradigm shaped the logic behind the belief that fixed-pitch propellers possessed a performance advantage over variable-pitch propellers. The interplay between the American military and Reed and Curtiss on the fixed versus variable-pitch debates highlights an environment of indeterminacy and ambiguity in propeller design during the 1920s.

Persistence and Conflict in the Propeller Community

The obstacle to the complete implementation of metal detachable-blade propellers were the realities of the Air Service's logistical situation in the early 1920s. First, the service's primary aircraft engine was the 400-horsepower Liberty V-12 left over from the World War I aviation production program and there were a large number of war surplus wood propellers to go with them. The gradual incorporation of metal propellers into the operational fleet as the number of wood propellers dwindled was a time-consuming process. Second, engines that were being phased out, such as the 300-horsepower Hispano-Suiza, were still plentiful. It was illogical to go through a significant outlay of engineering resources to design a ground-adjustable-pitch propeller for obsolete power plants.[73] The realities of paltry budgets and the logistics of introducing a new technology dictated the continued use of wood propellers.

[72] J. F. Curry to F. H. Russell, November 1, 1924, Folder "Heath [Reed], 1922–1925," Box 5582, RD 3143, RG 342, NARA.

[73] Ibid.

Russell, who was trying to sell the Army the fixed-pitch duralumin Reed propeller, disagreed. He emphasized that it was superior to the ground-adjustable-pitch propeller. The commitment to one efficient operating regime over flexibility was well apparent. Russell believed that adjustable-pitch was only suitable for experimental purposes, once flight tests confirmed the correct blade setting, the blades should be made as "immovable as possible" to maintain "all-round efficiency." The D-type gave efficiency from the hub to the tip and the ability to repair rather than replace after crashes. The traditional wood-type hub offered interchangeability. To prove the point, he suggested the Army conduct comparative tests of a Reed and "the other metal propeller," the Standard Steel design, which would highlight the superiority of his company's product. Recognizing the importance of government support, Russell emphasized that "we must have some duralumin propeller orders that we may continue this development."[74]

Reed also adamantly believed that detachable-blade propellers were inferior. He argued that the "single-piece style will in the long run be found superior to any style with multiple parts." He alleged that because "so much depends upon the root anchorage," the connection of the retaining shoulder to the hub was, by design, structurally weak. Any imperfection in the manufacturing process resulted in a "dangerous concentration of load" that caused the assembly to fail.[75]

The arguments put forth by Russell and Reed centered on their belief that the ground-adjustable-pitch propeller was not an original contribution to aeronautical technology. They were also well aware of the special relationship between Caldwell and the propeller unit and Standard Steel. The Reed Propeller Company filed suit in May 1924 against Standard Steel in the United States District Court for the Western District of Pennsylvania for infringement of Reed's first patent. Acknowledging the Engineering Division's involvement, Reed filed a second suit against the US government for infringement of both patents in January 1925.[76] Despite the obvious differences in the design and construction, Reed felt his propeller was a foundation technology in much the same way the Wright brothers viewed their airplane.

74 F. H. Russell to J. F. Curry, November 4, 1924, Folder "Heath [Reed], 1922–1925," Box 5582, RD 3143, RG 342, NARA.

75 Reed, "Technical Development of the Reed Metal Propeller," 58–59, 61.

76 *The Reed Propeller Company, Inc. v. The United States*, USCC No. 42133, in James A. Hoyt, *Cases Decided in the Court of Claims of the United States, December 1, 1941, to March 31, 1942*, Vol. 95 (Washington, DC: Government Printing Office, 1942), 264.

Curry's response to the Curtiss-Reed arguments clearly outlined the direction the Army was going. Operators could adjust the pitch of a detachable-blade propeller to meet the requirements of specific flight conditions, primarily climb, top speed, and cruise, for various engine/airframe combinations. Testing proved that ground-adjustable-pitch propellers were 2 to 3 percent more efficient than fixed-pitch propellers designed for the same performance regime. Both Reed and detachable-blade propellers could be repaired in the event of an accident, but the latter only required the replacement of a damaged blade by a mechanic. A damaged Reed propeller had to be replaced entirely, which limited operations in the field. Adjustable-pitch propellers were more conducive to bulk production because Standard Steel finished them in drop forging dies without subsequent machining. Curry emphasized that "our practical experience" brought those facts to light during years of propeller research and development.[77]

Curry suggested that the Reed propeller be used on moderate-size engines such as the Curtiss K-6 while detachable-blade propellers were reserved for use on larger engines. The Navy was working toward that division as well. Curry believed that allowing Reed to specialize alleviated their production problems and sustained two viable metal propeller manufacturers. Curry reasoned that "your company is entitled to a fair share of the metal propeller business along with your competitor."[78]

By 1926, the Army and the Navy considered the Standard Steel detachable-blade propeller superior to the R-type propeller. When given a choice of the two for the Curtiss P-1 Hawk and the Boeing PW-9 pursuit aircraft, the Office of the Chief of the Air Service decided that "in view of all the circumstances, it is believed preferable to procure the adjustable-pitch type."[79] Service tests at Selfridge Field, Michigan, conducted to determine a standard propeller for the Army's P-1 Hawk pursuit biplane with D-12 engine in March 1927 solidified that belief. The goal was to ascertain the existence of any "mechanical trouble" in a detachable-blade propeller and the D- and R-type propellers. To the test pilots, the detachable-blade propeller permitted a faster takeoff, generated more

[77] R. W. Dichman to Chief, Production Section, "Metal Propellers," March 7, 1927, Folder "Propellers–January through July, 1927," Box 5663, RD 3163, RG 342, NARA; J. F. Curry to F. H. Russell, November 20, 1924, Folder "Heath [Reed], 1922–1925," Box 5582, RD 3143, RG 342, NARA.

[78] Curry to Russell, November 20, 1924.

[79] Office of COAS to Chief, Engineering Division, "Allotment Table Fiscal Year 1927–Item No. 1, Class O1-H, Curtiss Reed Propellers," July 26, 1926, Folder "Reed Propellers, 1926–1929, 1932," Box 5867, RD 3216, RG 342, NARA.

thrust in the air, and had no tendency to flutter. They recommended that it be adopted as the standard propeller for the P-1 Hawk.[80] Based on a similar evaluation in August 1927, BuAer requested the replacement of all D-type propellers with detachable-blade-type propellers, preferably the Standard Steel nine-foot propeller.[81]

The continued failures of the D-type propeller led the Army to consider the complete ban of the design from service use in August 1927. The D-type propeller on an aircraft flown by Assistant Chief of the Air Corps, Brig. Gen. James E. Fechet, and Maj. Walter G. Kilner failed, causing the two officers to crash-land in the Chesapeake Bay. The Chief of the Air Corps, General Patrick ordered an investigation, but no conclusive evidence could be found for the exact reason why the propellers failed. After a meeting with Caldwell, Kilner, and Henry W. Harms of the Air Corps Executive, Patrick suggested that the Army replace the D-type with detachable-blade ground-adjustable-pitch propellers and "possibly wooden propellers."[82]

The Materiel Division did not entirely agree with Patrick's suggestion. The D-type propeller had been in disfavor for two to three years and the Army was buying only the much-improved R-type propellers. There was success with the Drawing 070717 Standard Steel propeller for the Liberty V-12, but variants for use with Curtiss engines were not available. It would be a "considerable expense" for the Army to change entirely over to detachable-blade propellers.[83]

The realities of the Army's procurement situation dictated the implementation of the R-type on the Army's Curtiss Hawk and Falcon aircraft. One R-type was cheaper at $360 while the Standard Steel ground-adjustable-pitch propeller cost $490.[84] In a time of tight budgets and fiscal expediency, the Army's use of the R-type was appropriate for those aircraft so long as adequate inspection and maintenance procedures were

[80] L. MacDill to Materiel Division Service Test Representative, Selfridge Field, March 5, 1927; Vincent B. Dixon to Chief, Engineering Division, Materiel Division, April 28, 1927, Folder "Propellers – January through July, 1927," Box 5663, RD 3163, RG 342, NARA.

[81] R. D. Thomas to Chief, BuAer, "Curtiss-Reed Slab Type Metal Propellers – Replacement of," August 16, 1927, Folder "F23, Volume 12," Box 1956, RG 72, NARA.

[82] H. W. Harms to Executive, Materiel Division, "Curtiss Reed Metal Propellers," August 19, 1927; W. G. Kilner to Chief, Materiel Division, "Propellers," August 19, 1927, Folder "Reed Propellers, 1926–1929, 1932," Box 5867, RD 3216, RG 342, NARA.

[83] J. E. Fickel to M. Patrick, "Propellers – Curtiss-Reed Type," August 19, 1927, Folder "Reed Propellers, 1926–1929, 1932," Box 5867, RD 3216, RG 342, NARA.

[84] F. H. Russell to W. E. Gillmore, "McCook Field Proposal E-27128 for Metal Propellers for D-12 and Liberty Motors," March 3, 1927, Folder "Propellers – January through July, 1927," Box 5663, RD 3163, RG 342, NARA.

TABLE 2 *Reed propeller production*

Year	Civilian	Army	Navy	Total
1923	9	4	6	19
1924	72	21	6	99
1925	177	76	95	348
1926	254	166	187	604
1927	111	135	210	456
1928	210	265	42	517
1929	415	4	2	421
1930	51	0	0	51
1931	23	0	0	23
1932	52	0	0	52
1933	27	0	0	27
				2,617

Note: "Curtiss Single Piece Propeller Production, 1923 to 1933," n.d. [1933], Box 4509, RG 123, NARA.

followed. Those logistical realities meant the Reed propeller would be around for much longer than it was wanted.

The Reed propeller had fallen into disfavor with the aeronautical community. After the successful air racing years, production and sales dwindled rapidly after 1929. Curtiss manufactured a total of 2,617 D- and R-type propellers for the Air Mail Service and other civilian operators, the Army, and the Navy (Table 2).

The decline in propeller sales did not mean that there was no other source of income for Reed. The patent infringement litigation against Standard Steel that began in 1924 dragged on for over four years. Standard Steel was unable to prove the Reed patent invalid in court as it applied to three specific propeller designs, including the pivotal specification No. 1519 for the Wright J-5 Whirlwind engine.[85] President Harry A. Kraeling acquired a manufacturing license in November 1928 and agreed to pay a small indemnity for past use as well as patent royalties. Within four years, the Reed Propeller Company received $159,547.51 in royalties from that license alone.[86]

Reed felt it was time for another improved propeller. He submitted to the Materiel Division his S-5 chrome molybdenum steel propeller design

[85] S. A. Stewart to W. E. Valk, 2 January 1936, in S. H. Philbin, "Production of Information in Compliance with Order of this Court," January 9, 1936, *Reed Propeller Company v. United States*, USCC No. 42,133, Folder GJ42133(3), Box 4504, RG 123, NARA.

[86] Plaintiff's Exhibit No. 8, USCC No. 42,133, "Royalties Paid the Reed Propeller Company by its Licensees Curtiss Aeroplane and Motor Company and Hamilton Standard Propeller

for the Pratt & Whitney Wasp engine for whirl testing in February 1932. The Army initially believed that Reed was making a donation to the Wright Field Museum. The chief of the engineering section at Wright Field, Maj. Clinton W. Howard, curtly informed Reed that the Army "practically abandoned the purchase of single piece propellers a number of years ago, and their use has been discontinued as rapidly as circumstances would permit." Considering the meager funds available for whirl testing, the Army could not justify testing an antiquated propeller design.[87] After that final rebuff, Reed retired from active propeller invention.

Nevertheless, Reed was mindful of his legacy in aeronautics. He believed that he deserved the technical recognition, and the accompanying financial compensation, that his design was the point-of-departure for all metal propellers. The number one threat to that legacy was the continued infringement of his patents by the US government in the early 1930s. Reed dismissed his earlier 1925 suit without prejudice in January 1929 as part of the license agreement with Standard Steel.[88] The deal also included offering the government a license, but the Justice Department rejected the offer. At the time, Reed and his lawyers were satisfied since the main manufacturers of aluminum alloy propellers in the United States either possessed a license or were defendants in an ongoing patent infringement case filed by the Reed Propeller Company.[89]

The specific issue was the American military's search for additional suppliers of ground-adjustable-pitch propellers in 1930. The Army Air Corps contracted the J. A. Fay and Egan Company of Cincinnati and Hartzell Propellers of Piqua, Ohio, to construct experimental aluminum alloy blades. Neither of those companies possessed a license from the Reed Propeller Company. The larger issue concerned what company patent attorney William Valk called the "inequity of the government

Corporation," n.d. [1933], Box 4509, RG 123, NARA; *Reed Propeller Company v. the United States*, *Cases Decided*, 264–265.

[87] S. A. Reed to C. W. Howard, February 11, 1932; C. W. Howard to S. A. Reed, February 16, 1932, Folder "Reed Propellers, 1926–1929, 1932," Box 5867, RD 3216, RG 342, NARA.

[88] "Judgments Without Opinions: The Reed Propeller Company, Inc. v. The United States, USCC No. E-144," in Ewart W. Hobbs, *Cases Decided in the Court of Claims of the United States, June 1, 1941, to (In Part) February 4, 1929*, Vol. 66 (Washington, DC: Government Printing Office, 1929), 745.

[89] S. H. Philbin, "Petition to the Honorable, the Judges of the Court of Claims of the United States," December 23, 1932, *Reed Propeller Company v. United States*, USCC No. 42,133, Folder GJ42133(1), Box 4504, RG 123, NARA.

deliberately and intentionally setting out to break down an established patent-protected industry by created subsidized and unnecessary competition." Letters to the secretaries of the Army and Navy in January 1931 threatened "possibly disastrous litigation" if a license was not secured.[90]

The resultant case in the US Court of Claims, *Reed Propeller Company v. United States*, started in 1932. It followed a well-worn pattern found in the large number of patent infringement suits concerning aeronautical, chemical, electrical, and optical equipment in the 1920s and 1930s. Reed and his associates, as the plaintiff, felt they held a monopoly on all airplane propellers made from aluminum alloy that incorporated centrifugal force for structural integrity. The defendant, the US government, assumed the role of the independent acting in violation of those patents and bore the burden of proving their invalidity in court. Typically, corporate plaintiffs could outspend the independents in terms of retaining patent lawyers and surviving protracted litigation as in Reed's suit against Standard Steel.[91] In this particular case, the Reed Propeller Company opposed the Justice Department and the vast legal, financial, political, and technical network of the federal government.

At the center of the alleged infringement was the "Government Propeller," the ground-adjustable-pitch propeller perfected at McCook Field and in Pittsburgh by Frank Caldwell and Thomas A. Dicks. The Justice Department's lawyers attacked the validity of Reed's patents through the use of hundreds of hours of testimony from an impressive list of expert witnesses including propeller pioneers Stuart Bastow, Victor W. Pagé, Henry Ivan Stengel, Spencer Heath, and Frank Caldwell. Artifacts from the Smithsonian Institution, including the Bastow-Pagé and Aeromarine ground-adjustable-pitch propellers, served as exhibits for the defense. Regarding Reed's first patent, they argued that the blades of the government propeller did not rely upon centrifugal force for structural rigidity. There were numerous examples of prior art, knowledge, and public use that predated Reed's second patent, primarily Alphonse Pénaud and Paul Gauchot's 1876 airplane patent and the flight of the Bastow-Pagé propeller on the *Rhode Island* in 1910. To the judges, it was well-established that the aeronautical community was "moving rapidly

[90] W. E. Valk to Secretary of War, "Reed Aeronautical Propeller Patents," January 23, 1931, Box 4509, RG 123, NARA; W. E. Valk to Secretary of Navy, "Reed Aeronautical Propeller Patents," May 21, 1931, Box 4509, RG 123, NARA.

[91] Joseph Borkin, "The Patent Infringement Suit: Ordeal by Trial," *The University of Chicago Law Review* 17 (Summer 1950): 636.

toward the adoption of the metal propeller" well before Reed even began to ponder turning his foghorn into a propeller. After ten years of active litigation, they ruled in favor of the defendant on January 5, 1942, and rendered the Reed patents invalid.[92]

Reed did not live to see the conclusion of *Reed Propeller Company v. United States*. He died on October 1, 1935, at the age of eighty-one at his home in New York City firmly believing his propeller was a central contribution to aeronautical progress. At the time, C. G. Grey generously regarded Reed as "one of the best dressed and best liked men in New York."[93] Reed's heirs donated his prototype D-1 propeller to the Smithsonian Institution shortly after the conclusion of the patent case. Associate Director John E. Graf assured Reed's family that the inventor's legacy was the introduction of the first aluminum alloy airplane propeller put into widespread use after the wood propeller.[94]

As the protracted legal fight against the government began to gain momentum, Reed made what would become his most long-lasting contribution to aeronautics. In December 1934, he created an annual award, administered by the Institute of the Aeronautical Sciences (IAS – a precursor to the American Institute for Aeronautics and Astronautics), that recognized individuals whose experimental and theoretical research had made a valuable, direct, and practical contribution to flight technology. As a founding member and a trustee of the institute in October 1932, Reed was, in Lester D. Gardner's words, "one of the most distinguished members and best friends" of the organization. As a result, the IAS named the award after him not only in recognition of his generosity, but also for his propeller. Reed personally wrote a $250 check for the first award for 1934 to meteorology pioneers Carl-Gustaf Rossby and Hurd C. Willett.[95] Still awarded today, the Sylvanus Albert Reed Award became a prestigious indicator of individual excellence in aerospace research and technology over the course of the twentieth century.

[92] *Reed Propeller Company v. the United States, Cases Decided*, 262, 314.

[93] Grey, "Sylvanus Reed," 461–462.

[94] L. G. Reed to J. E. Graf, June 12, 1942; J. E. Graf to L. G. Reed, August 17, 1942, "Reed D-1 Propeller," File A19430003000, RO, NASM; Grey, "Sylvanus Reed," 461–462.

[95] The IAS would receive a $10,000 endowment after Reed's death. "Fourth Annual Dinner," *Journal of the Aeronautical Sciences* 4 (February 1936): 114; Tom D. Crouch, *Rocketeers and Gentlemen Engineers: A History of the American Institute of Aeronautics and Astronautics ... and What Came Before* (Reston, VA: American Institute of Aeronautics and Astronautics, 2006), 78, 264.

Lessons from Failure

The story of Reed and his propeller adds layers of understanding to the role of failure and the importance of alternate pathways in technical development during the Aeronautical Revolution. There is the contradictory nuance to the argument that patents played no major role in the development of aircraft. Under the auspices of the 1917 cross-licensing agreement held by the consortium of airframe manufacturers making up the MAA, that is true.[96] Propeller patents, which existed outside the confines of the MAA, were important. If Reed was successful in court, he had the potential of impeding the economic futures of the true innovators in propeller design through licensing agreements. He would have also had the prestige of being recognized as the original inventor of the metal airplane propeller, which he desired just as much as financial restitution. That was not the case, but the overall eighteen-year litigation campaign was the type of serious distraction that is often overlooked in the history of aviation and technology.

The rise and fall of the Reed propeller also highlights the indeterminate synergy inherent in aeronautical invention, innovation, and use. The stunning performance of his invention in spectacular high-profile air races introduced the metal propeller to the world aeronautical community and to the public in general. In return, the greater aeronautical community celebrated Reed's invention as a technology representing "Progress," which included awarding him the 1925 Collier Trophy, several patents, and an important segment of the aviation market. The excitement over the performance of the light weight and simple fixed design on the part of the broader aeronautical community does reflect the existence of what one historian called a "progress ideology of metal" since the use of duralumin was its central design facet.[97] In the case of the Reed propeller, that ideology combined with the allure of spectacle reinforced the perception that the technology represented modernity and fueled the mainstream enthusiasm of the aeronautical community, but they did not shape Reed's decisions to choose duralumin.

Extensive evaluation, test results, and failure in operational use, in the end, however, revealed that Reed's propellers were a technological dead

[96] Alex Roland, "Pools of Invention: The Role of Patents in the Development of American Aircraft, 1917–1997," in *Atmospheric Flight in the Twentieth Century*, ed. Peter Galison and Alex Roland (Boston. MA: Kluwer Academic Publishers, 2000), 338.

[97] Eric M. Schatzberg, *Wings of Wood, Wings of Metal: Culture and Technical Choice in American Airplane Materials, 1914–1945* (Princeton, NJ: Princeton University Press, 1998), 4.

end. The D- and R-types were important intermediate developments over wood propellers. Caldwell and his specialist colleagues ignored the larger aeronautical community's excitement for them for important practical reasons. They developed the safer and equally efficient detachable-blade ground-adjustable-pitch propeller in parallel.

The inability of the Reed propeller to enhance the overall performance of the rapidly evolving modern airplane safely reveals the ambiguity and complexity in the development of propeller technology during the 1920s. Ultimately, the Reed diversion illustrated that there was no clear consensus within the aeronautical community on what constituted the ultimate form of the aerial propeller. In other words, the definition of "Progress" was not agreed upon universally. The Reed propeller, which appeared to be an alternate pathway during the period 1917–1930, was a complete failure.

Nevertheless, both the Reed and detachable-blade, ground-adjustable propellers were fixed-pitch in flight, which meant that neither would be the ultimate solution in propeller innovation as it pertained to enabling aircraft to fly higher, faster, and farther. The main difference was the design and who developed the technology. The ground-adjustable propeller, the product of an intense government-industry partnership, incorporated the potential of being developed into a practical variable-pitch mechanism with its multipiece construction. It remained to be seen if that same partnership, or another type of collaboration altogether, would lead to the creation of a propeller capable of shifting gears in the air.

7

No. 1 Propeller Company

Just as the American aviation industry started to grow in the wake of ground-breaking government legislation and Lindbergh's transatlantic flight, the stock market crash of October 1929 and the Great Depression threatened its continued existence. The newly formed Hamilton Standard Propeller Corporation, suffering from dwindling military and commercial contracts, scrambled for a new product to sustain itself. Chief engineer, Frank Caldwell, designed a hydraulic two-position, controllable-pitch, or hydro-controllable, propeller that promised to increase Hamilton Standard's place in the aviation marketplace. His new propeller used the engine's oil supply and centrifugal force exerted by counterweights to keep the blades at the desired pitch during flight. Eugene E. Wilson, the president of Hamilton Standard, knew instantly the propeller was an innovation the corporation desired and was looking for, which in his words, was "the answer to a maiden's prayer." He enthusiastically supported the development of the two-position, controllable-counterweight design, which was an extreme financial undertaking during the Great Depression.[1] The innovation was both an investment in the further improvement of the airplane and the financial fortunes of the company.

The creation of an industrial propeller community within the context of the emergence of the modern aviation corporation in the late 1920s made the variable-pitch propeller possible. Overall, from the end of World War I to the mid-1920s, the foundation of the American aviation

[1] Eugene E. Wilson, *Slipstream: The Autobiography of an Air Craftsman* (Palm Beach, FL: Literary Investment Guild, 1967), 165–166; George Rosen, *Thrusting Forward: A History of the Propeller* (Windsor Locks, CT: United Technologies Corporation, 1984), 42.

industry changed from one that traced back to individual pioneers such as Wilbur and Orville Wright and Glenn Curtiss to one based on a mainstream American corporate model characterized by heavy Wall Street involvement.[2] The modern variable-pitch propeller required the necessary managerial, financial, technical, and personnel infrastructure that only a corporate environment could provide, but that in turn meant the innovation had to have the potential for financial success. An understanding of that vital interrelationship is crucial to a broader understanding of the development of aeronautical technology in the 1920s and 1930s.

The momentum begun by government aviation legislation and the technical and symbolic success of Lindbergh's transatlantic flight contributed to the creation of modern aviation corporations in the United States in 1929. The struggling aviation industry, a collection of small businesses catering to a limited specialty market, underwent a period of rapid growth marked by consolidation into corporations with impressive financial, technical, and managerial resources. Wall Street analysts recognized the existence of a "big five" of large holding companies that represented a comprehensive approach to manufacturing, selling, and transportation within the entire aviation industry. The two oldest groups – the Curtiss Aeroplane and Motor Company, headed by Clement M. Keys, and the Wright Aeronautical Corporation controlled by Richard Hoyt – were in operation in the immediate post-World War I period. The merger of the two in July 1929 created the powerful Curtiss-Wright Corporation. The Detroit Aircraft Corporation represented interests from the General Motors Corporation, and the Harriman-Lehman-Pynchon banking group sponsored the Aviation Corporation (AVCO) of Delaware. By far, the "largest complete unit," was the United Aircraft and Transport Corporation (UATC).[3]

An Industrial-Corporate Community

UATC was the creation of aviation entrepreneurs Frederick B. Rentschler, Edward A. Deeds, William E. Boeing, and Chance Vought in 1929. Coming from a prosperous industrial family in Hamilton, Ohio, Rentschler was

[2] Dominick A. Pisano, "The Decade of Incorporation: Aviation, 1917–1926," unpublished manuscript, December 2001, 4–5.

[3] J. Roy Prosser and Company, "Present and Future Trends of Aeronautical Companies: Mid-Year Revision," June 1929, Folder "Resumé of Aeronautical Companies, 1929," Box 1, Series I, Accession 1629, HFM; Roger E. Bilstein, *The American Aerospace Industry: From Workshop to Global Enterprise* (New York: Twayne Publishers, 1996), 33.

an officer of the Wright Aeronautical Corporation. Disenchanted with the corporation's management, Rentschler left with fellow engineer George J. Mead to form a company dedicated to the production of radial, air-cooled engines. Rentschler sought assistance from an old family friend in Dayton, Col. Edward A. Deeds, a director of the National City Bank of New York, who had a controlling interest in Niles-Bement-Pond, a conglomeration of machine tool companies that dominated American and European industry. Deeds, a major player in the American World War I aviation production program, gave Rentschler and Mead control of one of the company's subsidiaries, New England precision tool and instrument manufacturer Pratt & Whitney. The pair moved to Hartford, Connecticut, added "Aircraft" to the name of the company, and began their pioneering work on the 400-horsepower nine-cylinder Wasp radial engine.[4]

At the same time the Wasp proved its value to the expansion of military and commercial aviation in the mid- to late 1920s, Congress enacted the Air Mail Act of 1925. The resulting privatization of airmail operations with the option of carrying fare-paying passengers stimulated the growth of American aviation. On the West Coast, the Boeing Airplane Company expanded its successful military aircraft manufacturing activities into carrying mail for the US government. Seattle lumber magnate William E. Boeing started producing aircraft in 1916 and soon became one of the leading military aircraft contractors. By 1928, the Boeing Airplane and Transport Corporation (BATC) included the Boeing factory in Seattle, the Boeing Air Transport transcontinental airline operating the Chicago–San Francisco mail run, and a 75 percent controlling interest in Pacific Air Transport, which operated the Los Angeles–San Francisco–Seattle mail run. The majority of Boeing's military and commercial aircraft used Pratt & Whitney radial engines, the top-selling power plants of the period.[5]

Realizing that Pratt & Whitney engines and Boeing aircraft had a common future, Rentschler and Boeing decided to integrate their operations. They joined with another important aviation entrepreneur, Chance Vought, a successful manufacturer of Navy aircraft, to form UATC. Deeds arranged for BATC's purchase of Pratt & Whitney Aircraft from Niles-Bement-Pond. Rentschler's brother, Gordon S. Rentschler, a Wall Street financier and soon-to-be president of the powerful National City Bank,

[4] "The United Aircraft and Transport," *U.S. Air Services* 14 (January 1929): 66; "No. 1 Airplane Company," *Fortune* 5 (April 1932): 47–48.

[5] "No. 1 Airplane Company," 47, 49–50.

arranged for floating the preferred and common stock on the public market. The new corporation's capitalization at $25 million made it, according to *Fortune* magazine, "a small titan, as American titans go" with a strong market share of radial engine-powered military and commercial aircraft. After the incorporation of UATC in Delaware on January 19, 1929, the new corporation was the world's "No. 1 airplane company."[6]

Rentschler and Boeing intended to operate UATC as a vertically integrated corporation – a holding company that controlled subsidiaries representing the major sectors of the aviation industry from airframes to engines to the airlines that used the finished product – to dominate the American and world aeronautical markets. One historian equated the corporation as an "integrated aerial octopus."[7] Wall Street analysts recognized that the key to the success of the American aviation holding companies was the acquisition and consolidation of the "pioneers of the industry" that had the "valuable experience" that made them important financial commodities.[8]

To that end, UATC acquired companies that complemented Pratt & Whitney engines and Boeing and Vought aircraft, including the Sikorsky Aviation Corporation, Northrop Aircraft Corporation, and the Stearman Aircraft Company. By late 1929, UATC oversaw fourteen different operations in the areas of manufacturing, transport, education, export sales, and airports.[9] UATC leadership, operating through its executive committee, planned to retain a decentralized administration where the member companies operated as separate entities whenever possible. UATC existed to guide the overall policies of the member companies and to provide financing when necessary.[10]

[6] Prosser and Company, "Present and Future Trends of Aeronautical Companies"; "No. 1 Airplane Company," 50; Ronald Fernandez, *Excess Profits: The Rise of United Technologies* (Reading, MA: Addison-Wesley Publishing, 1983), 50–51.

[7] Bilstein, *American Aerospace Industry*, 34.

[8] Prosser and Company, "Present and Future Trends of Aeronautical Companies."

[9] The manufacturing companies included: Pratt & Whitney Aircraft Company, the Boeing Airplane Company, Chance Vought Corporation, the Stearman Aircraft Company, the Northrop Aircraft Corporation, Sikorsky Aviation Corporation, and the Hamilton Standard Propeller Corporation. The transport companies were Boeing Air Transport, Pacific Air Transport, and Stout Air Services. The other UATC operations were the Boeing School of Aeronautics, United Aircraft Exports, United Airports Company of California, and the United Airports of Connecticut. Both Boeing and Pratt & Whitney operated subsidiaries in Canada.

[10] United Aircraft and Transport Corporation, *First Annual Report to Stockholders*, December 31, 1929, Hamilton Sundstrand Community Relations Division (hereafter cited as HSCRD), 7, 10; "The United Aircraft and Transport," 68.

Merger of Two Worlds: The Creation of Hamilton Standard

As part of the vertical integration, UATC needed a manufacturer of detachable-blade ground-adjustable-pitch propellers along with the all-important design patents for the corporate family. Boeing enticed Tom Hamilton of Hamilton Aero Manufacturing in Milwaukee to join in return for an exchange of the new corporation's stock. Hamilton bought the propeller business from Matthews Brothers and formed the Hamilton Aero Manufacturing Company in 1919. Hamilton Aero had made the most of the opportunity in the mid-1920s to become a secondary source of supply to the US government and the airlines by switching from wood propeller production to the manufacture of the metal ground-adjustable propeller pioneered by Standard Steel and the Army. Hamilton owned two aeronautical companies based in Milwaukee. His other enterprise, Hamilton Metalplane Company, experienced limited gains in a market dominated by the Ford Tri-Motor. Hamilton received 50,000 shares of UATC stock for selling both.[11]

Rentschler and Boeing endeavored to acquire the best manufacturers for the holding company's operations, but the personal connection between Hamilton and Boeing was at the root of the acquisition in 1929. Much like the corporation's overall beginnings, the acquisition of Hamilton Aero was characteristically nepotistic. Hamilton and Boeing were acquaintances in Seattle's upper-class society and shared its early fascination with aviation during the early flight period.[12]

On the promise of continued commercial success in propellers, UATC acquired Hamilton Aero. Rentschler and Boeing did not understand that Hamilton Aero was not an innovator in metal propeller technology. Hamilton Aero possessed no important patents and had no pioneering design knowledge to contribute. Instead, the company was manufacturing aluminum alloy propellers by using government specifications as the guide. Hamilton Aero could only replicate the detachable-blade ground-adjustable-pitch propeller pioneered by the Army and Standard Steel, a direct competitor.[13]

[11] "Thomas Foster Hamilton," *Who's Who in American Aeronautics* (New York: Aviation Publishing Corporation, 1928), 48; Prosser and Company, "Present and Future Trends of Aeronautical Companies"; Fernandez, *Excess Profits*, 60.

[12] "Rites Set for T.F. Hamilton, Flying Pioneer," *Los Angeles Times*, August 13, 1969, section 2, p. 3.

[13] "No. 1 Airplane Company," 50; Fernandez, *Excess Profits*, 60.

The experience of Ray Hegy, who joined Hamilton Aero as a "rougher" in 1925, illustrated the level of experience of the factory personnel and how the company found an empirical way into the lucrative metal propeller market. He joined the company after answering a Hamilton Aero Manufacturing advertisement in the *Milwaukee Journal* searching for expert cabinetmakers needed to make propellers. Hegy used the traditional woodcrafter's tools – planes, shaves, a protractor, and a laminated bench – and a flexible-shaft grinder to turn glued mahogany, oak, and birch blanks into a general propeller form. The mahogany blanks were surplus material left over from Matthews Brothers at the end of World War I, which many aircraft manufacturers, including Ryan Air Lines of San Diego, preferred. The rough propeller went from Hegy's bench to the finishing and tipping departments before being shipped out to a customer. Hegy and his coworkers were some of the best woodcrafters in aviation.[14]

Hamilton Aero was a profitable wood propeller business in 1925. The small company designed and manufactured propellers for Army PT-1 and Navy VE-7 trainers, Ryan M-1 mail planes, recreational Waco and American Eagle biplanes, and pioneering aircraft such as the Fokker F-VII-3m trimotor *Josephine Ford*, used by Navy Cdr. Richard E. Byrd, Jr., to explore the Arctic in May 1926. The Hamilton Aero propellers on the Navy's ill-fated airship, USS *Shenandoah*, were the largest ever made at eighteen-feet in diameter. Business was good and Tom Hamilton flamboyantly drove a Rolls-Royce automobile to the relatively new factory, having just been built for a defunct shoe manufacturer, on Keefe Avenue.

Nevertheless, Hamilton Aero needed to make the transition to metal if it were to remain a primary contractor to the US government. Early 1920s memories of late paychecks and the personal pleas of Hamilton's wife, Ethel, to not cash them until a government payment cleared the bank, and scrounging for wood scraps to heat the factory, were not that distant. The metal propeller was, in Hegy's words, "knocking at the door," and with it was the promise of continued government production contracts that forced the wood-based manufacturer to enter into the world of metal aircraft.[15]

[14] Ray Hegy, "That Little Old Prop-Maker ... Me," *Sport Aviation* 14 (July 1964): 4–6.

[15] Hegy, "That Little Old Prop-Maker," 4; Russell Trotman, "Turning Thirty Years," *The Bee-Hive* 24 (January 1949): 22; Fred E. Weick, Interview by Harvey H. Lippincott, March 4, 1981, Folder "Interview: United Aircraft and Hamilton, 1929-1930," Box 3, FWP, NASM, 76.

Hegy began to notice lathes, milling machines, and stacks of heavy aluminum alloy forgings accumulate in one side of the factory during the fall of 1925. A new shop superintendent, Arvid Nelson, arrived soon after. Hamilton Aero received an Army Air Corps contract to manufacture a detachable-blade ground-adjustable-pitch propeller for the Wright J-4 Whirlwind engine. Presented with a Standard Steel split hub, Hamilton Aero, as one of the more successful wood propeller manufacturers, was to make the duralumin blades.[16]

Hegy was given the job. He initially refused, but a salary increase of $0.75 an hour changed his mind. Hegy first "chopped" a rectangular shape out of the solid slab forgings. One of his coworkers installed a metal-cutting blade in the factory band saw and cut out the planform and marked the general twist along the length of the blade at intervals called stations. Another woodcrafter-turned-metalworker shaped the blade root, or shank, on one of the new lathes. Hegy used a pneumatic jack hammer with a ¼-inch chisel installed to cut grooves at each station to within ¼-inch of the finished surface. It took the strength of Hegy and Bill Pekowsky of the tipping department to handle Hegy's flexible-shaft grinder with a milling cutter mounted to exert enough pressure to remove the rest of the duralumin. Hegy remembered that the work was "slow and rather exhausting" and painful when the red-hot aluminum alloy chips found their way down shirt collars, up shirt sleeves, and in their faces. Eventually, they sourced handkerchiefs, gloves, and goggles to protect themselves. The replacement of the milling cutter with a sanding disc allowed Hegy to finish the propeller in much the same way his colleagues would have completed a wood propeller. He then applied a coat of black shellac for low glare and durability and placed the gold Hamilton Aero airfoil-shaped logo at the center of each blade.[17]

The process of making Hamilton Aero's first metal propeller blades was crude and unsophisticated to say the least. Ultimately, the Army rejected the propeller due to an errant forging mark on a blade shank and returned it to Milwaukee. Tom Hamilton sold the propeller to Chicago aircraft builder Mattie Laird, who used it for many years.

Hamilton Aero went on to adopt Pittsburgh- and Detroit-style methods of production. Duralumin supplier Alcoa put the blanks through

[16] General Inspector of Naval Aircraft, McCook Field, telegram to BuAer, October 27, 1925, Folder "F23, Volume 9," Box 1954, RG 72, NARA; Hegy, "That Little Old Prop-Maker," 6, 11.

[17] Ray Hegy to Charles W. Hitz, September 29, 1999, Courtesy of Charles W. Hitz; Hegy, "That Little Old Prop-Maker," 11–12.

proprietary forging dies and an ever-growing machine shop expanded to include split hub fabrication. The new blanks required the turning of the shank in the lathe and minimal grinding with a motorized sanding disc. Hegy and the majority of his coworkers had become "finishers," and with the title came a silvery haze of duralumin dust choking their lungs until the company installed a filtration system in the factory. Nelson's adoption of a time clock also ensured that the 100 workers kept production on schedule.[18]

UATC's acquisition of Hamilton Aero put the corporation in a precarious position. The other major American aviation combine, the recently formed Curtiss-Wright Corporation, controlled the Reed Propeller Company and Dr. Reed's patents for his aluminum alloy design. After a protracted patent infringement suit, Hamilton Aero's rival, Standard Steel, acquired what amounted to an exclusive license for manufacturing aluminum alloy propellers in November 1928. In return, Reed agreed to sue all patent infringers, primarily Hamilton Aero, which had been ignoring Reed's claims. Tom Hamilton and his lawyers were just beginning to fight Reed in the US District Court for the Eastern District of Wisconsin. Rentschler shrewdly defused a potentially volatile economic and legal situation by acquiring Standard Steel and with it the Reed license to manufacture aluminum alloy propellers in August 1929.[19]

Acquiring Standard Steel was an ingenious move. The company was the acknowledged industry pioneer in propeller research, development, and manufacturing, especially because of its new employee Frank Caldwell.[20] He had been directing the Army's propeller research and development program at McCook and Wright fields since 1917. His experiences with the Hart and Eustis and other variable-pitch propeller designs led to his decision to develop a new device privately and without the sponsorship of the Army. He left government service in August 1928 and joined Standard Steel in June 1929 as its chief engineer.[21] The addition of the leading American expert in propellers was no small feat for the small

[18] Ray Hegy to Charles W. Hitz, September 29, 1999, Courtesy of Charles W. Hitz; Hegy, "That Little Old Prop-Maker," 12–13, 16.

[19] "Dr. Reed Settles Suit of Standard," *Aviation* 26 (February 2, 1929): 326; *Pittsburgh Post-Gazette*, August 10, 1929; "No. 1 Airplane Company," 50; Wilson, *Slipstream*, 155–156; Weick Oral Interview, 26; Rosen, *Thrusting Forward*, 38, 40; Fernandez, *Excess Profits*, 65.

[20] C. M. Keys to E. G. McCauley, November 8, 1928, Folder 15, Box 10, CMK, NASM.

[21] "Frank Walker Caldwell," *Collier's Encyclopedia*, 4 (New York: Collier, 1957), 322–323; Ronald Miller and David Sawers, *The Technical Development of Modern Aviation* (New York: Praeger, 1970), 73; Rosen, *Thrusting Forward*, 42.

company. Caldwell's son, Walter, remembered hearing as a small boy, the "boss" at Standard Steel, Harry A. Kraeling, say, "If Caldwell stays two years, then we have it made."[22]

UATC got the patents and licenses it needed.[23] Caldwell's pioneering ground-adjustable and mechanically actuated variable-pitch propeller patents were the property of Standard Steel. As a result, Caldwell became an important contributing member to the most powerful aviation corporation in the United States as its propeller specialist.

Standard Steel's West Homestead factory produced the other half of the ground-adjustable-pitch propellers manufactured in the United States as well as for a growing international market. Despite Hamilton Aero's parity regarding production, it was clear that UATC saw Standard Steel as the dominant propeller company. UATC planned to make Hamilton Aero a subsidiary division of Standard Steel in September 1929.[24]

UATC altered that intention and merged Standard Steel with Hamilton Aero Manufacturing in November 1929 to form the Hamilton Standard Propeller Corporation, which became the world's number one propeller company with a virtual monopoly on high-performance propellers in the United States. Involving $14 million, the merger required approval by 75 percent of Standard Steel's stockholders, which included interests led by John J. Hillman, Jr., the chairman of the board of the Peoples-Pittsburgh Trust Company, and the president and general manager, Harry A. Kraeling. Hamilton Standard's officers represented the two former firms with Tom Hamilton as chairman and Harry A. Kraeling as president. There was also an outsider. Eugene E. Wilson, a former naval aviator and technical specialist in BuAer, became vice-president and general manager. Frank Caldwell became the chief engineer. The company concentrated its manufacturing activities in Pittsburgh and Milwaukee, established a service center at the Burbank Airport near Los Angeles for the benefit of the West Coast aircraft industry, and maintained speed

[22] Walter H. Caldwell, telephone interview by author, July 23, 1999.

[23] Clement M. Keys, president of the Curtiss-Wright Corporation remarked that, "It seems to me that everybody at Standard Steel has patents on things used by that company, which appears to be somewhat naïve and child-like in its implicit faith in the theory that the world was its oyster and had no guardian." C. M. Keys to E. G. McCauley, November 8, 1928, Folder 15, Box 10, CMK, NASM.

[24] "Aviation Merger Planned: United Aircraft and Standard Steel Propeller in $14,000,000 Deal," *New York Times*, August 10, 1929, p. 19; *Pittsburgh Post-Gazette*, August 10, 1929; G. S. Wheat to P. G. Johnson, September 5, 1929, Folder 15, Box 1, United Aircraft Files, BCA.

testing courses in Milwaukee and Burbank.[25] Casting a shadow over the merger was the stock market crash on Wall Street earlier in October, but it was a sound financial move.

The union of two formerly aggressive competitors was not easy. Hamilton and Kraeling frequently collided in executive meetings. The sales staff for both former companies harbored bitter memories of the lengths each would go to for an advantage in the marketplace. Carl Schory, who as a sales representative for Hamilton Aero before the merger, taunted a Standard Steel representative at an aircraft trade show, "Listen you, I'm going to undersell you five dollars on a propeller. I don't give a damn if I give them away." A five dollar discount on a ground-adjustable-pitch propeller that cost $300 was significant when it came down to which company sold more products. Schory found himself now allied with the same people he worked so hard to outwit and outsell in the marketplace.[26]

Hamilton Standard required new leadership to guide it within UATC's corporate family and through the Great Depression. Rentschler instituted changes that affected the original leadership of the two companies based on his estimation of their abilities. He terminated the chairman of the board position and forced Tom Hamilton out altogether. More importantly, Rentschler replaced Kraeling with Wilson as the president of Hamilton Standard during the spring of 1930 as a counterpoint to the contentious relationship between the two presidents of the formerly independent companies.[27]

Wilson was the perfect candidate for leading Hamilton Standard. As the influential head of BuAer's airplane design and engine sections, he was well versed in aircraft design, aeronautical propulsion systems, and government procurement processes. He supervised the development of the first generation of seagoing fighting, scouting, bombing, and torpedo aircraft to go aboard America's first carriers, *Langley*, *Lexington*, and *Saratoga*. Wilson contracted the first production orders for Wright and Pratt & Whitney radial engines and Standard Steel ground-adjustable-pitch propellers for the Navy. Despite that success, he faced an unpromising future as a naval aviator. Wilson had spent too many years on land

[25] "Aviation Merger Planned," *New York Times*, 19; *Pittsburgh Post-Gazette*, August 10, 1929; "UAT Propeller Firms Now Hamilton Standard," *Aviation* 27 (November 30, 1929): 1080; United Aircraft and Transport Corporation, *First Annual Report to Stockholders*, 5.

[26] Schory in Weick Oral Interview, 86–87, 103.

[27] Weick Oral Interview, 87.

working toward the technical future of naval aviation to garner promotions and more pay within the traditional sea-going framework of the Navy. He resigned in November 1929 and joined Hamilton Standard at Rentschler's personal invitation.[28] Wilson had walked through the "revolving door" in much the same way Caldwell had the previous year.

A natural executive, Wilson quickly turned Hamilton Standard into a unified company capable of maintaining its lead in the American propeller industry. The administration of two different factories hundreds of miles apart during economic hard times was not easy. The Army-Navy five-year plans offered financial solvency only if all manufacturing and design facilities were in one place. After a careful evaluation of both factories in Milwaukee and Pittsburgh, Wilson ordered the old Hamilton Aero factory closed in February 1930 and the move of all operations to Pittsburgh by April.[29] Pittsburgh's advantageous location on the main east-west rail lines made it more conducive to consistent manufacturing and delivery; furthermore, there was no heavy corporate income tax as there was in Wisconsin.

Hamilton Aero did have something to contribute to the new company. Arvid Nelson's layout of the Milwaukee factory impressed Wilson. The last lot of wood propellers had just left the city by train, which marked the final transition of the company to metal propeller production. Nelson took the position of assistant treasurer and factory manager for Hamilton Standard's new purpose-built 21,000 square foot brick facility. Approximately forty workers from Milwaukee relocated to Pittsburgh. Already gone and enjoying life in southern California, Tom Hamilton learned about the closing of "his" factory in Milwaukee from former employees after the move.[30]

Wilson appointed former Army officer Raycroft Walsh as the vice-president and director of Hamilton Standard in April 1930 to direct the propeller maker's sales effort from Pittsburgh. Walsh earned his aviator's wings at the outbreak of World War I. During the war, he commanded aviation training units at Kelly Field, Texas, and Rockwell Field, San Diego. As the Air Officer of the Panama Canal Department, which included the Sixth Composite Group and France Field, he commanded Air Service units during two grand war maneuvers in cooperation with

[28] "Wilson Quits Post as Expert Aviator," *Washington Post*, November 24, 1929, p. R5; Wilson, *Slipstream*, 156–157; Fernandez, *Excess Profits*, 66–67.

[29] "Propeller Company to Leave City," *Milwaukee Sentinel*, February 16, 1930, p. 14.

[30] Fernandez, *Excess Profits*, 67; Trotman, "Turning Thirty Years," 22; Weick Oral Interview, 77–78, 100.

the Navy. Walsh went on to hold command and administrative positions in the Panama Canal Zone from 1921 to 1924. During his tenure in the Canal Zone, he was also the commanding officer and pilot of the Central American Flight, which pioneered airways and provided the foundation for the expansion of air mail to Central America. Walsh left the Army in 1926 and worked for the McGraw Hill Publishing Company and the silk manufacturer, Cheney Brothers, before joining Hamilton Standard.[31]

Like Wilson, Walsh was especially well connected in Washington political and bureaucratic circles. Walsh first came to Washington to serve as Assistant Executive Officer of the Chief of Air Service. From 1920 to 1921, he was the Administrative Executive and Assistant Chief of the Supply Group of the Air Service. Most important, Walsh was the fiscal officer from 1924 to 1926 where he prepared financial estimates for the Bureau of the Budget and congressional committees. He also served as the contact officer for the House of Representatives, where he helped present military aviation issues before congressional committees, including the Morrow Board, which recommended the reorganization of the Air Service into the Air Corps.[32] Walsh possessed the qualifications and connections that made him an ideal person to represent Hamilton Standard's interests during military contract negotiations. After rising to the head of the company in 1931, he excelled at long-term planning and setting corporate policy.[33]

Hamilton Standard was ready to dominate the world propeller market with executive leadership knowledgeable of the intricacies of Wall Street and government procurement, an experienced factory planner capable of organizing production, and the world's leading propeller expert as chief engineer in charge of technical development. There was an important addition to the engineering team. Erle Martin became Caldwell's assistant in June 1930. Born in Tullahoma, Tennessee, in July 1907, Martin grew up in Kentucky and Pennsylvania. He received his bachelor's degree in electrical engineering from Pennsylvania State University in 1929. Martin joined the Kreider-Reisner Aircraft Company of Hagerstown, Maryland, a division of the Fairchild Airplane and Engine Corporation, as a draftsman before transferring three months later to Air Propellers, Inc., to work on a hollow-steel propeller design. Walsh and Caldwell discovered the young engineer when Hamilton Standard acquired Air Propellers for

[31] "Major Walsh Joins Hamilton Standard Propeller," *U.S. Air Services* 15 (April 1930): 55.

[32] Ibid., 55.

[33] Trotman, "Turning Thirty Years," 23.

the experimental blade late in 1930. Martin arrived in Pittsburgh and proved to be an asset to the engineering team.[34]

Hamilton Standard held an important financial position within UATC's corporate family. Every military and commercial UATC aircraft carried one or more Hamilton Standard ground-adjustable-pitch propellers, and many aircraft from other manufacturers did as well. Of the estimated $1.3 million in total sales for UATC in 1931, Hamilton Standard ranked third at $200,000 behind Pratt & Whitney's $850,000 and Boeing Airplane's $450,000. Chance Vought earned much less at $50,000 and Stearman, Northrop, and Sikorsky operated at considerable losses.[35]

Ground-adjustable-pitch propeller sales provided the revenue that kept the company going during the dark financial days of the Depression. There were only two competitors to the Hamilton Standard product – the Reed propeller produced by Curtiss and the Micarta propeller marketed by Westinghouse – and they were both structurally obsolete. In April 1930, Wilson announced that Hamilton Standard had received its largest single commercial order for 300 propellers from the Stinson Aircraft Corporation of Detroit, which the aviation community regarded as the first committed attempt at the quantity production of private aircraft. The order was fortuitous for Hamilton Standard, because the corporation had just moved to its new factory in Pittsburgh. With the addition of other production orders as well as replacement orders for current customers, the Pittsburgh facility was operating at normal capacity.[36] The immediate future of Hamilton Standard appeared bright.

The success of the newly consolidated Hamilton Standard, however, was not guaranteed. The new corporation was the first UATC member to feel the effects of the Great Depression. During the period March 1929-August 1930, which included the merger, ground-adjustable-pitch propeller sales amounted to $3.5 million with a net profit of $485,000. Those figures dropped to approximately half, $1.7 million and $224,397 respectively, from January 1931 to May 1932. The crucial period of June 1932 to April 1933 witnessed sales again reduced by nearly half to $664,000 and profits slashed by three-quarters to $38,500. On paper,

[34] Frank J. Delear, "He Decided on Aviation," *The Bee-Hive* 30 (Spring 1955): 8–9; Trotman, "Turning Thirty Years," 22–23; Weick Oral Interview, 89; "Erle Martin, Inventor, Dead; Improved Aircraft in the 30s," *New York Times*, December 14, 1981, File CM-1333000-01, BF, NASM.

[35] "No. 1 Airplane Company," 126.

[36] "Record Order for Propellers," *U.S. Air Services* 15 (April 1930): n.p.

Hamilton Standard was unable to sustain itself with dwindling sales and profits.[37]

Besides the contentious relationships in the front office and dwindling profits, Hamilton Standard faced labor unrest from the newly combined factory personnel. They went on strike in late August 1931. Within a week, Wilson followed through on an earlier threat to the union leaders and ordered the West Homestead factory at Pittsburgh permanently closed. Many of the employees felt that the Wilson, who kept his main office in East Hartford to be near the UATC leadership, and Rentschler had already decided to close the factory for financial reasons and used the strike as a convenient excuse.[38] The official reason announced in the *Wall Street Journal* was that consolidation allowed "closer contact" in the evaluation of the corporation's propellers, engines, and airframes.[39]

Hamilton Standard's engineering, sales, and service staff and tons of factory equipment and machinery moved to Connecticut between September 1931 and June 1932. Sixty employees from the West Homestead plant, including a sizable portion that had started with Hamilton Aero in Milwaukee, journeyed to East Hartford. They joined about 140 Pratt & Whitney factory personnel. They manufactured propellers in the unused portion of the Pratt & Whitney plant. It was a shrewd business move to consolidate operations at the UATC level, but it destroyed the already fragmented Hamilton Standard manufacturing community. The dislocation also cemented the bonds between the surviving original Hamilton Aero and Standard Steel workers and the company. Fifty-seven of those sixty itinerant propeller makers, including longtime Standard Steel employee, Alexander Mannella, still worked for Hamilton Standard at the end of the 1940s.[40]

As the Great Depression lingered, the American military attempted to offset the UATC-Hamilton Standard monopoly. Both the Army and Navy argued that a second source of supply was necessary, especially in times of war. The government encouraged new companies to manufacture propellers and to bid competitively against Hamilton Standard. This effort to deter the domination of an individual contractor's pricing policies

[37] For the eighteen-month period from March 1929 to August 1930, Hamilton Standard averaged $194,444 in sales and a $26,944 profit per month. For the sixteen-month period from January 1931 to May 1932, sales averaged $106,250 and profits $14,021. For the crucial eleven-month period from June 1932 to April 1933, the company averaged $60,363 in sales and $3,500 profit each month. Fernandez, *Excess Profits*, 79.

[38] Schory in Weick Oral Interview, 88, 91–92.

[39] "United Aircraft Plant," *Wall Street Journal*, September 10, 1931, p. 7.

[40] Trotman, "Turning Thirty Years," 24; Fernandez, *Excess Profits*, 79.

created a procurement system that accommodated competing sources of supply. One of those companies, the J. A. Fay and Egan Company, was a woodworking machinery manufacturer that attempted to get into the metal propeller business. Fay and Egan's inability to succeed in the venture, especially after being involved in patent litigation from both the Reed Propeller Company and Hamilton Standard, led to the closing of the eighty-three year-old company.[41]

Echoing the larger trend in the aviation industry, Hamilton Standard engineered new products while small firms underbid the company for the military production contracts. Since the propellers were intended for the US government, there were no issues related to patent licensing and royalties. This situation left Hamilton Standard and its UATC corporate leadership to subsidize new developments while smaller firms could profit from production contracts.[42]

Nevertheless, UATC was not a financially weak conglomeration, and it could well afford to finance the development of a variable-pitch propeller within its corporate family. UATC saw the airplane "as a means of transportation or of national defense, patronized by the public and a billion-dollar government." The editors at *Fortune* believed the financial success of UATC was no accident, but instead could be attributed to the "foresight" of its leadership. That meant UATC produced only aircraft capable of reliably and economically carrying military weapons, mail, or passengers. With its primary customer the US government through military aircraft contracts and airmail subsidies, UATC was "refusing to look too far ahead" of that strategy in its development of new innovations. The editors of *Fortune* approved: UATC aircraft, incorporating the products of the corporate family, were to be the result of entrepreneurial "reality rather than a dream." Two years into the Great Depression, UATC was on a firm financial footing. Its cash and marketable securities increased from $5 million in 1928 to $17 million in 1932. *Fortune* felt there was potential for even more growth.[43]

A Committee of Specialists

To ensure the interrelated technological and financial success of the corporation, UATC's executive committee created a Technical Advisory

[41] Folder "Fay and Egan, 1929–1935," Box 6035, RD 3264, RG 342, NARA.
[42] Fernandez, *Excess Profits*, 79.
[43] "No. 1 Airplane Company," 46, 47.

Committee (TAC) to facilitate cooperation among the member companies. The TAC, which held its first meeting in Hartford in May 1929, was to avoid duplication in research and development, share common technological interests, let the members know of the latest innovations, and overall maximize the engineering resources of the corporation. It achieved this by meeting biannually where representatives of the member companies reported on their products, their market competition, and their experimental programs. The committee issued a highly confidential report detailing the conclusions made during the meeting for distribution among the members and UATC's executive committee. The TAC also served as the primary conduit for research conducted by the soon-to-be created UATC Research Division located in Hartford.[44]

The TAC membership consisted of the design and engineering leadership from UATC's member companies (Figure 15). They were a team of equals, a veritable "Who's Who" in the larger aeronautical community, where each specialist was a leader in his field. It included the flying boat builder Igor I. Sikorsky; the intuitive engineer John K. "Jack" Northrop; general aviation designer Mac Short from Stearman; engine pioneers George Mead (and TAC chair) from Pratt & Whitney; Boeing designers Claire L. Egtvedt and Charles N. Monteith; and Charles J. McCarthy from Chance Vought. Charles H. Chatfield, working as an aeronautical engineer for Pratt & Whitney, coordinated research activities. Tom Hamilton, the only truly non-engineer in the group, and former NACA researcher Fred E. Weick represented the corporation's propeller interests.[45]

The second meeting of the TAC in Seattle in December 1929 was the first to include Frank Caldwell, who George Mead characterized as the "number one propeller man in the world."[46] In his overview report, Caldwell discussed Hamilton Standard's efforts to make ground-adjustable-pitch propellers stronger and lighter with magnesium alloy and hollow-steel blades; the reduction of propeller noise; the need for a propeller series designed for a variety of ratings ranging from 60- to 560-horsepower; and the correct placement of the propeller in an aircraft design. Because UATC assigned Hamilton Standard the task of advancing

[44] Report of the Second Meeting of the Technical Advisory Committee of the United Aircraft and Transport Corporation at Seattle, Washington, December 2-6, 1929, CE, BCA, 1–5; United Aircraft and Transport Corporation, *First Annual Report to Stockholders*, 7–8, 10.

[45] United Aircraft and Transport Corporation, *First Annual Report to Stockholders*, 10.

[46] George J. Mead, quoted in Louis M. Lyons, "Europe in 24 Hours Right in the Cards," *Daily Boston Globe*, September 8, 1935, p. B6.

FIGURE 15 An industrial community of aeronautical specialists: the UATC Technical Advisory Committee, December 1929. Front row, left to right: Charles H. Chatfield, Frank W. Caldwell, Thomas F. Hamilton, Mac Short, A. Kenneth Humphries, Claire L. Egtvedt; second row, left to right: Fred E. Weick, Charles N. Monteith, Igor I. Sikorsky, D.B. Colyer, John K. Northrop, George J. Mead. Boeing.

the state-of-the-art in propeller technology in response to the needs of the military and airlines, Caldwell reported that a variable-pitch propeller offered a "substantial improvement" in efficiency at takeoff and a "slight improvement" for cruise.[47]

The TAC discussed how Caldwell and Hamilton Standard should approach the development of a variable-pitch propeller. In an enthusiastic discussion among Caldwell, Egtvedt, Hamilton, McCarthy, Mead, and Sikorsky, the committee recommended the development of two different types of variable-pitch propellers: the two-position, controllable-pitch propeller for commercial use and a constant-speed propeller for military aircraft, specifically for pursuit and fighter applications. Mead thought it was in "United's best interests" to pursue both developments.

[47] Report of the Second Meeting of the Technical Advisory Committee, 3, 327, 391, 483.

The two-position propeller, then under development by Caldwell in Pittsburgh, met immediate performance requirements, while the constant-speed propeller would be the best in the long term. The key to the success of the Hamilton Standard constant-speed propeller program, according to Mead, was to get the government to bear the expense of development. Nevertheless, when asked which was more immediately important to UATC – lighter ground-adjustable-pitch propellers or the variable-pitch propeller – the TAC unanimously agreed that the former was more desirable.[48]

Frank W. Caldwell Reinvents the Propeller

With more than ten years of experience in the field, Caldwell knew that making his propeller design practical would be challenging, expensive, and time-consuming. His experiences with the Army's variable-pitch propeller programs influenced his decision to reject mechanical actuation. The ball bearings and linkages characteristic of the Army-sponsored designs were indicative of his educational background in mechanical engineering, but they proved to be inadequate. Caldwell began to look for other methods to alter pitch after he left government service. In November 1928, while eating lunch, he sketched on a tablecloth his concept of a hydraulically actuated variable-pitch propeller design.[49] Caldwell filed his patent application for the design on May 25, 1929, which became the foundation for his work in industry. As he had done with Standard Steel, he assigned the patent to Hamilton Standard.[50]

Caldwell's new propeller reflected his approach to innovation. His experience working for the Army taught him to build upon successive designs that served as the foundation for the next level of development. He achieved the first level with the metal ground-adjustable-pitch propeller. The knowledge of how to design metal propeller blades that were both strong and safe allowed the propeller community to concentrate its efforts on developing a practical variable-pitch mechanism. Caldwell reasoned that only a small amount of pitch variation was required to alleviate the design compromise of the fixed-pitch propeller, so he began with a two-position design with settings for takeoff and cruise. He did not

[48] Ibid., 399, 403, 405, 506.

[49] A copy of the original sketch is in the Frank W. Caldwell File, HSCRD.

[50] F. W. Caldwell, "Propeller," US Patent No. 1,893,612, January 10, 1933.

attempt to design the more complicated constant-speed propeller until he completed that first crucial step: a controllable-pitch propeller.

The timing of Caldwell's idea for a hydraulically actuated variable-pitch propeller came at a fortuitous moment for Hamilton Standard. Only the resources of the UATC would enable the research and development required to introduce a practical variable-pitch propeller. UATC recognized the synergistic nature of the airplane, but innovations had to be made workable and had to contribute to the profitability of the corporation's products. If variable-pitch propellers made an airplane fly higher, faster, and farther, which for UATC's leadership meant dominating the market, then it would invest its money into that project. When Wilson came to Caldwell looking for a new and lucrative innovation for the company to develop, the engineer was ready.

After the December 1929 TAC meeting in Seattle, UATC's executive committee authorized the development of Caldwell's controllable-pitch propeller. Rentschler and UATC's corporate leadership absorbed the research, development, and operating costs of Hamilton Standard until Caldwell made the propeller ready and the aeronautical community accepted it.[51] Nonetheless, UATC was a business. When he saw that research and development expenditures exceeded production profits, Rentschler warned Wilson that the new propeller "had better pan out or it will be just too bad for you."[52]

Caldwell hid himself away in a wood test shack only large enough for him, his experimental propeller, and the low-powered electric motor required to rotate it until he developed a successful device. A West Homestead city water line provided the hydraulic pressure. The experimental prototype propeller was ready in 1930. Hamilton Standard-UATC engineers tested it on an airplane with a 150-horsepower engine in front of Caldwell, Walsh, Nelson, and UATC's corporate leadership. The test pilot, Halsey R. "Hal" Bazley, who was the manager of the Curtiss Flying Service for Pittsburgh, reported that when he set the control to the specified pitch, the blades would move to the opposite setting, meaning if he wanted low pitch for takeoff, it would actually move into high pitch position and vice versa. The embarrassment of what appeared to be a major technical blunder dissipated once Hamilton Standard mechanics

[51] "UAT Propeller Firms Now Hamilton Standard," 1080; Report of the Third Meeting of the Technical Advisory Committee of the United Aircraft and Transport Corporation at Hartford, Connecticut, May 19–23, 1930, Folder 5, Box 5778, CE, BCA, 1; Rosen, *Thrusting Forward*, 42.

[52] Wilson, *Slipstream*, 155, 165–166, 167.

found that they had mistakenly installed a crucial three-way valve backwards. Their reinstallation of the part resulted in a flawless demonstration flight.[53]

The new propeller weighed fifty-six pounds as compared to the forty-two pounds of the standard ground-adjustable-pitch propeller. Keeping in mind the prospect of "profitable future business," Caldwell noted that Hamilton Standard's efforts were directed toward producing a "light, economical, and simple propeller" that was "relatively free from maintenance." He felt there was not much chance to "develop a profitable business in the manufacture of these propellers" until he and his design team achieved that goal.[54] Caldwell was following his characteristic process of cautious development. Knowing that his technical career, the financial fortunes of the corporation, and the lives of aircraft operators and passengers were at stake, he was not going to move hastily toward production.

The design, called the two-position, controllable-counterweight propeller, or "hydro-controllable," was an "extremely simple device" that offered only two pitch settings for climb and cruise. Engine oil pressure actuated low pitch and centrifugally operated counterweights set the blades to high pitch. All actuation depended on an oil control mechanism consisting of a three-way valve leading from the engine's regular oil pressure system. Each blade had a flange attached to its base that had an arm extending through the hub barrel. The arm had a curved slot cut in it at an angle. A sliding cylinder with a projecting lug carrying a ball-bearing roller, forming the center of the hub, engaged in the slot cut in the arm. To actuate low pitch, the pilot moved a lever in the cockpit that introduced engine oil pressure through the hollow crankshaft into the cylinder. The cylinder moved forward and the roller on the projecting lug (as it rolled in the curved slot) forced the blade flange arm to rotate. The propeller control had to be made an integral part of the engine. Pratt & Whitney engineers designed a modification package to fit any engine their company offered.[55]

When the pilot moved the control lever in the other direction, the oil pressure was shut off. The blade flange arms acted as counterweights

[53] "Halsey R. Bazley," in *Who's Who in Aviation, 1942–1943* (New York: Ziff-Davis, 1943), 30; Trotman, "Turning Thirty Years," 23.

[54] F. W. Caldwell, "The Controllable-Pitch Propeller," Appendix II, Report of the Third Meeting of the Technical Advisory Committee, 1–3.

[55] "Hamilton Standard Announces Improved Type Controllable-Pitch Propeller," *The Bee-Hive* 7 (March 1933): 5–6, 9.

FIGURE 16 Bernt Balchen inspecting the early model Hamilton Standard controllable-counterweight propeller installed on the Northrop Gamma *Polar Star* in 1932. Smithsonian National Air and Space Museum (NASM 95–8737).

and the centrifugal force of the spinning crankshaft made the weights seek a wider arc and rotate the blade to its original high pitch position. A valve shut off the oil pressure into the crankshaft and opened a drain to force the oil out of the sliding cylinder and to reenter the oil sump. Provision was made to regulate the amount of blade rotation, which determined the degree of pitch change possible. The blade flange could be readily indexed around the blade to secure any desired pitch change not to exceed six degrees.[56]

The exposed components of the first Hamilton Standard two position controllable-counterweight propeller design facilitated ease of

[56] Ibid., 5–6, 9.

maintenance and inspection, but it was expensive to manufacture. The biggest problem with the design was the shape and placement of the counterweights. Caldwell hinged the long and narrow counterweights behind the blade at the center of the hub, which looked like large ice tongs. The end of the counterweights reached toward the end of the piston at the front of the propeller in the low-pitch setting. At the high-pitch setting, the counterweights extended well out from the crankshaft at the forward end at almost a forty-five-degree angle from the hub. The area of movement for the counterweights prohibited the installation of a propeller spinner for aerodynamic and aesthetic reasons.[57]

Hamilton Standard began evaluating controllable-counterweight designs for a variety of engines beginning in 1930. The first production propeller was for the 500-horsepower Wasp radial engine built by Pratt & Whitney. Hamilton Standard was quick to put the propeller into the public eye. Hamilton Standard service manager Carl Schory and Sikorsky pilot George Meissner entered a Sikorsky S-39 single-engine amphibian equipped with the first propeller into the Curtiss Marine Trophy competition for seaplanes at the Miami Air Races in January 1930. Models designated for various engines rated up to 700 horsepower appeared over the next two years (Figure 16).[58]

Corporate Competition in the Aeronautical Marketplace

The UATC's main rival in the aggressive aeronautical marketplace, the Curtiss-Wright Corporation, was also in the process of acquiring and developing its own variable-pitch propeller. Wallace R. Turnbull, encouraged by the successful tests at Camp Borden, decided the next step was to develop his electrically actuated propeller as a commercial product. He sent his propeller to the Garden City, Long Island, factory of the Curtiss Aeroplane and Motor Company for evaluation in April 1928.[59] Turnbull wanted $100,000 for his invention and thirteen years of private development work. After months of negotiation and compromise, the inventor awarded the exclusive license for his design in the United States to the Reed Propeller Company, now a Curtiss-Wright subsidiary, on December

[57] Ibid., 5.

[58] Weick Oral Interview, 66; Miller and Sawers, *Technical Development*, 74; Report of the Third Meeting of the Technical Advisory Committee, 29.

[59] W. R. Turnbull to CAMC, April 3, 1928, Folder 5, Box 22, CWC, NASM.

3, 1929 for forty shares of corporate stock valued at $5,000 and a share of the sales royalties.[60]

The Curtiss-Wright deal also included Turnbull remaining as a consultant until 1931 at the Garden City factory and then at the Kenmore facility after the move of all operations to Buffalo. He worked with his Curtiss colleagues, including chief engineer Theodore P. Wright and the thirty-five members of the propeller department led by Werner J. "Pete" Blanchard. A 1924 graduate of Kansas State College, Blanchard used his mechanical engineering degree to teach high school science and mathematics before joining Curtiss as a draftsman in early 1928. Recognized as a bright and talented engineer, he was responsible for all propeller development within the corporation within a year. Turnbull, Wright, and Blanchard transformed the primarily wood No. 2 into a commercially viable product with duralumin blades and a steel hub mechanism. The resultant No. 3C propeller was for the 200-horsepower range and weighed 128 pounds.[61] Blanchard filed a patent claim for his reinterpretation of the Turnbull electrically actuated hub on September 16, 1930, which became the foundation for the new design, and assigned the rights to Curtiss.[62]

The research and development program for the new series of Curtiss controllable-pitch propellers concluded just in time for the spring Detroit Aircraft Show in May 1931. Curtiss marketed it as part of a complete aeronautical product line, which included Whirlwind and Cyclone radial engines, the Curtiss Anti-Drag Ring Cowling, and complete aircraft such as the T-32 Condor airliner and the F8C-5 Helldiver fighter/dive-bomber.[63]

[60] R. F. Hoyt, "Turnbull Patent License," November 22, 1929; "License Agreement," December 3, 1929; W. R. Turnbull to Executive Committee, CAMC, January 7, 1930, Folder 5, Box 22, CWC, NASM. Turnbull's designs are expressed in US Patent numbers 1,793,652 and 1,793,653 dated February 24, 1931 and 1,828,303 and 1,828,348 dated October 20, 1931.

[61] "1932 Accomplishments of the Curtiss-Wright Corporation," n.d. [1932], "Curtiss-Wright" File, NASAHO; J. H. Parkin, "Wallace Rupert Turnbull, 1870–1954: Canadian Pioneer of Scientific Aviation," Canadian Aeronautical Journal 2 (January–February 1956): 41–42; Mac Trueman, "On a Wing and a Prayer: Wallace Rupert Turnbull," *The New Brunswick Reader* 4 (December 7, 1996):10; "Werner J. Blanchard," *Who's Who in Aviation, 1942–1943* (Chicago, IL: Ziff-Davis Publishing Company, 1942), 44; Pete Blanchard to the Men of My Family, April 1928, and Juliet Blanchard to the Family, June 27, 1929, in Juliet Blanchard, *A Man Wants Wings: Werner J. Blanchard, Adventures in Aviation*, 1986, www.margaretpoethig.com/family_friends/pete/wings_bio/index.html (Accessed May 24, 2012), 21–22, 24–25.

[62] Werner J. Blanchard, "Variable-Pitch Propeller," US Patent No. 1,951,320, March 13, 1934.

[63] "Variable-Pitch Propeller," *Curtiss-Wright Review* 2 (May 1931): 16.

The Army first tested the Curtiss Electric propeller in 1931 and within three years, both services had placed orders for the innovation.[64]

Curtiss basically had to start from scratch when the Reed propeller fell out of favor with the aeronautical community. Turnbull's invention and Blanchard's development work made them competitive again in the aeronautical marketplace. As a reward, Curtiss expanded Blanchard's responsibilities as project engineer, without a pay increase, to include the design of all power plant installations on the corporation's aircraft.[65]

Hamilton Standard and Curtiss were poised to become industry leaders worldwide, but they could not have reached that point without the resources of their parent corporations. The interrelationship between technology and business, as seen through the creation of the modern variable-pitch propeller, illustrates the nature of aviation during the late 1920s and early 1930s. The business community controlling the aviation industry saw the technical innovations of the Aeronautical Revolution and the individuals that created them as commodities that ensured economic and financial survival. The variable-pitch propeller was vital to the success of the modern airplane and the specialists who created it were crucial to the survival of the propeller companies and their corporate leaders. The new technology had yet to be proved in the air, but they gambled that it would in the near future.

[64] CAMC, "Curtiss Controllable-Pitch Propeller," n.d. [1934], File B5-260080-01, PTF, NASM; Propeller Division, Curtiss-Wright Corporation, "Curtiss Propellers: Thirty Years of Development," September 4, 1945, File B5-260000–01, PTF, NASM.

[65] Pete Blanchard to the Men of My Family, September 1931, in Blanchard, *A Man Wants Wings*, 28–31, 97.

8

A Gear Shift for the Airplane

In October 1934, American-designed and built Douglas DC-2 and Boeing Model 247-D commercial airliners finished second and third behind a special-purpose British de Havilland DH.88 Comet air racer in the 11,300-mile London-to-Melbourne MacRobertson Trophy Race. The *New York Times* hailed the performance of the two revolutionary aircraft as a victory for American aeronautical technology and its superiority over European technologies. The diminutive British entry, equipped with two propellers capable of only changing their pitch once in flight, carried only its pilot and copilot. Hamilton Standard controllable-pitch propellers enhanced the overall performance of the much larger transports, especially Dutch KLM's DC-2 *Uiver*, as they carried passengers and mail much like they would have in regular commercial operations (Figure 17). The crews of the DC-2 and 247-D lost the race and the resulting $50,000 prize, but won the international competition for aeronautical technological superiority.[1] Introduced to the world through the MacRobertson Race, the variable-pitch propeller was a critical technology to the Aeronautical Revolution of the late 1920s and early 1930s in the United States and Great Britain.

Aircraft designers drew upon an impressive listing of aerodynamic, propulsive, and material innovations as they worked to meet the specifications issued by their military and commercial clients. Streamlined design involved the removal or redesign of drag-inducing structures such

[1] "Big Air Race Viewed As A Victory for U.S.," *New York Times*, October 29, 1934, pg. 9; "MacRobertson Score Sheet," *Aviation* (November 1934): 365; "London-Melbourne Race," *Aircraft Yearbook for 1935* 17 (1935): 163–168.

FIGURE 17 The arrival of Dutch KLM's DC-2 *Uiver* at Melbourne brought American aeronautical technology, including the variable-pitch propeller, to the world stage. Smithsonian National Air and Space Museum (NASM 2002–4238).

as fixed landing gear, engines and radiators, struts and wires, open cockpits, and armament to increase overall performance. Flaps mounted on the trailing edge of the wings changed the aerodynamic shape of the wing to create more lift at takeoff and landing. Lightweight air-cooled radial engines offered increased horsepower while the NACA's cowling reduced drag and improved cooling at the same time. Structural improvements in the form of all-metal construction, internally supported cantilever wings, and retractable landing gear resulted in further efficiency and operational durability. How those innovations were incorporated into a particular design and in what combination was the key to realizing the modern airplane.

Despite the promise of increased performance offered by the variable-pitch propeller, the aeronautical community did not immediately welcome its introduction. The engineers and companies developing this new technology had to overcome much opposition. Besides the economic and technical arguments that the new propeller's cost, weight, and complexity of operation were too great, there was a larger reason for its limited introduction in the early 1930s. Conservative design logic guided the path of the integration of the variable-pitch propeller into the technical system of the airplane. The propeller community had to convince the larger aeronautical community of the importance of the innovation to the overall development of the airplane.

The question of whether or not to incorporate a variable-pitch propeller into a new aircraft design reveals the nature of engineering design logic in the late 1920s. The evolution of technology gravitated from the simplistic to the complex. Each of the component technologies that make up an airplane became more sophisticated, which increased performance and was the result of significant research and development on the part of specialists.

The nature of engineering design logic expressed by the larger aeronautical community can be characterized as performance-specific. This type of design hinged upon considering all of the variables – intended cruising altitude, weight, range, power plant, and number of passengers – and could tailor the project to meet those criteria. Within that format, a fixed-pitch propeller fit well within the myriad of design choices and introduced simplicity rather than complexity into the design process. The weakness of a performance-specific design was that the limitations of individual components degraded overall performance once a pilot took an airplane beyond the intended operating environment, which could be disastrous from a military or commercial standpoint.

A critical event to the success of the Aeronautical Revolution was the emergence of performance-general design logic. This process greatly facilitated the introduction of new synergistic aeronautical innovations such as the variable-pitch propeller that offered performance flexibility. An airplane able to deviate from a flight plan and fly safely in all conditions, at unexpected altitudes, and from airfields of varying sizes, ensured the continuous development of commercial and military aviation.[2]

Resistance to Innovation in Great Britain

Work continued on the Hele-Shaw Beacham constant-speed propeller at Gloster Aircraft through the 1920s. The first successful design was a nine-foot ten-inch-diameter propeller designed for the supercharged nine-cylinder Bristol Jupiter VI and fourteen-cylinder Armstrong-Siddeley Jaguar

[2] Walter G. Vincenti, "The Retractable Landing Gear and the Northrop 'Anomaly': Variation-Selection and the Shaping of Technology," *Technology and Culture* 35 (January 1994): 22–26.

IV air-cooled radial engines and test flown on Gloster Grebe, Gamecock, and Guan fighter aircraft from 1926 to 1928. The propeller featured hollow-steel and drop-forged duralumin blades and a pitch range of twelve degrees. A Hele-Shaw Beacham–equipped Grebe achieved 165 mph in level flight at 10,000 feet and could climb to 20,000 feet in sixteen minutes. Another Grebe exhibited an increased rate of climb of 200 feet per minute. Overall, these early flight trials totaled 250 hours and illustrated the potential of the Hele-Shaw Beacham design because the propeller regulated its pitch automatically. The RAF and Bristol pilots were able to focus on aerobatic maneuvers – climbing, leveling out, looping, and diving – and operating their machine guns in the biplane fighters without worrying about propeller settings or a change in engine rpm.[3]

Encouraged by the initial tests of their new product, Burroughes, Hele-Shaw, and Beacham went to the British aeronautical community searching for development contracts. The British power plant community rejected their overtures because it believed specific engine developments provided the extra power needed at takeoff and that the hydraulic constant-speed mechanism was underdeveloped. Despite its initial interest, the Air Ministry was equally suspicious of the merits of the new Gloster propeller and promised minimal development contracts. Gloster granted a license to the Okura Company of Japan in 1928 to defray the mounting costs of propeller development program due to the lack of support at home.[4]

The international aeronautical community periodically met and discussed the merits of the variable-pitch propeller in the open forum of professional meetings in Great Britain and the United States. Hele-Shaw and Beacham presented their research to the RAeS in London on April 12, 1928.[5] Their address was the first public discussion of their hydraulically actuated, constant-speed mechanism and the merits of variable pitch overall. The opportunity to present the initial findings of a research program turned into a heated discussion over which technologies would be appropriate in future aircraft designs.

[3] H. S. Hele-Shaw and T. E. Beacham, "The Variable-Pitch Airscrew: With a Description of a New System of Hydraulic Control," *Journal of the Royal Aeronautical Society* 32 (July 1928): 530; Derek N. James, *Gloster Aircraft Since 1917* (London: Putnam, 1971), 16, 103–104, 106, 116, 134; C. G. Grey, *A History of the Air Ministry* (London: George Allen and Unwin, 1940), 64; Bruce Stait, *Rotol: The History of An Airscrew Company, 1937–1960* (Stroud, Gloster, England: Alan Sutton Publishing, 1990), 2.

[4] James, *Gloster Aircraft*, 16–17.

[5] Hele-Shaw and Beacham, "The Variable-Pitch Airscrew," 525–554.

The ensuing discussion after the presentation indicated the existence of two propeller factions within the British aeronautical community. Despite his participation in the 1917–1918 variable-pitch propeller tests at the RAE, E. J. H. Lynam joined with Dr. Henry C. Watts to lead a self-styled "anti-VP school," that represented the staunch design conservatism toward the variable-pitch propeller and highlighted several points to negate its advantages. Primarily, they argued that the issue of weight prevented the variable-pitch propeller from being a practical design choice. The Hele-Shaw Beacham design was 80 percent heavier than a fixed-pitch propeller, which meant it would bring the performance of an airplane down to levels below those attained with fixed-pitch propellers.[6]

Lynam directed the tests of the Hele-Shaw Beacham propeller at the RAE in 1928. Even though the propeller had been flown for more than thirty hours without fault, he stated that the RAE was still "not satisfied" with its operation. Overall, he maintained that "it was ... very difficult to make a really strong case for fitting a variable-pitch airscrew," and recommended the design emphasis be placed upon engine improvements and the fixed-pitch propeller. Lynam added that "such devices must be costly and complicated, and therefore less reliable than the simple fixed-pitch screw." Speaking as the British government's representative authority on the design of propellers, he believed that a fixed-pitch propeller designed for a definite and well-known performance regime was better, meaning more practical and simplistic, than a heavy, complicated, and unreliable variable-pitch propeller.[7]

If a variable-pitch propeller were to be used on an airplane, the anti-VP school had specific ideas about their application and the design of the mechanism. Constant-speed operation with automatic pitch adjustment was appropriate for fighter aircraft with supercharged engines where the pilot would be too busy fighting the enemy to manually adjust propeller pitch in the midst of a dogfight. On military transport and commercial aircraft, they believed the propeller should be controllable by the pilot to adjust pitch variation. Lynam and his group vehemently disagreed with the use of a hydraulic system and favored a mechanical system such as the device developed by the RAE during World War I.[8] The overall tone of their response showed a contempt for Hele-Shaw and Beacham. They were "newcomers" to the propeller community with no real experience.

[6] E. J. H. Lynam, in Hele-Shaw and Beacham, "The Variable-Pitch Airscrew," 538–541.
[7] Lynam, et al., in Hele-Shaw and Beacham, "The Variable-Pitch Airscrew," 538–547.
[8] Ibid.

Most of the responses to Hele-Shaw and Beacham's lecture were positive, which reflected the existence of an equally strong "pro-VP school" led by those who had the most to gain from the propeller's introduction: Pop Milner and Hugh Burroughes of Gloster Aircraft. This faction believed that the theoretical advantages of variable pitch could be made into reality after further development work that resulted in a much lighter, reliable, and inexpensive apparatus. These engineers emphasized that the increased margin of safety, increase in power, and shorter takeoff run justified the extra weight of the variable-pitch mechanism, especially on heavily loaded seaplanes, transports, and bombers with geared engines and slow-turning propellers.[9]

The pro-VP faction was emphatic in its support of the variable-pitch propeller. Society Treasurer Maj. D. H. Kennedy of the RAF predicted, "I foresee a time, not far distant, when the commercial airliner will have engines equipped with variable-pitch airscrews."[10] P. A. Ralli of Supermarine Aviation Works Ltd. designed the fixed-pitch Fairey-Reed propeller for the 1927 Schneider Trophy-winning S.6 racer. His comments highlighted the indeterminacy of design in the 1920s and how that applied to the incorporation of variable-pitch propellers into new aircraft designs. He commented that high-speed, heavily loaded aircraft operating under difficult takeoff conditions warranted their use. Aircraft with limited speed range, or with high speed and light wing loading with no takeoff restrictions such as an air racer, favored the use of fixed-pitch propellers. Aircraft designers had to keep these types of aircraft in mind when debating the merits of variable-pitch in the application of a design.[11]

To the president of the RAeS, Sir William Forbes-Sempill, the resistance exhibited by the anti-VP school was decidedly unprogressive. He recognized that the "wave of pessimism" that characterized the opening of the discussion was "unnecessarily severe" because the variable-pitch propeller was in the early stages of development. The "optimistic" tone that encouraged further experimentation found in previous RAeS presentations was obviously lacking.[12] The editor of *Flight*, Stanley Spooner,

[9] H. L. Milner and H. Burroughes, in Hele-Shaw and Beacham, "The Variable-Pitch Airscrew," 542–550.

[10] D. H. Kennedy, in Hele-Shaw and Beacham, "The Variable-Pitch Airscrew," 547.

[11] P. A. Ralli, in Hele-Shaw and Beacham, "The Variable-Pitch Airscrew," 550; "Schneider Trophy Machine Design," *Flight* 20 (February 2, 1928): 66.

[12] William Forbes-Sempill, in Hele-Shaw and Beacham, "The Variable-Pitch Airscrew," 550–551.

asserted that he did not "share the pessimism that was so freely expressed at the lecture" because the variable-pitch propeller was "well worth developing."[13] Major Mayo of the RAF declared, "The variable-pitch propeller seemed to have been judged before it had a proper trial."[14] The pro-VP school wanted the British aeronautical community to recognize the promise of variable-pitch and what it could offer to overall aircraft performance after continued experimentation.

Gloster attempted to garner interest in its propeller after the RAeS presentation by displaying it at Great Britain's major aeronautical trade fair, the International Aero Exhibition held at the Olympia in London, the following June. Frank Caldwell, still working at McCook Field, and Tom Hamilton of Hamilton Aero Manufacturing, were two of the notable members of the propeller community in attendance.[15] Caldwell witnessed a flight demonstration at the Gloster factory at Hucclecote later during the visit. He would not sketch his idea for a hydraulic controllable-pitch propeller until November 1928, so his visit to England certainly informed his design choices.

It was not until January 1929 that the company received any real interest from the British government. The Air Ministry issued a contract for the design and construction of twelve propellers for tests on Armstrong-Whitworth, Bristol and Fairey biplane fighters. These aircraft flew in May and September 1929, and February 1930 respectively. Gloster installed propellers on Gamecock aircraft for continued flight-testing in October 1930. From November 1930 to April 1931, Gloster conducted a fifty-hour flight test program on those propellers. The company flew an additional twenty-five hours in October 1931 to evaluate modifications that resulted from the earlier program.[16]

Gloster's Hele-Shaw Beacham constant-speed propeller development program foundered in the early 1930s. There were well-known individuals within the British aeronautical community who supported the innovation, primarily Roy Fedden, chief engineer of the Bristol Aeroplane Company Ltd., the designer of the famed Jupiter engine. He recognized the synergy of propulsion system design and believed the absence of a

[13] Stanley Spooner, "Editorial Comment: Variable-Pitch Propellers," *Flight* 20 (April 19, 1928): 257–258; "The Variable-Pitch Airscrew: With a Description of a New System of Hydraulic Control," *Flight* 20 (April 19, 1928): 269–270.

[14] Mayo, in Hele-Shaw and Beacham, "The Variable-Pitch Airscrew," 545–546.

[15] R. K. Smith to author, October 13, 1999; James, *Gloster Aircraft*, 17–18; Thomas Foxworth, *The Speed Seekers* (New York: Doubleday, 1975), 74–75.

[16] "The 'Bristol' Constant-Speed Controllable-Pitch Airscrew," *The Bristol Review* (June 1937): 42; James, *Gloster Aircraft*, 17, 118, 122; Stait, *Rotol*, 2.

practical variable-pitch propeller was an obstacle to the further development of British aircraft and engines. Fedden's enthusiasm surprised and even annoyed his colleagues and peers. He collaborated with Gloster during the early 1930s flight tests, but he soon became dissatisfied with their progress. Fedden secured an Air Ministry contract, took over the program, and brought Pop Milner to Bristol in 1934.[17]

The American Debate over a New Idea

Across the Atlantic, the general belief that variable-pitch propellers were heavy, unreliable, and overall impractical also existed. Nevertheless, there was an impetus within the corporate propeller community to go ahead with development. Before the creation of the IAS in 1932, aeronautical engineers interacted professionally through closely related professional societies, primarily the American Society of Mechanical Engineers (ASME) and the Society of Automotive Engineers (SAE). At the August 1929 SAE meeting in Cleveland, Ohio, held concurrently with the National Air Races, two papers, one delivered by Frank Caldwell and another jointly by Curtiss-Wright chief engineer Theodore P. Wright and Wallace R. Turnbull, represented a significant turning point in the development of the variable-pitch propeller during the Aeronautical Revolution.

Both papers offered convincing arguments for the use of variable-pitch propellers. Caldwell's presentation was full of detailed analysis, to-the-point, and based on his decade-and-a-half of experience working in propellers. His test curves of a nine-foot diameter propeller demonstrated that an ability to change blade angle from twenty-two degrees, an ideal setting for cruising, to fourteen degrees would increase thrust by 40 percent at takeoff and rate of climb by 60 percent.[18]

Wright and Turnbull revealed that an airplane with a supercharged engine and a fixed-pitch propeller could climb to 20,000 feet in thirty-two minutes. With the addition of a variable-pitch propeller, the same airplane would make the flight in 27 ½ minutes, a gain of 17 percent. They added that the "desirability of the controllable-pitch propeller, recognized in the early days of aviation, is rapidly becoming a necessity if the airplane

[17] E. M. Nixon, "Aircraft Engine Developments," *Aeronautical Journal* 70 (January 1966): 156; Stait, *Rotol*, 4; Bill Gunston, *Fedden: The Life of Sir Roy Fedden*, Rolls-Royce Heritage Trust Historical Series No. 26 (Derby, England: Rolls-Royce Heritage Trust, 1998), 135, 195–196.

[18] Frank W. Caldwell, "Variable-Pitch Propellers," *SAE Transactions* 24 (1929): 475.

is to develop its full possibilities as an established means of transportation." Wright and Turnbull confidently predicted that "extensive use of the controllable-pitch propeller is inevitable in the next few years" because the new technology could potentially save the military and commercial airlines hundreds of thousands of dollars a year in operating costs.[19]

Edward P. Warner, the editor of the influential journal, *Aviation*, recognized that the Cleveland presentations emphasized real-world application over discussing hypothetical designs. To him, this clearly indicated that the propeller community was preparing itself for the introduction of a practical variable-pitch propeller. The immediacy of the technological moment fueled by financial and performance requirements would keep the community focused on practical applications. Unfortunately, attendance could have been better. Only fifty people were present at the important session because it had to compete with the extensive flying program of the National Air Races for attendance on a sunny Monday afternoon.[20]

The performance deficiencies of the new low-wing monoplanes with high wing loadings being introduced in the United States and incorporating some of the advances associated with the Aeronautical Revolution illustrated the need for the variable-pitch propeller. The transitional Boeing Model 200 Monomail of 1930 was a mail and cargo airplane that featured a single open cockpit, cantilever monoplane wings, all-metal stressed-skin semi-monocoque construction, retractable landing gear, a single 575-horsepower Pratt & Whitney R-1860-B Hornet engine enclosed in a ring cowling, and a two-blade ground-adjustable-pitch propeller. With the propeller set for cruise, the high pitch setting ensured that a significant portion of the blade stalled at low aircraft speeds, meaning there was a significant loss of thrust. The Monomail could not even move forward and take off under its own power for its first test flight with a full payload in May 1930 at Boeing Field in Seattle. The Boeing company test pilot, Eddie Allen, used every inch of runway to get off the ground after company engineers opted for a compromise propeller setting that did not generate full performance in either regime.[21] The Materiel Division at

[19] Theodore P. Wright and Wallace R. Turnbull, "Controllable-Pitch Propeller," *SAE Transactions* 24 (1929): 478, 481.

[20] Edward P. Warner, "The SAE Propeller and Power Plant Sessions," *Aviation* 27 (August 31, 1929): 469.

[21] Peter M. Bowers, *Boeing Aircraft Since 1916* (Annapolis, MD: Naval Institute Press, 1989), 198–200; G. W. Carr to C. N. Monteith, May 7, 1930; G. W. Carr to P. G. Johnson, May 7, 1930, Folder 2, Box 181, Model 200 Files, BCA; Harold Mansfield, *Vision: The Story of Boeing* (New York: Popular Library, 1966), 42, 44, 46.

Wright Field evaluated the Boeing Model 200 Monomail as a potential Air Corps transport in July 1930. The test pilots complained of slow takeoff and climb. Overall, they believed a "better propeller setting" would enhance engine performance.[22]

The poor performance of the Model 200 led Boeing to modify it into a new airplane, the Model 221A, which United Air Lines (UAL) used on its Cheyenne-to-Chicago route. The UATC Research Division sponsored a joint development program conducted by Boeing, Hamilton Standard, and Pratt & Whitney at Chicago, Cheyenne, and Hartford from July to November 1931 to investigate the value of the controllable-pitch propeller to the performance of the Model 221A. The program evaluated the takeoff and climb performance of three three-blade and six two-blade controllable-pitch propellers that varied in diameter from nine feet three inches to eleven feet. United Aircraft researchers concluded that the "chief reason for using a hydro-controllable propeller on the Monomail is to improve takeoff and climb," which the test results clearly indicated.[23]

The comparison tests of fixed- and controllable-pitch propellers conducted at Rentschler Field in Hartford illustrated the importance of the variable-pitch propeller. Specifically, the addition of a low-pitch setting to the performance equation made an important difference. For the Model 221A, the takeoff run, the distance covered by the airplane before lifting off the ground, decreased from the 1,085 feet with a fixed pitch propeller designed for cruise to 825 feet for the controllable-pitch propeller at its low-pitch setting, which was an overall reduction of 24 percent. The calculated rate of climb for the fixed-pitch propeller was 690 feet/minute. The controllable-pitch propeller climbed at 845 feet/minute, a gain of 22 percent. In terms of safety, it was important that a fully loaded airplane be able to clear an obstruction at takeoff. At the end of a 2,000 feet test run, the Monomail with the fixed-pitch propeller climbed to an altitude of thirty-seven feet. With the controllable-pitch propeller, the height nearly doubled to seventy-two feet. Researcher Charles H. Chatfield emphasized that, "these improvements in takeoff and in climb were obtained at little if any sacrifice in maximum and cruising speed" because

[22] Air Corps Materiel Division, "Report of Test on Boeing Transport Model 200 (Monomail)," July 18, 1930, Folder 15, Box 1033, Model 200 Files, BCA.

[23] D. S. Harvey, "Performance Tests of the Boeing Monomail at Chicago, Cheyenne, and Hartford, 1931," UATC Research Division Report No. T-51, February 25, 1932, Folder 10, Box 185, Model 221A Files, BCA.

the controllable-pitch propeller in the high pitch setting matched the performance of the best fixed-pitch propeller.[24]

While the UATC conducted its tests, American aviators, interested in breaking records and exploration, were quick to install the new propeller on their aircraft. Ruth Nichols flew a Lockheed Vega owned by the Crosley Radio Corporation and equipped with one of Hamilton Standard's first controllable counterweight propellers. In the early 1930s, she set several altitude, speed, and distance records across the United States, including an official flight of 1,977 miles for women pilots between Oakland, California, and Louisville, Kentucky, on October 24–25, 1931.[25]

The success of the Model 221A tests and Ruth Nichols' flight notwithstanding, Hamilton Standard's new product faced design conservatism throughout the American aeronautical community. According to Eugene Wilson, the Army, Navy, industry, and even the UATC corporate leadership rejected outright the idea of a variable-pitch propeller, especially Caldwell's hydraulic design, in the early 1930s. He contended many years later that aircraft designers and engineers believed hydraulic control was excessively heavy, expensive, and was the most complicated method of pitch actuation.[26]

For the Materiel Division, the government organization most responsible for propeller development, it was not a blind reluctance to adopt new innovations. The primary reasons were the perceived application of the new device and the nature and cost of its engineering development. They saw the controllable-pitch propeller as the best device to increase airplane performance, but it had no official specifications for the technology at the end of 1930. Engineering section chief, Maj. Clinton W. Howard, simply asserted that the "development of this article has not progressed sufficiently to enable any one to set up even a general specification." Fundamentally, the requirements were for it to be simple and easy to operate, as light as possible, as strong as a ground-adjustable-pitch propeller, and to be reliable and longstanding in service.[27]

[24] C. H. Chatfield to R. Walsh, November 18, 1931; D. S. Harvey, "Flight Test of Boeing Model 221 with Hamilton Standard Hydro-Controllable Propeller," November 5, 1931, Folder 9, Box 185, Model 221A Files, BCA.

[25] "Hamilton Standard Controllable-Pitch Propeller – Ruth Nichols," File A19400027000, RO, NASM.

[26] Eugene Wilson, *Slipstream: An Autobiography of an Air Craftsman* (Palm Beach, FL: Literary Investment Guild, 1967); Mansfield, *Vision*, 44.

[27] C. W. Howard to Smith Engineering Company, December 31, 1930, Folder "Propellers – Smith Engineering Company-Lycoming Manufacturing Company, 1931–1934," Box 5977, RD 3247, RG 342, NARA.

Air Corps engineers saw value in using a variable-pitch propeller on a supercharged engine. The power plant unit at Wright Field pursued work on two types of superchargers: the geared centrifugal supercharger for radial engines and the exhaust-driven centrifugal turbo supercharger for V-type engines. Combined with a variable-pitch propeller, superchargers increased power output at cruise altitudes and improved overall low-level performance. Before the Army specified variable-pitch propellers for any of those aircraft, supercharged or otherwise, they had to provide the anticipated performance, safety, and reliability that only exhaustive ground and flight testing by the propeller unit could confirm.[28]

To that end, the Materiel Division continued to evaluate controllable-pitch designs during the spring of 1931. The designs submitted by Curtiss and the Southern Aircraft Corporation of Birmingham, Alabama, completed preliminary testing. The Stimson-Patton propeller project became inactive due to the financial difficulties of its owners. The Materiel Division had its own design, sponsored by Adam Dickey and Lt. Orval R. Cook, and intended for the obsolete Liberty V-12 engine. One of those propellers had successfully flown for 500 hours without a malfunction.[29]

The most promising design was the Smith propeller. The Smith Engineering Company of Cleveland submitted its drawings for its experimental, mechanically actuated, controllable-pitch propeller with hollow-steel blades to the Army earlier in January.[30] S. Harold Smith, president of the company and designer of the propeller, used the motive power supplied by the rotation of the hub around a special control gear worm, mounted in the thrust bearing cover plate, to actuate pitch. The control was a flexible cable that was uncomplicated, light, and desirable for an aeronautical community interested in simplicity. The control had three settings: increase pitch, decrease pitch, and a neutral stop to lock the blade angle in flight. The pilot could set the blade angle at an infinite number of settings in accordance with the air and engine speeds for maximum efficiency. Maximum and minimum pitch stops kept the blades within the correct blade angle range. Smith manufactured both the hubs and blades from heat-treated chrome vanadium steel. Installation modifications

[28] Materiel Division, "Current Experimental and Development Projects," May 5, 1931, Folder 400.112, "Experimental Projects and Developments, 1931," Box 5812, RD 3202, RG 342, NARA; A. J. Lyon to R. Walsh, January 21, 1931, Folder "Propellers – January-June 1931," Box 5822, RD 3205, RG 342, NARA.

[29] Materiel Division, "Current Experimental and Development Projects," May 5, 1931.

[30] L. G. Craft to D. A. Dickey, January 21, 1931, Folder "Propellers – Smith Engineering Company – Lycoming Manufacturing Company, 1931–1934," Box 5977, RD 3247, RG 342, NARA.

required only the replacement of the engine thrust bearing cover plate with one that incorporated the control assembly.[31] The Materiel Division found no "impractical features" and invited Smith to submit a propeller for further testing.[32]

During the period 1928–1931, the Army gave Hamilton Standard one experimental controllable-pitch propeller contract amounting to $4,500 for the Douglas BT-2B basic trainer. Raycroft Walsh pleaded with Brig. Gen. Henry C. Pratt, Chief of the Materiel Division, for $65,000 in experimental and service test orders that would garner the Army "outstanding achievements in propeller development" in 1932. Hamilton Standard had propellers ready for the Pratt & Whitney R-1340 Wasp and R-1860 Hornet radial engines used on the Air Corps' new Boeing P-12 pursuit and YB-9 bomber aircraft.[33]

On behalf of the Materiel Division, Maj. Henry H. Arnold explained the realities of the pursuit of new technologies during the "present period of financial depression" to Walsh. The Air Corps was unable to allocate funds for experimental and service test contracts for a "progressive development program" with any manufacturer because it had to encourage simultaneous innovation in engines, carburetion, ignition, radios and communication, navigational and all-weather flying instruments. He went on:

> The Air Corps, of course, would like to be of greater assistance to the aircraft manufacturers during periods of depressions, but it would require a large amount of money for us to undertake, by means of development programs, to support all of the manufacturers who are carrying out experimental work in which we are interested. Our appropriations for this purpose have not been increased for a number of years, and under present conditions we are not likely to receive any additional funds for development work.[34]

[31] Smith Engineering Company, "Smith Controllable Pitch Propeller," n.d. [1931], File B5-807000-01, PTF, NASM; E. B. Schaefer, "The Smith Controllable-Pitch Propeller," *Aviation Engineering* 7 (December 1932): 16; D. Adam Dickey and O. R. Cook, "Controllable and Automatic Aircraft Propellers," *SAE Transactions* 30 (March 1932): 108; Aeronautical Chamber of Commerce of America, *Aircraft Year Book for 1933* (New York: D. Van Nostrand, 1933), 372.

[32] A. J. Lyon to L. G. Craft, January 30, 1931, Folder "Propellers – Smith Engineering Company-Lycoming Manufacturing Company, 1931–1934," Box 5977, RD 3247, RG 342, NARA.

[33] R. Walsh to H. C. Pratt, June 11, 1931; "Recommended Program of Propeller Tests," [April 1931], Folder "Propellers – January–June 1931," Box 5822, RD 3205, RG 342, NARA; James C. Fahey, *U.S. Army Aircraft, 1908–1946* (New York: Ships and Aircraft, 1946), 22, 24,32.

[34] H. H. Arnold to R. Walsh, June 17, 1931, Folder "Propellers – January–June 1931," Box 5822, RD 3205, RG 342, NARA.

The Materiel Division was not ignorant of the value of variable-pitch; it simply could not afford to promote its development within the realities of austere engineering budgets in the early 1930s.

Moments of Acceptance

A well-documented example of the industry's design conservatism toward variable-pitch propellers concerned the development of the Boeing 247. Boeing designed and built the new monoplane for United Air Lines, its corporate partner, for use on its newly opened transcontinental routes. Before the 247's introduction, the thirty-one-hour westward transcontinental flight provided by the United's "Coast to Coast Limited" service in 1931 consisted of three legs over three separate mountain ranges where the altitude above sea level fluctuated dramatically. The first leg took travelers from sixty feet at Newark, New Jersey, over the Appalachians to Chicago at 600 feet in a Ford Tri-Motor. The second leg left in a Boeing Model 80 airliner headed for Cheyenne, Wyoming. While over the Great Plains of Nebraska, the three-engine biplane began a five-hour 4,888 feet climb to reach Cheyenne at 6,058 feet high in the foothills of the Rocky Mountains. The Model 80 used ground-adjustable-pitch propellers. The best its crew could do was to follow the terrain as they gradually flew higher and higher. The third leg continued in another Model 80 through the Rockies to Rock Springs, Wyoming, the highest point during the journey at 6,261 feet, and to Salt Lake City at 4,251 feet. With the Rockies behind it, the airliner still had to clear the final hurdle, the Sierras. The airliner descended from Reno at 4,600 feet to Oakland at eighteen feet above sea level.[35] Nothing illustrated the need for variable-pitch propellers more than the extremes United's airliners faced as they flew the transcontinental route from Newark to Oakland, which, at the time, was the fastest way to travel across the United States.

The 247 was Boeing's answer to the mediocre performance of the Tri-Motor and the Model 80. One 247 could make the entire transcontinental flight in twenty-one hours. The airplane featured the latest innovations of the Aeronautical Revolution: a streamlined aerodynamic shape including nacelles correctly placed in the wing's leading edge according to NACA data; all-metal semi-monocoque fuselage and cantilever monoplane wing construction; retractable landing gear; advanced instrumentation; trailing edge flaps to help slow the sleek airliner for landing; and

[35] "No. 1 Airplane Company," *Fortune* 5 (April 1932): 46, 51.

two supercharged 550-horsepower Pratt & Whitney S1D1 Wasp radial engines.[36]

Boeing engineers specifically rejected variable-pitch propellers because they felt they were too heavy and that other innovations in aerodynamics, structures, and power plants provided the desired performance. Much like the rest of the aeronautical industry, Boeing emphasized the prevailing design paradigm of light weight and simplicity, which favored the ground-adjustable-pitch propeller. Charles N. Monteith, chief engineer of the Boeing Airplane Company, stressed that "the controllable-pitch propeller is another item which will add some weight to the installation and should be avoided if possible."[37] He was aware of the potential value of using a variable-pitch propeller. Ten years earlier in 1921, he had recommended their development while serving as the chief of the Airplane Section at McCook Field. Monteith simply chose to follow the well-established paradigm of pursuing the lighter alternative.

The UATC technological leadership emphasized that characteristic as well. George Mead, speaking before UATC's TAC, stated, "I don't believe there is any question as to wanting a lighter propeller, and I don't believe any of us care much how we get it as long as we get it, I think that is our job."[38] As a result, Boeing designers equipped the original 247 with two Hamilton Standard three-blade ground-adjustable-pitch propellers. The description of the 247's propulsion system in the Pratt & Whitney company magazine, *The Bee-Hive*, mentioned the propellers as an afterthought.[39]

The first test flight of the 247 at Boeing Field in late February 1933 yielded impressive results. The cruising speed at 5,000 feet was 165 mph. The takeoff run was 770 feet, and the rate of climb to 5,000 feet was 830 feet per minute. Believing that the 247 had proven its capabilities, Boeing manufactured the first aircraft for UAL and intended to have at least eighteen ready for a May 1, 1933 delivery.[40]

[36] F. Robert van der Linden, *The Boeing 247: The First Modern Airliner* (Seattle: University of Washington Press, 1991), 75; Bowers, *Boeing Aircraft*, 207, 213.

[37] C. N. Monteith to C. L. Egtvedt, "Mr. Rentschler's Memo of October 16, 1931," October 22, 1931, Folder 18, Box 5, Model 247 Files, BCA.

[38] Report of the Second Meeting of the Technical Advisory Committee of the United Aircraft and Transport Corporation at Seattle, Washington, December 2-6, 1929, CE, BCA, 25–26.

[39] "New Boeing Commercial Transport Surpasses Expectations in Test Flight," *U.S. Air Services* 18 (April 1933): 13, 15; "Wasp-Powered Twin Engine Boeing Transport Shows Remarkable Performance in Initial Tests," *The Bee-Hive* 7 (April 1933): 10.

[40] "New Boeing Commercial Transport Surpasses Expectations in Test Flight," 14–15.

Oddly enough, while the 247 first took to the air with ground-adjustable-pitch propellers, the propulsion community recognized that Caldwell and Hamilton Standard's variable-pitch propeller was a crucial part of the overall technical synergy of the modern airplane. *The Bee-Hive* remarked in March 1933 that, "Every airplane is the result of many compromises." Designers, using new performance-enhancing innovations were taking some of the compromise out of aircraft design. Certainly, one of those innovations was the two-position, controllable-counterweight propeller introduced by Pratt & Whitney's corporate partner, Hamilton Standard. While the idea was not new, *The Bee-Hive* saw the "successful application of a simply [*sic*] operating mechanism" that was "dependable and light" as a "new step forward." Caldwell and his fellow engineers at Hamilton Standard had "removed one more compromise from the field of modern aircraft design."[41] The appearance of the propeller indicated that changes in the prevailing attitudes of aircraft design were taking place slowly, especially from the vantage point of the specialists who created it. But the change had yet to occur within the UATC's aircraft design community.

Progress-oriented rhetoric and optimism characterized the initial reactions to the 247. *U.S. Air Services* announced the dawning of "a new era in commercial air transport" after the first test flight at Seattle. The 247's "high performance, exceptional strength, profitable payload, close attention to the comfort of the air traveler, and of particular interest – its operating economy" promised to usher in a new age of American technological superiority – and profitability – in aeronautics. In the larger sense, the 247 helped alter humankind's perceptions of time and space while adding "greatly to the laurels of American ingenuity and progress." Hailed as the fastest multiengine commercial airplane in the world, the 247 was the product of experience gleaned from UAL in flying transcontinental routes and Boeing's sixteen years of manufacturing aircraft.[42]

Ironically, the 247 was also a reflection of a design conservatism that ruled out a variable-pitch propeller. Boeing's decision to use ground-adjustable-pitch propellers on the 247 almost proved to be the design's undoing. The new airliners performed poorly during their UAL acceptance tests in the spring of 1933. UAL's main technical facility was at

[41] "Hamilton Standard Announces Improved Type Controllable-Pitch Propeller," *The Bee-Hive* 7 (March 1933): 5.

[42] "New Boeing Commercial Transport Surpasses Expectations in Test Flight," 13–15; "Wasp-Powered Twin Engine Boeing Transport Shows Remarkable Performance in Initial Tests," 8.

Cheyenne. The aircraft could barely climb to 6,000 feet and did not exhibit the impressive performance characteristics demonstrated near sea level at Boeing Field.[43] The Boeing engineers quickly learned that all of the innovations of the Aeronautical Revolution, including variable-pitch propellers, needed to act in synergy before their new design could make it over the Rockies safely.

Hamilton Standard, eager to prove the value of its new product, sent Caldwell to Wyoming to investigate the problem. Dissatisfied with his first design, Caldwell and his team in Hartford had just finished work on a refined two-position, controllable-counterweight design. Even though it retained the same internal mechanism as the earlier propeller, it was, as *The Bee-Hive* described it, "more rugged and compact" and significantly cheaper to manufacture. The primary difference was the "clean and compact counterweights." Caldwell placed them just ahead of the propeller to reduce the amount of travel to achieve the same pitch setting and so that they would lie close to the crankshaft in both high and low pitch settings. The refined design allowed the use of a spinner for a more streamlined aerodynamic effect. The new propeller was ready for production for both military and commercial aircraft by early 1933.[44]

Caldwell's tests concluded that the use of variable-pitch propellers reduced the 247's takeoff run by 20 percent and increased the rate of climb by 22 percent and cruising speed by 5.5 percent from 165 to 171 mph. In short, his invention maximized the performance of an already revolutionary design. As a result, Boeing placed the first production order for the Hamilton Standard two-position, controllable-pitch propellers for use on all of its transport aircraft, not just the refined Model 247-D. The new propellers started arriving in Seattle in June 1933. Additional tests conducted at Rentschler Field in Hartford supported Caldwell's findings.[45]

[43] Wilson, *Slipstream*, 168–170.

[44] "Hamilton Standard Announces Improved Type Controllable-Pitch Propeller," 5; Ronald Miller and David Sawers, *The Technical Development of Modern Aviation* (New York: Praeger, 1970), 74.

[45] "Transport Cruises 171 M.P.H. Using New Type Props," *Boeing News* 4 (May 1933): 1; Charles H. Chatfield, "Controllable-Pitch Propellers in Transport Service," *Aviation* 32 (June 1933): 180; C. L. Egtvedt to P. G. Johnson, "Report," June 5, 1933, Folder 1, Box 10, General Management Files, Boeing Airplane Company, 1916–1934, BCA; Miller and Sawers, *Technical Development*, 74–75; Roger E. Bilstein, *Flight Patterns: Trends in Aeronautical Development in the United States, 1918–1929* (Athens, GA: University of Georgia Press, 1983), 105; van der Linden, *The Boeing 247*, 46; D. S. Harvey, "Flight Tests on the Boeing 247 Transport with Two Types of Propellers," UATC Research Division Report No. T-70, May 8, 1933, Folder 4, Box 1038, Model 247 Files, BCA.

The original 247 was, in essence, a transitional aircraft that was a product of performance-specific design. The addition of the variable-pitch propeller along with two other advances, the NACA engine cowling and a rearward slanting windscreen, after the fact, made the 247-D "modern" and a flexible performance-general design.

The story behind the inclusion of the variable-pitch propeller in the design of the even more revolutionary Douglas DC transports reveals how an engineering community chose the new technology at the start of the design process. Jack Frye, vice president for operations of Transcontinental and Western Air (TWA), wanted to replace the airline's wood wing Fokker trimotor aircraft after the fatal March 1931 crash of an airliner carrying the famed Notre Dame football coach, Knute Rockne. Unable to buy Model 247 airliners initially due to the direct corporate connection between UAL and Boeing in the UATC family, he asked the Douglas Company and other aircraft manufacturers, including Consolidated, Curtiss, General Aviation, and Martin, to submit proposals for a new all-metal three-engine monoplane airliner in August 1932. A primary requirement for the new airliner was that it be able to fly safely on two of its three engines with a full payload.[46]

The Douglas Company responded with a totally original, streamlined, all-metal, cantilever-wing, twin-engine monoplane with retractable landing gear that was more than competitive with the Model 247. The two high-output 700-horsepower Wright Cyclone SR-1820-F-3 radial engines, enclosed in NACA cowlings, eliminated the need for the third center engine. The Douglas engineers were confident that the airliner could go one step farther than Frye's requirement by being able to operate on only one engine in the event of power plant failure.[47]

Initially, the Douglas Company designed the DC-1 with three-blade fixed-pitch propellers. Early photographs of the TWA DC-1 at the time of its first flight in July 1933 featured this anomaly.[48] Frye, impressed by the initial performance estimates of the new twin-engine transport, was reluctant to use variable-pitch propellers. He asked the Douglas chief engineer, James H. Kindelberger, "Do you secure this [performance] without the use of any devices, such as variable-pitch propellers, etc., that have not

[46] J. Frye to Douglas Aircraft Corporation, "Attention: Mr. Donald Douglas," August 2, 1932, The James S. McDonnell Prologue Room, BCA.

[47] René J. Francillon, *McDonnell Douglas Aircraft since 1920* (London: Putnam, 1979), 165–173.

[48] Peter W. Brooks, *The Modern Airliner: Its Origins and Development* (Manhattan, KS: Sunflower University Press, 1961), 74.

to date been thoroughly service tested and proven?"[49] Douglas first contacted Caldwell at Hamilton Standard in September 1932 about propellers for its new transport for TWA. In the detailed listing of specifications, Kindelberger stated that "there is a possibility of using controllable-pitch propellers on this airplane at a later date, but they are not in the picture at the present time."[50]

By February 1933, the ever-increasing weight and resultant need for higher performance of the DC prototype led the Douglas design team to incorporate variable-pitch propellers. Performance analyses revealed that the new technology increased the original cruising speed by 25 mph. With the weight of the wing and fuselage and the increase in speed, the Douglas engineers learned fixed-pitch propellers made TWA's rigorous one-engine operation requirement impossible. The airline's technical advisor, Charles A. Lindbergh, required single-engine operation to increase the safety margin of the new airplane. In a Douglas memorandum, engineer Arthur E. Raymond revealed that, "fixed pitch propellers are out of the question for we can obtain no ceiling whatever on one engine when using them," while "controllable-pitch propellers, on the other hand, show a much greater benefit than they did before." Raymond went on to elaborate on the merits of variable-pitch for the DC design:

> I discussed the matter with Doug [Donald W. Douglas] this morning and he agreed with me that we should purchase and install on the first airplane controllable propellers if available. If TWA insists that the tests be run with fixed-pitch, all that will be necessary is for them to keep the propellers at one setting. However, the fact that the controllable feature is incorporated and can be used at any time and that with it the single-engine guarantee can be met (with luck), together with the fact that the airplane will be so much faster than expected, should go far towards selling the airplane.[51]

This memo reflects the Douglas engineering team in transition from performance-specific to performance-general design logic. The Douglas Company incorporated Hamilton Standard controllable-pitch propellers into the design of their revolutionary DC-1 transport before its first flight on July 1, 1933.[52]

[49] J. Frye to J. H. Kindelberger, August 15, 1932, File MO15127, DC-2 Files, BCA.

[50] J. H. Kindelberger to F. W. Caldwell, September 23, 1932, File MO15127, DC-2 Files, BCA.

[51] A. E. Raymond to J. H. Kindelberger, February 13, 1933, File MO15127, DC-2 Files, BCA.

[52] *Development of the Douglas Transport* (Santa Monica, CA: Douglas Aircraft Company, n.d. [1933]), n.p., File MO15127, DC-2 Files, BCA; Miller and Sawers, *Technical Development*, 74–75; George Rosen, *Thrusting Forward: A History of the Propeller* (Windsor Locks, CT: United Technologies Corporation, 1984), 42.

While the Douglas engineers were realizing the value of the variable-pitch propeller, the development of the DC coincided with Hamilton Standard's final development tests and federal certification of the two-position, controllable-pitch propeller. One last obstacle to the introduction of the propeller was the need to adapt it to the oil system and reduction gearing of the Wright Cyclone engine. The hydro-controllable propeller operated with the engine's oil supply connected through the engine crankshaft. A selling point of the Hamilton Standard propeller for use with Pratt & Whitney engines was that the corporate partners had already collaborated on the design of the necessary modifications required for hydraulic operation. Wright Aeronautical had to make the necessary modifications to make the propeller available for the DC project.[53] Those modifications forced the Douglas design team to wait for the availability of the Hamilton Standard controllable-pitch propellers before they could be installed on the DC-1 or its immediate successor, the DC-2.

The famous "one engine out" tests at Winslow, Arizona, in November 1933 justified the incorporation of controllable propellers into the DC-2 design. To comply with Lindbergh's directive, the DC-2 had to be able to safely operate on one engine under a normal or heavy load. Winslow was TWA's main maintenance and repair facility for the western leg of its transcontinental routes. At 4,878 feet above sea level, the location offered TWA the same type of environment as United's Cheyenne facility for evaluating the high-altitude takeoff performance of its airliners. As the fully loaded DC-2 traveled down the runway, the pilot shut down one motor after 2,200 feet, took off, and climbed to 9,000 feet at a cruising speed of 120 mph. From Winslow, the DC-2 flew the 240 miles to Albuquerque, New Mexico, by way of the Continental Divide at an altitude of 7,243 feet, all on one engine equipped with the Hamilton Standard propeller.[54]

The DC-2 entered commercial service with TWA in May 1934 equipped with eleven-foot-six inch-diameter Hamilton Standard Model 45C2 three-blade propellers that weighed 155 pounds each.[55] In recounting the

[53] S. A. Stewart to A. E. Raymond, December 13, 1932; B. G. Leighton to T. A. Morgan, December 22, 1932, File M015127, DC-2 Files, BCA; A. C. Foulk to Chief, Engineering Section, "Modification of Engines to Accommodate Hamilton Standard Controllable Propeller," July 11, 1932, Folder "Propellers – General, 1932," Box 5867, RD 3216, RG 342, NARA; Francillon, *McDonnell Douglas Aircraft*, 170.

[54] Douglas Aircraft, *Douglas Transport: DC-2* (n.d.), File M015130, DC-2 Files, BCA; "New TWA Douglas 'Airliner' at Rest and at Work," *U.S. Air Services* 18 (November 1933): 20; "Up and Over on One Motor," *U.S. Air Services* 18 (November 1933): n.p.

[55] F. W. Caldwell to J. H. Kindelberger, October 3, 1932, File M015127, DC-2 Files, BCA; *Development of the Douglas Transport*, n.p.

story of the development of the original DC, *Fortune* magazine noted that "the most dramatic development to be made available to the Douglas engineers while their ship was abuilding [*sic*] was the controllable-pitch propeller" introduced by Hamilton Standard.[56]

One of the first production aircraft designed from the ground up to benefit from the use of variable-pitch propellers was the transoceanic Sikorsky S-42 flying boat. Weighing in at nineteen tons, the S-42 was the largest of its kind. Pan American Airways called them Flying Clippers as they cruised at 170 mph while carrying thirty-seven passengers at a maximum range of 1,200 miles. Starting in 1934, S-42s, piloted by daring airline captains such as Edwin C. Musik, pioneered Pan Am's Caribbean, Atlantic, and Pacific routes. According to Raycroft Walsh, Sikorsky engineers planned from the beginning to "utilize to the utmost" controllable-pitch propellers. Each of the behemoth's four 700-horsepower Pratt & Whitney Hornet radial engines had a Hamilton Standard three-blade propeller.[57] The S-42 represented a revolution in design for flying boats the same way the Boeing 247-D and Douglas DC-2 represented new advances for land-based aircraft and was a symbol of progress through technology.

Alternative Technologies

Hamilton Standard was not Douglas's only option for the DC airliners. The company's engineers also contacted Curtiss and a new competitor, the Smith Engineering Company, in late 1932 to inquire about their respective variable-pitch designs.[58] Smith introduced its mechanically actuated propeller to the public at the Detroit Aircraft Show in September 1931.[59] Quickly, the propeller entered the spotlight of aerial spectacle that characterized the 1920s and 1930s. At the end of August 1932, Jimmy Doolittle broke the world speed record for landplanes at 296 mph in front of a crowd of thousands at the National Air Races in

56 "Success in Santa Monica," *Fortune* 11 (May 1935): 175.

57 Raycroft Walsh, "Hamilton Standard Grants Licenses for Controllables in Three Countries," *The Bee-Hive* 8 (September 1934): 8; "Hornet-Powered Sikorsky Blazes Transpacific Trail to Hawaii," *The Bee-Hive* 9 (May 1935): 3–5.

58 S. A. Stewart to A. E. Raymond, December 13, 1932; B. G. Leighton to T. A. Morgan, December 22, 1932, File M015127, DC-2 Files, BCA.

59 S. H. Smith to R. Leyes, March 30, 1992, File B5-807000-01, PTF, NASM; P. F. Hackethal to O. R. Cook, January 25, 1932, Folder "Propellers – Smith Engineering Company-Lycoming Manufacturing Company, 1931–1934," Box 5977, RD 3247, RG 342, NARA.

Cleveland. Days after the record flight, he won the Thompson Trophy race at an average of 252 mph where he exceeded 302 mph during one lap. His diminutive monoplane racer, the Granville Brothers Gee Bee R-1 Super-Sportster, was purpose-built for speed with a radically streamlined fuselage, an 800-horsepower Pratt & Whitney Wasp Senior engine, and a Smith propeller.[60] Doolittle remarked that the propeller "functioned perfectly" as it contributed to the R-1's overall performance.[61] Lauren D. Lyman noted in the *New York Times* that it was "very well liked" by other racers due to the eight degrees of pitch adjustment.[62]

Smith Engineering had an opportunity to enter the world spotlight in July 1933. Wiley Post prepared his Lockheed Vega 5C *Winnie Mae* for his solo round-the-world flight during July 15–23, 1933. He consciously incorporated "additional flying help" through the use of a two-blade Smith propeller to complement the 550-horsepower Pratt & Whitney Wasp engine and a Sperry automatic pilot.[63] Until the destruction of the propeller during a failed takeoff at Flat, Alaska, on July 20, the Smith propeller performed flawlessly. Post made a point to send a telegram to the company:

> The Smith Controllable-Pitch Propeller certainly worked beautifully. Official observers of my New York takeoff tell me my takeoff and distance were [*sic*] by far shorter than any other transatlantic flyers. The various pitch settings obtainable on your propeller certainly proved a godsend as I was forced to take off from emergency fields with heavy loads. Short takeoff and steep climb are great and I had to use many different pitch settings because of both takeoff and flight at different altitudes. This multiplicity of pitch changes is absolutely necessary. I was surprised and gratified at increase in cruising speed and fuel saving.[64]

The landing accident at Flat was unfortunate for Smith. When Post arrived at Floyd Bennett Field, the *Winnie Mae* had a Hamilton Standard ground-adjustable-pitch propeller installed.[65] The opportunity to demonstrate and sell Smith propellers in the wake of Post's flight was lost.

[60] Lauren D. Lyman, "Doolittle Flies 302 Miles An Hour," *New York Times*, September 2, 1932, p. 2; Terry Gwynn-Jones, *Farther and Faster: Aviation's Adventuring Years, 1909–1939* (Washington, DC: Smithsonian Institution Press, 1991), 166–168, 285, 302.

[61] J. H. Doolittle to L. P. Zinke, September 7, 1932, File B5-807000-01, PTF, NASM.

[62] Lauren D. Lyman, "Air Races Aid Service," *New York Times*, September 4, 1932, p. XX6.

[63] "Post Plane to Use Novel Devices," *New York Times*, July 16, 1933, p. 2; Gwynn-Jones, *Farther and Faster*, 247–250.

[64] W. Post to Smith Engineering Company, July 24, 1933, File B5-807000-01, PTF, NASM.

[65] Stanley R. Mohler and Bobby H. Johnson, *Wiley Post, His Winnie Mae, and the World's First Pressure Suit* (Washington, DC: Smithsonian Institution Press, 1971), 57–59.

The stellar performance of the Smith propeller in the Doolittle and Post flights made it a valuable commodity as the period of rapid acquisition and merger in the aviation industry began to wane. The UATC and Curtiss-Wright had owned the Caldwell and Turnbull variable-pitch propeller designs since 1929. The Cord Corporation of Chicago acquired Smith Engineering and made it part of the Aero Engine Division of the Lycoming Manufacturing Company in Williamsport, Pennsylvania, in October 1933. Cord's aviation interests operated through AVCO.[66] Like its corporate competitors, AVCO marketed an aeronautical product line that featured Stinson airliners and private airplanes, Lycoming engines, and Smith propellers.[67]

Under the leadership of McCook Field alumnus Glen T. Lampton, Lycoming engineers developed an electric pitch shift control for the basic Smith design and added counterweights for more precise pitch actuation on multiengine aircraft. Electrical solenoids activated the mechanically actuated hub in lieu of the manually operated flexible cable.[68]

The primary market for the new propeller was the American military. The Air Corps was interested in installing a three-blade controllable Smith propeller on aircraft equipped with the inline Curtiss Conqueror engine as well as various radial engines including the Wright R-1820 Cyclone and the Pratt & Whitney R-1690 Hornet.[69] The Airplane Section at Wright Field installed a Smith propeller on a Boeing P-12C biplane fighter for preliminary tests in February 1932, and flight testing of a Curtiss XP-6E Hawk biplane fighter began in September 1932.[70] The Navy installed a

[66] "Cord Corporation Buys Propeller Concern, Widening Interest in Aviation Industry," *New York Times*, September 1, 1933, p. 23; F. M. Bender to Materiel Division, October 16, 1933, Folder "Propellers – Smith Engineering Company-Lycoming Manufacturing Company, 1931–1934," Box 5977, RD 3247, RG 342, NARA.

[67] Lycoming Division, Aviation Manufacturing Corporation, "Lycoming Controllable Propellers," n.d. [1937], File B5-569000–06, PTF, NASM.

[68] G. T. Lampton to Materiel Division, December 19, 1933, Folder "Propellers–Smith Engineering Company-Lycoming Manufacturing Company, 1931–1934," Box 5977, RD 3247, RG 342, NARA; "Lycoming Develops Electric Control and Pitch Indicator for Controllable Propellers," June 23, 1934, File B5-569000–20, PTF, NASM.

[69] C. W. Howard to P. F. Hackethal, February 27, 1932; G. J. Brew to Chief, Materiel Division, March 26, 1934, Folder "Propellers–Smith Engineering Company-Lycoming Manufacturing Company, 1931–1934," Box 5977, RD 3247, RG 342, NARA.

[70] C. W. Howard to Chief, Flying Branch, "Airplane Assignment," September 7, 1932, Folder "Propellers–Smith Engineering Company-Lycoming Manufacturing Company, 1931–1934," Box 5977, RD 3247, RG 342, NARA; "Smith Controllable Propeller Control," February 10, 1932, File B5-807000–20, PTF, NASM.

two-blade propeller on the Grumman XF2F-1 experimental fighter with a Pratt & Whitney R-1535 engine in December 1933.[71]

The Air Corps awarded Lycoming an initial contract for thirty-four of the new controllable-pitch propeller hubs in May 1934 for service testing.[72] Encouraged by the results, the Army ordered an additional 250 three-blade Smith propellers for the all-metal monoplane Martin B-10 bombers of the General Headquarters (GHQ) Air Force. The B-10, considered the first modern bomber because it incorporated the latest innovations of the Aeronautical Revolution in a synergistic design, garnered Martin Aircraft the 1932 Collier Trophy for its design. It was also the only military aircraft to employ the Smith propeller in service.[73] The Air Corps used the B-10 to demonstrate to the world the continent-ranging ability of strategic bombers during an epic long-distance mass flight from Washington, DC, to Fairbanks, Alaska, during July–August 1934.[74]

As the Smith Engineering Company enjoyed its moment in the sun, Pete Blanchard and his team at Curtiss introduced an improved Electric propeller series for the 200- to 800-horsepower range during the fall of 1933. The design relied upon a small twelve-volt, high-speed, one-quarter-horsepower electric motor geared to the blades by means of a high-ratio speed-reducing mechanism for pitch control. The speed reducer, consisting of a system of planetary gears, transmitted the energy of the electric motor to the blades through a bevel gear in the hub that was connected to bevel gears in the blade roots. Power from the battery reached the motor via a slip ring and contact brush assembly on the propeller shaft. The motor used the same amount of energy required to power a single landing light. Unlike the two-position Hamilton Standard design,

[71] R. P. McConnell to A. J. Lyon, "Vibration Test–Smith Controllable Pitch Propeller 8'6" on XF2F Airplane," December 28, 1933, Folder "Propellers–Smith Engineering Company-Lycoming Manufacturing Company, 1931–1934," Box 5977, RD 3247, RG 342, NARA.

[72] K. B. Wolfe to Lycoming, "Air Corps Inspection of Propeller Hubs," May 29, 1934, Folder "Propellers–Smith Engineering Company-Lycoming Manufacturing Company, 1931–1934," Box 5977, RD 3247, RG 342, NARA.

[73] O. P. Echols to Office of COAC, February 12, 1936, Folder "Propellers–Smith Engineering Company-Lycoming Manufacturing Company, 1935–1937," Box 6195, RD 3303, RG 342, NARA; Lycoming Division, Aviation Manufacturing Corporation, "Lycoming Controllable Propellers," n.d. [1937].

[74] "Army Planes Back From Alaska Trip," *New York Times*, August 21, 1934, p. 19; Maurer Maurer, *Aviation in the U.S. Army, 1919–1939* (Washington, DC: Office of Air Force History, 1987), 352–354.

the Curtiss Electric propeller was a multi-pitch propeller. The pilot could set the blade pitch to any angle between zero and ninety degrees.[75]

Gaining Momentum

The time for the variable-pitch propeller had arrived. The editors of *U.S. Air Services* recognized that "a great undertaking" was occurring in aviation. Commercial airliners carried mail and passengers across the continental United States at ever-increasing speeds. Complimentary developments in aerodynamics, propulsion systems, stability and control, materials, internal systems, structures, and manufacturing were converging into the modern airplane. *U.S. Air Services* asserted that the variable-pitch propeller was a crucial component of that momentum because it improved performance at takeoff and cruise.[76] Reginald M. Cleveland noted in his *New York Times* aviation column that "the development of the controllable angle propeller has always been desirable, but with the arrival of modern, high-speed aircraft it became absolutely necessary. Quick takeoff and single-engine flight by loaded twin-engined [*sic*] planes depend on them."[77]

The introduction of the variable-pitch propeller on the Boeing 247, the Douglas DC-series transports, the Sikorsky S-42, and the Martin B-10 paved the way for the expansion of commercial and military aviation and aircraft performance in the United States. Hamilton Standard was eager to make the most of its opportunities. Still facing paltry government contracts, the company sought licensing opportunities in foreign commercial markets that benefited the overall engineering development of the two-position, controllable-pitch propeller. Due to the fundamental nature of the variable-pitch propeller, the Army, Navy, and the State Department allowed Hamilton Standard to pursue those commercial avenues overseas.[78] Raycroft Walsh informed the Army that BuAer concluded that "in view of the number of controllable-pitch propellers under development

[75] "The Turnbull V.P. Propeller," *Flight* 24 (May 13, 1932): 419–420; "Curtiss Controllable-Pitch Propeller," *U.S. Air Services* 18 (November 1933): n.p.; CAMC, "Curtiss Controllable-Pitch Propeller," n.d. [1934], File B5-260080-01, PTF, NASM.

[76] "Notes of Men and Machines," *U.S. Air Services* 18 (July 1933): 36.

[77] Reginald M. Cleveland, "Contact," *New York Times*, June 3, 1934, p. 8.

[78] R. Walsh to BuAer, July 12, 1934; E. J. King to COAC, July 21, 1934; H. C. Pratt to COAC, "Woodward Type Control for Controllable-Pitch Propellers (Hamilton Standard)," August 9, 1934, Folder "Propellers–July through December, 1934," Box 5977, RD 3247, RG 342, NARA.

and their probable wide distribution, it appears impracticable to attempt to keep these developments secret or confidential."[79]

There was no immediate threat to national security regarding the sale of Hamilton Standard variable-pitch propellers to foreign interests. Both the military and Hamilton Standard agreed that the "time lag" between the signing of a licensing contract with a foreign manufacturer and the actual start of production ensured that no foreign power secured "any real and practical benefit from designs released to them." License agreements signed in 1934 did not result in license-built propellers until 1936, which, in an environment of unprecedented innovation, still gave the United States a two-to-three-year lead in innovation.[80]

As a result, Hamilton Standard sold the foreign rights to manufacture the propeller in Great Britain to the de Havilland Aircraft Company, to the Société Française Hispano-Suiza in France, to the Junkers-Flugzeugwerk A.G. in Germany, and to Sumitomo Metal Industries in Japan during the 1934–1935 period. Besides Great Britain, each of these countries had their own variable-pitch propeller development programs, which were in various stages of development. In terms of practicality, however, these enterprising companies found it advantageous to simply license a proven and available design. By December 1934, Hamilton Standard had manufactured 1,000 controllable counterweight propellers.[81]

The dramatic improvement in performance offered by the variable-pitch propeller won Caldwell and Hamilton Standard the 1933 Collier Trophy (Figure 18). President Franklin D. Roosevelt bestowed the honor on behalf of the NAA during a special ceremony at the White House on May 29, 1934.[82] Roosevelt's reading of the award citation highlighted the importance of their achievement:

> The success of [Caldwell's] propeller has revealed a new horizon of aeronautics and taken the limits off speed. Henceforth, our pace through the air will be as fast as the daring and imagination of the engineers. Without it we could never have taken advantage of the improvement in design and in motors which is behind the

[79] R. Walsh to Materiel Division, November 7, 1935, Folder "Propellers, 1935," Box 6035, RD 3264, RG 342, NARA.

[80] Ibid.

[81] "1,000th Controllable," *The Bee-Hive* (December 1934): 4.

[82] "Caldwell Wins Aviation Trophy," *New York Times*, May 29, 1934, p. 7; Carl B. Allen, "Hamilton Standard Wins Collier Trophy for Controllable-Pitch Propeller," *The Bee-Hive* 8 (June 1934): 1; "Collier Trophy Awarded," *New York Times*, May 30, 1934, p. 3; Bill Robie, *For the Greatest Achievement: A History of the Aero Club of America and the National Aeronautic Association* (Washington, DC: Smithsonian Institution Press, 1993), 230; Rosen, *Thrusting Forward*, 42.

FIGURE 18 Triumph of the propeller specialists: Hamilton Standard employees (left to right) Arvid Nelson, Erle Martin, Frank W. Caldwell, Raycroft Walsh, and sales manager Sidney A. Stewart with the Collier Trophy.
Courtesy of the Smithsonian Institution Libraries, Washington, DC. Reproduced by permission of United Technologies Corporation, Pratt & Whitney.

new era of high speeds, for with the old fixed-pitch propeller none of the new swift airliners could take off with safety from any but the largest airports. Now Caldwell's propeller shrinks the distance needed for the takeoff, and makes every field available for the modern airplane. It gives quicker takeoff and more rapid climb. It makes transatlantic operation possible on a profitable basis.[83]

The new innovation and its ability to shift gears in the air enhanced the new modern aircraft and engines that appeared during the rapidly emerging Aeronautical Revolution.

Soon, the "gearshift of the air" became another example of America's progress through technological achievement. Commentators quickly capitalized upon the automotive analogy in weekly columns and journal articles.[84] The "gearshift" analogy was a widespread and ingenious way to explain the value of the variable-pitch propeller to the public. Most

[83] Franklin D. Roosevelt, quoted in "Shifting Gears in the Air," *Technology Review* 36 (July 1933–1934): 354.

[84] Contemporary articles that used the "gearshift" analogy included: "Shifting Gears in the Air," *Technology Review*; Cleveland, "Contact," *New York Times*, June 3, 1934; "Plane 'Gear Shift' Wins Reed Award," *New York Times*, January 31, 1936, p. 14; "Gear Shift Aids Flight," *New York Times*, August 23, 1936, p. 13.

Americans, primarily young males, tinkered with automobiles and understood the concept and they could easily explain it to those who did not.

In the United States, the disparate efforts of industry, the airlines, and the military resulted in the integration of the variable-pitch propeller into the technical system of the airplane. It was not a deliberate achievement based on a singular belief that innovation should be embraced, but one based on the complementary ideologies of safety, engineering logic, and the quest for economic survival in a competitive environment – not to mention the need for greater altitude to fly over mountains rather than through them. At the core of this synergistic impulse was Frank Caldwell, who had conceived the nucleus of a practical hydraulic variable-pitch system. The American aviation industry learned through practical application the value of the new technology.

International military and commercial competition enabled the American aeronautical community to show the world its superior technology. Industrialist and philanthropist Sir MacPherson Robertson of Melbourne, Australia, announced his intention to sponsor an international air race from Mildenhall, England, to Melbourne to encourage the development of commercial aviation. To entice competition, he offered a £10,000 prize and a gold loving cup. Advertising on both three continents touted it as the "world's greatest air race." Slated to begin on October 20, 1934, the MacRobertson Race, which used the same combined name as Robertson's confectionary company, was actually two contests: one for all-out speed and another that placed a premium on cargo-carrying capability. The Royal Aero Club officiated while the International Commission on Aerial Navigation set the entry requirements. The main prerequisite was that every entry had to climb to a height of at least sixty feet at a point no more than 660 yards from the start of their takeoff run while fully loaded.[85]

The takeoff requirements for the MacRobertson Race and the desire to increase Britain's aeronautical prestige in the world forced the de Havilland Aircraft Company to incorporate variable-pitch propellers into the design of its entry, the DH.88 Comet. De Havilland engineer Lee Murray stated that had it not been for those requirements, their design would have employed fixed-pitch propellers.[86] Despite the efforts of Roy Fedden and Pop Milner at Bristol, a variable-pitch propeller designed

[85] "Air Race Will Be World's Greatest," *New York Times*, October 19, 1934, p. 2; Gwynn-Jones, *Farther and Faster*, 251–259.

[86] Lee Murray in Reginald M. Cleveland, "Contact," *New York Times*, November 11, 1934, p. 6.

in Britain was not available for incorporation into the DH.88 design. Desperate for the new technology, de Havilland's general manager, Frank T. Hearle, and chief designer, Arthur E. Hagg, visited Hamilton Standard in the United States. They acquired the license to manufacture and sell in Great Britain the two-position, controllable-pitch propeller and all subsequent designs in June 1934 for £18,000. De Havilland was alone in what appeared to be a risky proposition. Hearle and Hagg were unable to interest any other British aeronautical companies in a joint venture or receive a guarantee from the Air Ministry for orders.[87]

Two Hamilton Standard propellers arrived at Hatfield a few weeks later. A DH.88 flew with them installed on September 8, 1934, just three weeks before the start of the MacRobertson Race. The de Havilland test pilots reported that the Comet's two Gipsy Six R engines immediately overheated. Investigation revealed that the thick propeller blade roots, designed according to American practice for use with radial engines, did not adequately cool the British inline engines. After having invested so much to get the propellers, de Havilland was not able to redesign the blades in time for use in the race to Australia.[88]

The other propeller available for use by de Havilland was the Ratier design from France, which was a popular choice for small high-speed aircraft. The company's two-position pneumatically controlled propeller was essentially a "one-way" mechanism. Air pressure kept the blade at low pitch for takeoff. As the airplane climbed to cruising altitude air speed increased drag on a disc mounted at the front of the hub. The increasing air pressure pushed the disc rearward, which in turn released the air pressure in the cylinder that moved blade pitch to cruise. This operation could only change pitch once for each flight so landing did not benefit from low pitch. Before the next flight, a mechanic had to use a bicycle pump to recharge the cylinder.[89] The Ratier propeller was a desperate, but necessary, expediency for de Havilland.

The Comet was a successful airplane. Charles W. A. Scott and Tom Campbell Black in the scarlet DH.88 *Grosvenor House* flew the course in approximately seventy-one hours at 176 mph for 11,300 miles, winning the speed contest portion of the MacRobertson Race.[90] The

[87] C. Martin Sharp, *D. H.: A History of de Havilland* (Shrewsbury, England: Airlife, 1982), 138–139, 152–153, 155.

[88] "Enterprise in Airscrews: First Details of a Mighty New de Havilland Airscrew and the Story of 21 Years of Achievement," *Flight* 69 (March 2, 1956): 243.

[89] C. M. Poulsen, "Controllable-Pitch Airscrews, Part II," *Flight* 27 (May 9, 1935): 500; Sharp, *D.H.*, 142–143; Rosen, *Thrusting Forward*, 44.

[90] Grey, *A History of the Air Ministry*, 243.

incorporation of the Ratier propeller into the Comet's design represented the performance-specific design logic characteristically used by the British in approaching a design challenge. The second and third place winners, Dutch KLM's DC-2 *Uiver* and the Boeing 247-D *Warner Brothers Comet* flown by Roscoe Turner and Clyde Pangborn, reflected the newer performance-general design logic so heavily influenced by recent innovations of the Aeronautical Revolution. The differing approaches to solving an aeronautical design challenge were not lost on the international aviation press, which emphasized the importance of the American aircraft designs.

British air transport and aircraft technology came under scrutiny as a result of the MacRobertson Race. The editor of London's *Saturday Review* commented that "Britain has won the greatest air race in history; but she has yet to start on an even greater air race: a race in commercial and military supremacy."[91] As for KLM, the Dutch airline bought Hamilton Standard two-position, controllable-counterweight propellers for all its aircraft. One propeller, bought in September 1933, was in operation up to 2,000 hours on a Fokker F.XVIII airliner on the Amsterdam to Batavia, Dutch East Indies route two years later. By May 1935, there were 102 controllable propellers in operation that had flown over 2.5 million miles.[92] American aeronautical technology, and its growing influence on the conduct of international air commerce, was celebrated as a triumph of modern technological progress.

Practicality, Performance, and a Larger Role for the Specialists

The new technologies of the Aeronautical Revolution merged into the modern airplane during the 1930s. Still, the level of resistance to the introduction of the variable-pitch propeller in the United States illustrates the complexity and ambiguity inherent in technological development. Even though the new device offered unprecedented performance as expressed through professional presentations and publications, engineering groups who possessed conflicting ideologies about how to design an airplane had to be convinced of its value through practical application. If the idea of "Progress" was the overriding ideology driving aviation in

[91] Terry Gwynn-Jones, "Farther: The Quest for Distance," in *Milestones of Aviation*, ed. John T. Greenwood (New York: Hugh Lauter Levin Associates, 1995), 67.

[92] "KLM Uses Hamilton Standard Controllable Propellers," *Aviation* 34 (July 1935): n.p.; Brooks, *The Modern Airliner*, 91.

the 1920s and 1930s, there would have been no question whether or not to install variable-pitch propellers. The aeronautical community was not committed to the automatic and blind adoption of new innovations.

The most profound argument for variable-pitch, regardless of the enthusiasm generated by its proponents, would be the performance it generated when integrated into a new design. The 1934 MacRobertson Race symbolized that reality. American aircraft engineers learned quickly that all of the innovations of the Aeronautical Revolution, including variable-pitch propellers, had to act in synergy, which accommodated performance-general design logic. That the propeller community had to make the most of the technical opportunities to exploit its invention illustrates the hierarchical place of the propeller community in the larger aviation community.

The interplay between the individuals and groups responsible for aircraft design and the propeller community advocating innovation shaped the technological system of the modern airplane. The traditional ideologies of weight, reliability, cost, and "practical engineering" were components of a larger tension based on social status within the aeronautical community. The ultimate questions of who controlled the design of an aircraft and what technologies were to be appropriated for the new design were at the foundation of the resistance to the variable-pitch propeller. Once that opposition was overcome, the new device was celebrated as yet another example of humankind's technological prowess and considered a standard component of the modern, and distinctly American, high-performance airplane of the 1930s. What was left for the international aeronautical community was to move on to a new stage of technical refinement and use as nations used the airplane to meet their visions of the future.

9

Constant-Speed

Millionaire aviator Howard R. Hughes and four crew members took off from Floyd Bennett Field, New York, in a Lockheed 14 Super Electra twin-engine transport on July 10, 1938. Three days, nineteen hours, and fourteen minutes later, the silver monoplane, named the *New York World's Fair 1939*, had covered 14,824 miles in a round-the-world flight that symbolized the imminent arrival of the futuristic "World of Tomorrow" that the exposition celebrated. The NAA awarded Hughes and his crew the 1938 Collier Trophy for their well-executed long-range flight that highlighted innovations in navigation, communications, and engineering and illustrated overall the superiority of American aeronautical technology. Hughes "praised highly" the two Hamilton Standard Hydromatic constant-speed propellers, which had only become available just a few months before the flight.[1] Hughes's flight marked the arrival of a modern and refined airplane, propeller and all, in its most complete form.

Spectacular intercontinental flights such as Lindbergh's solo Atlantic crossing in 1927, the *Uiver* in the MacRobertson Race in 1934, and Hughes's around-the-world flight were indications that the aeronautical community expanded its conception of the synergistic system of the commercial and military airplane. Such spectacular uses of the airplane

[1] "Hamilton Standard Propellers Help Hughes Around World," *The Bee-Hive* 13 (July–August 1938): 1; "Congratulations, Mr. Hughes!," *Collier's* 121 (May 14, 1939): 78; "Howard Hughes' Lockheed 14 Monoplane Used for the Record-Breaking World Flight," *Aero Digest* 33 (August 1938): 47, 68; "Hughes Does It With the New Hydromatic Propellers!," *Aero Digest* 33 (August 1938): 77; Terry Gwynn-Jones, *Farther and Faster: Aviation's Adventuring Years, 1909–1939* (Washington, DC: Smithsonian Institution Press, 1991), 270–273.

led many in the United States to believe that a new era in history, an Air Age of peace, unlimited progress, and opportunity for humankind, was just around the corner.[2] The world's other aeronautical nations demonstrated the promise of aviation in different ways during the late 1930s. For Nazi Germany, the airplane was a potent technological symbol of fascism. Great Britain used commercial and military airplanes to connect its far-flung global empire.

The constant-speed propeller was a central technology in these modern airplanes. It was the ultimate form of a variable-pitch mechanism because it changed blade pitch automatically according to varying flight conditions while the engine speed remained the same, which maximized propeller, engine, and fuel economy and offered hands-off operation. Proposed first by Hele-Shaw and Beacham in 1924, not a single one had flown on an operational airplane in Europe or North America in the decade that followed. National aeronautical communities in the United States, Nazi Germany, and Great Britain approached this next step in different ways in the 1930s.

The United States: Refinement and Continued Growth

In the United States, Hamilton Standard and its corporate partners in the UATC were at the height of success in 1934. Excitement for the MacRobertson Race began to gain momentum across the United States and Europe during the summer. New Boeing 247-D transports with their Hamilton Standard propellers and Pratt & Whitney Wasp engines and operated exclusively by United Air Lines stood poised to dominate the domestic airline market flying over America at three miles a minute. Fearful of an aerial octopus reminiscent of the Gilded Age, Congress declared the UATC and other combines like it an illegal monopoly. The Air Mail Act of 1934 separated the corporation's manufacturing and transport operations into the United Aircraft Corporation (UAC) and UAL.

Nevertheless, business was still booming for the UAC. After his ouster from Hamilton Standard and a brief tenure as the president of United Airports of California in Burbank, Tom Hamilton moved to more exciting, and profitable, pursuits. He became the corporation's European sales representative for United Aircraft Exports Corporation in Paris. Hamilton negotiated the licensing rights of highly successful products

[2] Joseph J. Corn, *The Winged Gospel: America's Romance with Aviation, 1900–1950* (New York: Oxford University Press, 1983), 135.

such as the two-position controllable counterweight propeller to Junkers and the Pratt & Whitney Hornet D series engine to the Bavarian Motor Works (BMW), which led to a stronger relationship between the UAC and the mobilizing Nazi regime.[3] An estimated 78 percent of aviation orders for the rapidly rearming Nazi Germany in 1934 alone went to UAC with Hamilton Standard's products representing a significant percentage.[4]

Hamilton Standard, under Caldwell's leadership, continued to develop and refine the hydraulic variable-pitch propeller into a constant-speed propeller that changed blade pitch automatically through the use of a speed regulating governor. Caldwell originally intended his first propeller from 1930 to be constant-speed. The immediate demand for a variable-pitch propeller influenced his first developing the two-position controllable model. The success of making it over the initial hurdle of creating the controllable-counterweight mechanism led to the development of a dedicated constant-speed propeller starting in 1932. Caldwell quickly focused on designing a control responsive enough to frequent changes in engine speed, which was the key to making the new unit practical.[5]

Experiencing difficulty, Caldwell solicited the assistance of the Woodward Governor Company of Rockford, Illinois, a successful manufacturer of governing mechanisms for American industry since the 1870s. The company's founder, Amos W. Woodward, invented the first mechanical governor to control the speed of waterwheels, the source of power for factories and mills, in 1869. His son, Elmer E. Woodward, pioneered mechanical and hydraulic compensating governors in 1899 and 1917, which facilitated the widespread introduction of hydroelectric power across the United States and Canada. The advent of diesel engines for

[3] Ronald Fernandez, *Excess Profits: The Rise of United Technologies* (Reading, MA: Addison-Wesley Publishing, 1983), 66–67, 81–82. After the fall of France, Hamilton and his family returned to the United States via Spain. He worked as a "dollar-a-year man" after American entry into World War II overseeing the construction of Boeing B-17 engine nacelles. Hamilton pursued ventures related to resorts and land development in the western United States and Canada. He died in August 1969 in Los Angeles. "Thomas F. Hamilton: Early Plane Builder – Aviator–Propeller Manufacturer," from the *Flying Pioneers Biographies of Harold E. Morehouse*, n.d., MFRC; "Rites Set for T. F. Hamilton, Flying Pioneer," *Los Angeles Times*, August 13, 1969, sec. C5, p. 1.

[4] Raycroft Walsh, "Hamilton Standard Grants Licenses for Controllables in Three Countries," *The Bee-Hive* 8 (September 1934): 8; Eugene E. Wilson, *Slipstream: The Autobiography of an Air Craftsman* (Palm Beach, FL: Literary Investment Guild, 1967), 170; Ronald Miller and David Sawers, *The Technical Development of Modern Aviation* (New York: Praeger, 1970), 75; Fernandez, *Excess Profits*, 82–83.

[5] "The Hamilton Standard Constant-Speed Propeller," *The Bee-Hive* 10 (September 1936): 1–3; Miller and Sawers, *Technical Development*, 74–75.

electrical power generation in the early 1930s led Woodward to develop a hydraulic governor that became a mainstay for utility, industrial, and agricultural uses in 1933. Woodward and his company possessed an experience firmly rooted in the American industrial revolution of the late nineteenth and early twentieth century combined with the flexibility to successfully navigate new trends.

Woodward assigned the project to his younger engineers because of their enthusiasm for the airplane and its cultural position as the most exciting technology of the time. They failed to produce a suitable design after several weeks and, according to company lore, "the 'Old Man' quietly rolled up his sleeves and took over the job." The seventy-three-year-old Woodward dealt with Frank Caldwell, Erle Martin, and their engineering staff directly for two years. Unlike his other innovations that were the size of small houses, his hydraulic "baby" aircraft propeller governor design weighed two pounds and was about the size of a baseball. The first flight test occurred during the fall of 1934 on a Boeing 247-D. Woodward's company placed the first governor model, the PW-34, into production in late 1935.[6]

The foundation of the Hamilton Standard design was the commonality found in the early two-position model and its adaptability to constant-speed operation. The addition of the PW-34, a small engine driveshaft, and hydraulic pipes converted the proven two-position controllable-counterweight propeller into an automatic gearshift. The governor eliminated manual control by metering the flow of engine oil to-and-from the propeller automatically to adjust blade pitch. The governor kept the engine's rpm constant during climbing, diving, and level flight. Rather than using the throttle to change engine rpm, the pilot used the propeller pitch control. The ability to separate the control of engine speed from the engine throttle setting generated maximum engine efficiency under all flight conditions regardless of the airplane's altitude or forward speed. For pursuit and fighter aircraft, this arrangement enabled pilots to make violent maneuvers in combat without having to bother with adjusting both the propeller control and the engine throttle.[7] Reflecting the excitement generated by Wiley Post's use of an autopilot, named "Mechanical

[6] Martha Strolberg, *A Gentlemen Named Elmer* (Rockford, IL: Wilson-Hall, 1974), File CW-889000–01, BF, NASM, 10, 13, 16, 19, 20, 24. The first PW-34 survives in the collection of the Smithsonian National Air and Space Museum.

[7] "The Hamilton Standard Constant-Speed Propeller," 1–3; "How the Hamilton Standard Constant-Speed Propeller Works," *U.S. Air Services* 21 (September 1936): 19.

Mike," in the mid-1930s, American aviation journalists liked to call the new and futuristic constant-speed propeller a "robot" gearshift.[8]

The availability of the constant-speed, counterweight propeller in the mid-1930s facilitated the expansion of the American military and naval air arms. Believing strongly in the potential of the airplane in winning future wars, the US Army Air Corps' GHQ Air Force concentrated on developing the tactics and technology required to wage strategic aerial warfare. Those two areas came together in the concept of daylight, high-altitude precision bombing. The combination of a fast, high flying, multiengine delivery system that embodied the latest innovations of the Aeronautical Revolution, such as the four-engine Boeing B-17 Flying Fortress, with a new military weapon, the Norden bombsight, promised to destroy only strategic targets while at the same time easily out flying any enemy aircraft attempting to intercept them. The GHQ Air Force aimed to be a deterrent to large-scale conflict by attacking an enemy's ability to wage war through the destruction of its industrial production and infrastructure. Hamilton Standard received orders for over 400 constant-speed propellers for new high-performance military aircraft. Military planners specified and manufacturers designed aircraft such as the B-17 and the Navy's long-range twin-engine Consolidated PBY Catalina flying boat with the constant-speed propeller as an integral part of the new designs.[9]

The increased performance offered by the constant-speed propeller for military aircraft potentially derailed the sales efforts of Hamilton Standard in the commercial marketplace. Both the Army Air Corps and BuAer had the right to prevent the commercial sale and use of new innovations for one year if they were considered to be of sufficient military value. Frank Caldwell conducted a letter-writing campaign to the major airlines and aircraft and engine manufacturers soliciting support to have the constant-speed control released for commercial use. He argued that "the necessity for this device on modern high performance commercial air transport airplanes is even greater than in the case of military airplanes." The bottom line for these companies was that a converted propeller weighed only three to four pounds more than the two-position version, offered a "material improvement" in efficiency, and it was safe.[10] In the end, the government could not refuse Hamilton Standard the

8 "Robot Gearshift Stands High Among Aviation Refinements," *Christian Science Monitor*, January 7, 1936, p. 9.

9 "Over 400 Constant-Speed Propellers," *U.S. Air Services* 21 (July 1936): n.p.

10 F. W. Caldwell to Douglas Aircraft Company, September 7, 1934, File M015127, DC-2 Files, BCA.

opportunity to create revenue with a device it had developed by itself. Aircraft operators, primarily the airlines, bought the constant-speed governor and a modification kit to convert their two-position propellers or bought new ones outright.[11]

As military and commercial operators gobbled up production orders of the new Hamilton Standard propellers, Caldwell paused from his work at Hartford to travel to New York City to attend the Annual Dinner of the IAS at the Faculty Club of Columbia University on January 30, 1936. The forty-seven Fellows of the institute awarded Caldwell the Sylvanus Albert Reed Award for 1935. Reed had just passed away the previous October. Institute President Glenn L. Martin recognized Caldwell's success in "increasing the effectiveness of aircraft through development and improvement of controllable and constant-speed propellers." Caldwell noted in his remarks to the audience that he appreciated the award and its connection to Reed and interpreted it as further incentive to work toward increasing the speed and performance of aircraft.[12] Despite the accolade, he must have felt a tinge of irony during the ceremony since his difficult relationship with Reed was well into its second decade.

Hamilton Standard's competitor in the aeronautical marketplace, Curtiss-Wright, paralleled the incremental design development. Curtiss followed with an engine-driven constant-speed governor for use on aircraft with supercharged engines in early 1936. Unlike the mechanism developed in East Hartford, the Curtiss design was both controllable and constant-speed. The pilot had the choice of either manually controlling the propeller or setting it on automatic.[13]

Curtiss found a market niche with military pursuit, attack, and transport aircraft. The first aircraft to use the Curtiss Electric propeller was

[11] Russell Trotman, "Turning Thirty Years," *The Bee-Hive* 24 (January 1949): 26.

[12] Even after Reed's death, Caldwell devoted a considerable amount of time consulting and being deposed as an expert witness for the defense in Reed's final patent infringement suit against the United States that concluded in 1942. "Caldwell Will Get Aeronautics Award," *New York Times*, January 20, 1936, p. 21; "Plane 'Gear Shift' Wins Reed Award," *New York Times*, January 31, 1936, p. 14; "Frank Caldwell Gets Reed Prize in Aeronautics," *Washington Post*, February 9, 1936, p. 7; *Technology Review* 42 (March 1935–1936): 1; "Fourth Annual Dinner," *Journal of the Aeronautical Sciences* 4 (February 1936): 114–115.

[13] P. H. Schneck to Chief, Materiel Division, March 16, 1936, Folder "Curtiss Electric Propeller, 1934–1938," Box 6302, RD 3330, RG 342, NARA; "Curtiss Propellers," n.d. [1937], File B5-260000-01, PTF, NASM; "Curtiss Constant-Speed Propeller," *Aero Digest* 31 (December 1937): 60; Propeller Division, Curtiss-Wright Corporation, "Curtiss Propellers: Thirty Years of Development," September 4, 1945, File B5-260000-01, PTF, NASM.

the Consolidated PB-2A two-seat pursuit airplane. In 1935, the Air Corps considered it the ultimate high-altitude fighter with two forward- and one rearward-firing .30-caliber machine guns and a 700-horsepower turbosupercharged inline Curtiss Conqueror V-1570 engine. Curtiss products included the P-36A and P-37 pursuit aircraft and the twin-engine Y1A-18 attack airplane. The Douglas C-33, C-38, and C-39 transport aircraft, based on the revolutionary DC-series airliners, all used Curtiss Electric propellers.[14]

Constant-speed operation increased the efficiency of operational aircraft whether they were single-engine fighters or multiengine airliners and transports. Automatic pitch variation, which meant that there was a wider range of blade angles, also facilitated an important safety feature for multiengine airliner, bomber, and transport aircraft called blade feathering. In the event of engine failure during a flight, the slipstream flowing through the dead engine's propeller made it "windmill" or "run away." A windmilling propeller produced an enormous amount of drag and severely limited the speed and altitude performance of a stricken airplane, rendering it virtually uncontrollable. Severe windmilling resulted in the disintegration of the airplane from structural failure through extreme vibration. Both conditions threatened crew and passenger safety when the first major generation of Americans started to travel by air. From the financial standpoint, even if the airplane made it to the ground safely, the engine damage would be beyond repair.

Feathering set the blades to a position parallel to the slipstream. The analogy for this new type of propeller came from long established sport of rowing. To "feather the oar" means to turn the blade at the end horizontal as it leaves the water to keep it from hitting the water as the rower moves the oar forward to begin the next stroke. A feathered oar and a feathered propeller accomplish the same thing: both lessen wind resistance.[15] Furthermore, a feathered propeller does not disrupt the flow of air over the wing after engine failure. This improves the airplane's lift/drag coefficient, which makes it more controllable by the pilot even if the engine failure occurred during landing and takeoff. Tests conducted by American Airlines with a Douglas DC-3 powered by a single-engine

[14] F. O. Carroll to N. Heath McDowell, February 14, 1940; F. S. Borum, to CO, Fairfield Air Depot, "Propeller Installation–Douglas C-33 Airplane," May 11, 1940, Folder "Curtiss Propeller Division (Electric Propeller), 1940," Box 6588, RD 3403, RG 342, NARA.

[15] R. C. Lehmann, *Rowing* (New York: Edward Arnold, 1897), 14, 20, 21, 23–24; W. P. Keasbey, "Did You Ever Wonder What A 'Full-Feathering' Airplane Propeller Is?," *Christian Science Monitor*, September 24, 1941, p. 23.

revealed that a feathered propeller increased the altitude of the airliner by 300 feet.[16] In the Rockies and other high mountain ranges of the world, that was an important safety feature.

Curtiss was the first to introduce blade feathering on its electric controllable-pitch propeller. Due to its design, the mechanism already permitted a blade angle of 90 degrees parallel to the airstream. The company promised that the ability to feather the blades offered superior takeoff and climb characteristics, speed and maneuverability at cruising altitudes, and additional power during one-engine operation. Those qualities were vital to the safe operation of multiengine transports such as Curtiss-Wright's new T-32 Condor biplane airliner that began nighttime operations with American Airlines and Eastern Air Transport in 1933.[17] Curtiss won an important Army Air Corps contract for 240 electric feathering propellers at $2,134.42 each during the fall of 1937.[18] What amounted to a mere by-product of Turnbull's original design, the feathering capability gave Curtiss an advantage over competing propeller designs in the early 1930s.

Hamilton Standard was second to Curtiss in introducing the feathering propeller. The first Hamilton Standard constant-speed propeller still employed counterweights to vary pitch in one direction. Caldwell, promoted to engineering manager, and a team led by new chief engineer, Erle Martin, developed a new propeller that relied on hydraulic pressure for all pitch actuation and, as a result, utilized engine power even more efficiently. Introduced in early 1938, the Hydromatic propeller – the name was a combination of "hydraulic" and "automatic" – employed major improvements over the earlier variable-pitch designs. The propeller's "quick-acting" cams provided more responsive control of pitch variation, facilitated multiengine synchronization, and removed the risk of "over-speeding" the engine while diving. The Hydromatic was easier to maintain and operate while in service because its operating mechanism was sealed and operated under constant engine oil pressure (Figure 19).

[16] George Rosen, *Thrusting Forward: A History of the Propeller* (Windsor Locks, CT: United Technologies Corporation, 1984), 46, 49; O. P. Echols to Curtiss Aeroplane Division, February 12, 1938, Folder "Curtiss Electric Propeller, 1934–1938," Box 6302, RD 3330, RG 342, NARA.

[17] "Curtiss Controllable-Pitch Propeller," *U.S. Air Services* (November 1933): n.p.; Propeller Division, Curtiss-Wright Corporation, "Curtiss Propellers: Thirty Years of Development," PTF; K. M. Molson, "Some Historical Notes on the Development of the Variable Pitch Propeller," *Canadian Aeronautics and Space Journal* 11 (June 1965): 182.

[18] "Transport: Full Feathering," *Time* 30 (September 6, 1937), www.time.com/time/magazine/article/0,9171,758152,00.html (Accessed February 3, 2010).

FIGURE 19 Cutaway view of the fully hydraulic Hamilton Standard Hydromatic propeller. Note the governor mounted at the top of the nose casing. United Technologies Corporation Archive via Smithsonian National Air and Space Museum (NASM 9A07407).

To distinguish their design from Curtiss, described the Hydromatic as a "quick-feathering" propeller.[19]

Competition between Hamilton Standard and Curtiss stimulated the introduction of blade feathering, but it was not a new idea. Hele-Shaw and Beacham recognized the need for feathering in their original patents dating back to 1924.[20] The combination of constant-speed control and feathering was a growing ideal in the British aeronautical community that appeared first in the United States.

Despite being second to Curtiss, Hamilton Standard made the Hydromatic the industry standard for high-performance aircraft, primarily the highly successful "modern" airliner, the Douglas DC-3. C. R. Smith, president of American Airlines, recognized that the adoption of the DC-3 by American air carriers heralded a new chapter in the history of

[19] Frank W. Caldwell, "Hamilton Standard Hydromatic Propeller," *Aviation* 37 (July 1938): 28; and "A Review of the Hydromatic Propeller," *The Bee-Hive* 14 (July 1939): 3–5; "Hydromatic Aircraft Propeller," *Automotive Industries* (July 23, 1938): 114.

[20] H. S. Hele-Shaw and T .E. Beacham, "Feathering Screw Propellers," UK Patent No. 250,292, April 6, 1926; H. S. Hele-Shaw and T. E. Beacham, "Feathering Screw Propeller," US Patent No. 1,723,617, August 6, 1929.

commercial aviation. The twin-engine monoplane was the first airliner to make money carrying passengers, which released the airlines from their dependence on government subsidy through airmail contracts. Passenger miles on the airlines increased 600 percent from 1936 to 1941.[21]

The first large order for Hydromatics came from United Air Lines in December 1937. Rudy Schroeder, vice-president in charge of operations, was optimistic that the twenty-eight propellers would increase the performance of the airline's DC-3 airliners.[22] American Airlines installed a small number in March 1938 and ordered seventy more for its transcontinental DC-3s at a cost of $200,000 the following April.[23] A complete DC-3 cost $125,000 each.[24] The approximately $5,600 for two Hydromatics was a small price to pay for both performance and safety.

The American aeronautical community tacitly endorsed Hamilton Standard designs as superior, especially the Hydromatic propeller, but that did not mean they were the "best" or the most innovative. MIT professor of aeronautical engineering, Jerome C. Hunsaker, conducted an independent survey of Hamilton Standard's research program and its interrelationship with the marketplace at the request of George Mead of the UAC Technical Advisory Committee. Both engineers certainly knew each other. They had been longtime friends since attending MIT as students twenty-five years earlier. Hunsaker's July 1937 report reflected the latest trends in university-based aeronautical research and development. He offered suggestions that he felt would ensure Hamilton Standard's long-term success. For the Hydromatic, he suggested the addition of internal gears to achieve a greater range of pitch actuation and to simplify the feathering capability. In general, Hamilton Standard needed to investigate four-blade configurations, blade materials other than forged duralumin, compressibility at high tip speeds, and new hub and blade designs derived from wind tunnel testing.[25]

[21] C. R. Smith, quoted in Robert J. Serling, *Eagle: The Story of American Airlines* (New York: St. Martins, 1985), 110; US Department of Commerce Statistics, 1932–1941, referenced in Roger D. Launius and Janet R. Daly Bednarek, eds., *Reconsidering a Century of Flight* (Chapel Hill: University of North Carolina Press, 2003), 5.

[22] James V. Piersol, "Air Currents," *New York Times*, December 12, 1937, p. 217.

[23] "New Propeller in Use," *New York Times*, March 20, 1938, p. 166; "Airline Buys New Propellers," *New York Times*, April 1, 1938, p. 36; James V. Piersol, "Air Currents," *New York Times*, April 10, 1938, p. 164.

[24] "Four Planes for Braniff," *New York Times*, September 10, 1939, p. X15.

[25] J. C. Hunsaker, "Report on Propeller Developments," July 7, 1937, Folder 8, Box 26, Jerome C. Hunsaker Papers (hereafter cited as JCH), NASM.

Caldwell's response reflected his almost three decades of experience as a propeller specialist. While the Hydromatic propeller did not provide as full a range of pitch actuation as the Curtiss Electric propeller, it was more rugged and dependable in service. Caldwell was more concerned with mechanical and structural stability rather than highly refined features such as complicated contrarotating propellers and new and untested blade materials such as hollow-steel.[26] His awareness of Hamilton Standard's place in the technological landscape, which focused on applied practicality, illustrated the differences he saw in theoretically oriented university research and the ideas of other engineers. Caldwell knew his ideas of what constituted an ideal solution differed greatly from the visions of others. The main difference was that his ideas worked and were in practice around the world.

Within the international aviation community, Caldwell served as the leading propeller specialist. In October 1937, Caldwell joined a contingent of aeronautical engineers from the United States, Great Britain, Germany, and Italy, including Jerome Hunsaker, to address the newly formed Lilienthal-Gesellschaft für Luftfahrtforschung (Lilienthal Association for Aeronautical Research) in Munich. The association was part of the new aeronautical order sponsored by Nazi Germany and under the direct leadership of Luftwaffe chief, Hermann Göring. Adolf Baeumker, chief of research for the Reichluftfahrtministerium (Reich Aviation Ministry), and leading German aerodynamics pioneer, Ludwig Prandtl, were also driving forces. The high profile meeting included many well-known international aviation professionals, as among who were Charles A. Lindbergh and Il Duce's pilot son, Bruno Mussolini.[27]

It was not long until the first public demonstration of the full-feathering capability of the Hydromatic propeller took place. The assistant chief pilot of United Air Lines, George Grogan, took off in the United Air Lines *Mainliner* DST from Newark airport on the afternoon of Wednesday, April 6, 1938. As the airliner neared the Central Park area of New York City, he shut down the left engine and feathered the windmilling propeller. With one propeller motionless and one spinning, he

[26] F. W. Caldwell to G. J. Mead, August 3, 1937, Folder 8, Box 26, JCH, NASM; William F. Trimble, *Jerome C. Hunsaker and the Rise of American Aeronautics* (Washington, DC: Smithsonian Institution Press, 2002), 157–158.

[27] "Lindbergh Expected at Munich Congress," *New York Times*, October 9, 1937, p. 4; Helmuth Trischler, "Self Mobilization or Resistance? Aeronautical Research and National Socialism," in *Science, Technology and National Socialism*, ed. Monika Renneberg and Mark Walker (Cambridge: Cambridge University Press, 1994): 76–77.

flew through the clouds, climbed, and turned without difficulty. The spectacle for New Yorkers taking an afternoon walk was seeing an airliner fly so effortlessly with one engine.[28]

The New York demonstration took place a day before United inaugurated its fifteen-hour coast-to-coast transcontinental service with DSTs and DC-3s on April 7. United officials boasted of the new airliner's 205 mph cruising speed, unheard of at the time. They also made much of the fact that the Hydromatic propellers enhanced the performance of the thirteen ton airliners during takeoff, climb, and cruise.[29]

While the drive to develop the Hydromatic propeller was rooted in supplying commercial aviation, forward-thinking aviators once again took the technology "off-the-shelf," combined it with innovations catering to their specific flights, and showed the world the promise of flight. Howard Hughes and his team chose for the Lockheed 14 Super Electra two 1,100 horsepower Wright Cyclone nine-cylinder radials, the most powerful engines available at the time, installed extra fuel tanks, and used the most advanced navigation and safety equipment available to make the approximately 15,000 mile flight around the Northern Hemisphere. They even filled an empty fuselage compartment with thousands of ping-pong balls for extra flotation in case they had to land in the ocean. The modifications made the all-metal monoplane a global record-breaker. The incorporation of Hamilton Standard's latest constant-speed propeller into the *New York World's Fair 1939* was not a special modification. The two Hydromatics were a standard purchase from the manufacturer.[30]

Continued refinement of the airplane and its component systems contributed to one of the fundamental pursuits in aeronautics: safety. The aeronautical community recognized the longstanding importance of the variable-pitch propeller to the safe operation of aircraft. Aviation safety expert Jerome Lederer, an aeronautical engineer who pioneered aviation insurance in the late 1930s, commented on the relationship between technology and safety. Besides improvements in structural design, avionics, instrumentation, special-purpose blind-flying equipment, engine synchronization systems, and tricycle landing gear, Lederer saw the full-feathering

[28] Piersol, "Air Currents," April 10, 1938, 164; Hamilton Standard Propellers, *Wherever Man Flies* (Hartford, CT: United Aircraft Corporation, 1946); American Society of Mechanical Engineers, *Hamilton Standard Hydromatic Propeller: International Historic Engineering Landmark*, Book No. HH 10 90 (November 8, 1990), 3.

[29] "Fast Air Liners Go Into Service," *Los Angeles Times*, April 8, 1938, p. A2.

[30] "Howard Hughes' Lockheed 14 Monoplane Used for the Record-Breaking World Flight," 46–47, 68.

variable-pitch propeller as a vital component of the modern, and safe, airplane by 1939.[31] The Civil Aeronautics Authority (CAA – the forerunner of the Federal Aviation Administration) went so far as to recommend the use of Hydromatic propellers "for all planes [*sic*] operating on United States airlines."[32] Another indicator of the revolutionary nature of these new devices was their reliability and safety. From a documented number of forty-one in-flight propeller failures in 1932, the rate dropped to twenty-three in 1938 and to zero in 1940.[33]

The cash-strapped Army Air Corps found it difficult to justify the increased cost of full-feathering props for new twin and multiengine aircraft. Before placing an order with Hamilton Standard or Curtiss, the Materiel Division at Wright Field made an inquiry to three major airline operators, TWA, American, and UAL, in September 1938. The Air Corps wanted to know if the increased safety of full-feathering propellers justified the considerable extra expense. They requested specific instances of where the technology prevented a forced landing or any other type of accident for that matter as well as operating conditions that only the new type of propeller could create. In terms of competing designs, the Air Corps invited a comparison of the Hamilton Standard and Curtiss propellers as a basis of evaluation for an eventual contract purchase.[34]

By the fall of 1939, the Hydromatic was clearly a "shelf item" available to the world's airlines, and Hamilton Standard was clearly the leading propeller manufacturer. Approximately 75 percent of all propellers, from ground-adjustable to fully feathering constant-speed, used by American and international commercial airlines and the military in the late 1930s left the factory at East Hartford. Hamilton Standard sales personnel believed they were on the verge of establishing a monopoly.

The new constant-speed technology offered the aeronautical community an "automatic" gearshift of the air before everyday people on the ground had them in their automobiles. GM announced the introduction

[31] Jerome Lederer, *Safety in the Operation of Air Transportation* (Northfield, VT: Norwich University, April 20, 1939): iii, 15–30; National Safety Council, "Jerome Lederer: 1989 Inductee into the Safety and Health International Hall of Fame," 1989, www.nsc.org/insidenscshhofi/bioled.htm (Accessed July 1, 2002).

[32] "Four Planes for Braniff," X15.

[33] "Propeller Producers Widening Bottleneck despite Unusual Dependence on Hand Labor," *Wall Street Journal*, May 16, 1941, p. 28.

[34] J. P. Richter to Transcontinental and Western Air, September 1, 1938; J. P. Richter to American Airlines, September 1, 1938; and J. P. Richter to United Air Lines, September 1, 1938, Folder "Propellers–July through December, 1938," Box 6301, RD 3330, RG 342, NARA.

of the first production hydraulic automatic automobile transmission, called Hydra-Matic Drive, in September 1939. The corporation offered it on the 1940 Oldsmobile line of cars with the new innovation standard on the Custom 8 Cruiser family of sedans and convertibles. For observers of the automobile industry, it was single greatest mechanical improvement for 1939.[35]

As Hamilton Standard became the propeller maker most associated with aircraft in the late 1930s, the Curtiss-Wright Corporation increased its stake in the aeronautical marketplace. Curtiss-Wright solidified its position by establishing the independent Propeller Division in 1938.[36] Uncertainty in both Europe and Asia were the indicators the corporation needed to justify expansion. Due to world events, there would be an increased demand for propellers over the next few years.

The Propeller Division continued to innovate and seek out technical features Caldwell and his colleagues at Hamilton Standard believed to be impractical. The pitch-changing motor central to the design of the Curtiss Electric propeller exhibited a nearly unrestricted field of movement. The incorporation of a double field winding made it capable of reversible pitch, which meant the blades generated thrust in the opposite direction of flight. The new feature, long desired by the aeronautical community since World War I, aided flying boats in water maneuverability and served as a landing brake for large multiengine aircraft.[37]

Another innovation that had long eluded successful technical development was the hollow-steel blade. Since leaving Standard Steel in 1928, Thomas A. Dicks worked toward that goal at the Dicks Aeronautical Corporation. Dicks, engineer James H. McKee, and sales manager Hamilton Foley experienced considerable success innovating and selling experimental hollow-steel propeller blades. Dicks and McKee developed a fabrication method of arc-welding two steel sheets together to form a blade. They received considerable interest on the part of BuAer due to the increased resistance of steel to saltwater corrosion. The Pittsburgh Screw and Bolt Corporation, a manufacturer of industrial threaded fasteners, acquired Dicks's company in February 1929 and gave him a corner of its

[35] William C. Callahan, "No Clutch Is Needed: Automatic Transmission by Oldsmobile Out," *New York Times*, September 24, 1939, p. 141; General Motors, *Oldsmobile's Exclusive Hydra-Matic Drive* (n.p.: n.d. [Detroit: General Motors, 1940]), General Motors Heritage Center Archives.

[36] Propeller Division, Curtiss-Wright Corporation, "Curtiss Propellers: Thirty Years of Development," PTF.

[37] Ibid.

North Shore factory until moving the entire operation to Neville Island on the Ohio River. A successful test to 2,700-horsepower at the Wright Field propeller test facility led to an Air Corps order for seventy-five blades for installation on the ground-adjustable hubs for training aircraft.[38] There was also considerable interest in the early 1930s on the part of the Airplane Division of the Ford Motor Company and the growing commercial aviation industry, primarily American Airways, Northwestern Airways, Pan-American Airways, and Eastern Air Transport.[39]

The engineers at Curtiss had been watching the blade-making activities at Pittsburgh Screw and Bolt with great interest. Seeing an opportunity for a greater market share in aviation, the Curtiss-Wright Corporation acquired the company in 1939. President Guy W. Vaughan remarked the acquisition reflected the "increasing trend toward the use of hollow-steel blades, because of their lightness and greater durability" supplemented aluminum alloy blade production. More importantly, the purchase was a "logical and important addition" that resulted in the ideal product for the corporation: the Curtiss Electric propeller equipped with hollow-steel blades.[40]

The Curtiss Electric propeller became standard equipment on many of the new "modern" pursuit and fighter airplanes of the late 1930s. As a major contractor to the Army Air Corps and exporter to foreign air forces, Curtiss installed them on its aircraft as a matter of course, especially the groundbreaking aircraft of designer Donovan Berlin. The radial engine-powered P-36 Hawk and the more numerous Hawk 75 export version initiated the trend. The inline engine-powered P-40 Warhawk was faster at 350 mph. The unprecedented peacetime order for 524 Warhawks in April 1939 was a major boost for Curtiss-Wright and the Propeller Division. Other manufacturers chose the electric propeller for their pursuit airplanes as well. Lockheed chose two for its P-38 Lightning twin-engine interceptor designed by Hall Hibbard and Clarence "Kelly"

[38] "Work Started Year Ago on Curtiss-Wright Plant Here," *The* [Beaver and Rochester] *Daily Times*, April 21, 1942, p. 15; William F. Trimble, *High Frontier: A History of Aeronautics in Pennsylvania* (Pittsburgh, PA: University of Pittsburgh Press, 1982), 116; Rosen, *Thrusting Forward*, 44.

[39] Hamilton Foley to H. A. Hicks, "Flight Test No. 1012," June 17, 1932, Box 99, Series I, Accession 18, HFM.

[40] "Propeller Works to Curtiss-Wright," *New York Times*, October 27, 1939, p. 37. Near the outbreak of World War II, Dicks served as a consulting engineer for both Curtiss and Hamilton Standard. Dicks died on September 30, 1944, at his home in Pittsburgh. "Thomas A. Dicks Dies," *Aviation News* 2 (October 9, 1944): 13-14; "Hamilton's Thomas A. Dicks in Propeller Field Since 1917," *The Bee-Hive* 18 (June–July–August 1943): 14.

Johnson. Bell Aircraft ordered a special example with a hollow propeller shaft that accommodated a 37mm cannon for the P-39 Airacobra. Seversky Aircraft selected them for the follow-on aircraft to Alexander Kartveli's P-35, the P-43 Lancer.[41]

A final innovation from the Curtiss-Wright Propeller Division in 1939 reflected the increase in power available for small, high-speed pursuit airplanes and multiengine transport and bomber aircraft. Engines such as the inline Allison V-1710 and Pratt & Whitney R-1830 Twin Wasp produced in excess of 1,000 horsepower. To create a proportionate amount of thrust, propeller designers had to increase the blade area, which meant expanding the diameter of the blades. To do so on a three-blade propeller required taller and heavier landing gear on a single-engine airplane. For multiengine aircraft, that also meant the nacelles needed to be spaced further apart from the fuselage. Robert L. Earle, vice-president and general manager, announced in July that the addition of a fourth blade increased blade area without a substantial increase in structure and blade diameter. The development of a four-blade propeller was a cooperative effort between the Propeller Laboratory and the propulsion specialists of the Materiel Division at Wright Field. The prototype four-blade propeller flew first on a Curtiss P-36A pursuit airplane and incorporated features found as previous designs, including constant-speed, selective-pitch, and full-feathering.[42]

Curtiss-Wright's share of the aeronautical marketplace grew without the influence of its long-suffering head of the propeller and power plant design sections, Pete Blanchard. After the introduction of the feathering mechanism in 1933, he became increasingly frustrated with what he felt to be Curtiss-Wright's near-sighted policies on future propeller design, manufacturing, and marketing. In March 1935, he began to spend his free time sketching designs for a new variable-pitch propeller. If it proved promising, he would leave Curtiss-Wright. When president Ralph S. Damon denied Blanchard's requests for better engineering facilities, more competitive bidding for government contracts, and a raise during a heated meeting the following October, he immediately resigned.[43]

[41] Propeller Division, Curtiss-Wright Corporation, "Curtiss Propellers: Thirty Years of Development," PTF.

[42] "Army Planes to Get New Propeller of 4-Bladed 'Electric-Fan' Type," *New York Times*, July 24, 1939, p. 5.

[43] Pete Blanchard, Diary of Business Activity, October 1935, in Juliet Blanchard, *A Man Wants Wings: Werner J. Blanchard, Adventures in Aviation*, 1986, www.margaretpoethig.com/family_friends/pete/wings_bio/index.html (Accessed May 24, 2012), 32–34.

Blanchard became an independent consulting engineer and moved his family to Dayton, Ohio, to be closer to the Army Air Corps' Materiel Division at Wright Field. The officers of the propeller department were "mystified" by his abrupt departure from Curtiss-Wright during the Great Depression. Nevertheless, they sustained Blanchard with development contracts for propeller blades. One of his former colleagues, Charles S. J. MacNeil, Jr., joined him as a partner in February 1936. MacNeil received his engineering degree from MIT in 1933, worked for Curtiss under Blanchard, and took a brief position with the University of Michigan aeronautical engineering department before coming to Dayton. Blanchard and MacNeil formed Engineering Projects, Inc., on January 12, 1937. They worked toward the improved design of everyday technologies, which included roller skates, a home food canning system, and the new propeller featuring a constant-speed feathering hub and hollow-steel blades.[44]

Blanchard and MacNeil's propeller was a self-contained hydraulically actuated mechanism. It consisted of two major assemblies. The first was the one-piece forged hub, hollow-steel blades featuring internal ribs for extra strength, and a pitch changing mechanism mounted in the root of each blade. The second was the regulator assembly that contained an oil pump, pressure control valve, hydraulic fluid, and the governor. The application of hydraulic pressure from the regulator to the top and bottom of a piston housed in each pitch changing mechanism created an axial movement that enabled a change in pitch. The governor maintained the correct blade angle automatically. The pressure control valve ensured consistent operation and the pump supplied the necessary oil. Blanchard and MacNeil christened their design the Unimatic to reflect its "unit" construction and "automatic" constant-speed operation.[45]

Two unique features of the Unimatic proved to be attractive to the Army's propeller department at Wright Field, which had a voice throughout the design process. From the production and operational standpoint, its compact design made it possible for the propeller to be mounted on

[44] Pete Blanchard, Diary of Business Activity, October 1935, in Blanchard, *A Man Wants Wings*, 36; "Charles Seward Jadis MacNeil," *Who's Who in Aviation, 1942–1943* (Chicago, IL: Ziff-Davis Publishing Company, 1942), 269; University of Michigan, *A Century of Engineering Education* (Ann Arbor: University of Michigan Press, 1954), 1185.

[45] Werner J. Blanchard, "Propeller and Method of Making Same," US Patent No. 2,205,132, June 18, 1940; Werner J. Blanchard and Charles S.J. MacNeil, "Propeller Mechanism," US Patent No. 2,307,101, January 5, 1943; and "Propeller Mechanism," US Patent No. 2,307,102, January 5, 1943.

any engine in twenty minutes without special components or an outright conversion. The hollow center shaft of the propeller was a by-product of each blade having its own pitch changing mechanism. The open space made the installation of an aerial cannon possible, which reflected the growing trend of increasing the accuracy and lethality of aircraft armament.[46]

During the fall of 1938, Engineering Projects was ready to offer the Unimatic to a prospective buyer. Solid bids for the exclusive license to manufacture from Hamilton Standard and Curtiss-Wright never materialized. By December, Blanchard and MacNeil were increasingly open to an offer submitted by Ernest Breech, a vice-president of GM. Breech suggested that GM acquire Engineering Projects to serve as the nucleus of a new propeller division that would complement its Allison aircraft engine division manufacturing V-1710 power plants. As they negotiated the terms, Blanchard and MacNeil sustained their small company with continued development contracts from the Army. Their most important contract was for a contra-rotating propeller, which consisted of two propellers placed one before the other and spinning in opposite directions.[47]

Nazi Germany: Aerial Fascism and Innovation

While the American propeller industry rose to prominence, a potent new player in military aviation emerged in Europe in the late 1930s. Rising from the ashes of the Weimar disaster and ignoring the restrictions of the Versailles Treaty, Nazi Germany introduced to the world a revived air force, the Luftwaffe, and an aviation industry capable of producing state-of-the-art aircraft.[48] The Luftwaffe's modern airplanes served as potent aerial symbols of Adolph Hitler, National Socialism, and the technological prowess of the Deutsches Reich. The majority of those menacing warplanes featured VDM constant-speed propellers that gave them

[46] *Pilot's Information for Aeroprop Unimatic Constant-Speed Propellers* (Dayton, OH: Aeroproducts Division of General Motors Corporation, 1942), File BS-007000-55, PTF, NASM; H. M. McCoy, *The Aeroproducts Propeller: Engineering Information* (Dayton, OH: Aeroproducts Division, General Motors Corporation, n.d. [1943]), File BS-007000-02, PTF, NASM; Aeroproducts Division, General Motors Corporation, *Blades for Victory: The Story of the Aeroproducts Propeller and the Men and Women Who Build It* (Dayton, OH: Aeroproducts Division, General Motors Corporation, 1944), 12.

[47] Blanchard, *A Man Wants Wings*, 65–67.

[48] James S. Corum, *The Luftwaffe: Creating the Operational Air War, 1918–1940* (Lawrence: University Press of Kansas, 1997), 125, 225.

ultimate performance as Nazi Germany used them to intimidate the rest of Europe.

VDM stood for Vereinigte Deutsche Metallwerke, or United German Metalworks, which was a society of medium-sized family firms formed in August 1930 and centered on Frankfurt am Main, Heddernheim, and Hamburg, Gross-Borstel. VDM's main business was the sale of nonferrous metals, primarily copper and brass, aluminum, stainless steel, and nickel, to European industry. As the organization that controlled aluminum resources in Germany, it sought ways to create new products that used the material in sheet, extrusion, cast, and pressed form.[49]

One of VDM's member companies, the Heddernheimer Metal Company of Frankfort am Main, initiated the development of metal propellers in Germany by introducing a ground-adjustable propeller in the late 1920s similar in construction to the Standard Steel propeller.[50] German gliding pioneer and aeronautical engineer Arthur Martens joined Heddernheimer as the head of the sales department and contributed to the design of those early propellers. His one hour soaring flight over the Wasserkuppe at the 1922 Rhön gliding competition catapulted him to the status of a national hero, which generated public interest and visibility for the company.[51]

Despite Marten's fame and a growing market for ground-adjustable-pitch propellers, VDM struggled like the rest of Germany industry as the Great Depression and its stranglehold on the world economy intensified in the early 1930s. VDM's leadership, chairman Bernard Unholtz, technical manager W. Helbig, and Martens, recognized that the society's aluminum alloy blades would be even more desirable to the European aeronautical industry if they were connected to a variable-pitch hub. The society lured mechanical engineer Dr. Hans Ebert away from the Deutsche Versuchsanstalt für Luftfahrt (German Research Center for Aviation, or DVL) in Berlin in early 1933 just as the Nazi Party took over

[49] George C. McDonald, "Report on the Manufacture of Aircraft Propellers by Vereinigte Deutsches Metal Werke, Heddernheim, Frankfurt-Am-Man," May 23, 1945, B5-90000–01, PTF, NASM; *25 Jahre Vereinigte Deutsche Metallwerke A.G.* (Frankfurt-on-the-Main: Vereinigte Deutsche Metallwerke, 1955), 1–2. Passage translated by Peter Weil, June 23, 2011; Thyssen Krupp VDM, "History," 2011, www.thyssenkrupp-vdm.com/en/corporate-information/history/ (Accessed May 23, 2011).

[50] Jacob W. S. Wuest to Assistant Chief of Staff, G-2, War Department, "RS Light Propeller Hub," December 9, 1932, B5-900060-01, PTF, NASM.

[51] Peter Fritzsche, *A Nation of Fliers: German Aviation and the Popular Imagination* (Cambridge, MA: Harvard University Press, 1992), 108–109, 111.

the German government. The engineer's job was to head a research program that would result in a new variable-pitch propeller design.[52]

Ebert chose an ingenious and altogether original method of pitch actuation.[53] A small reversible electric motor mounted on the engine crankcase provided actuation via a flexible shaft connected to small primary drive reduction gearbox, which was in turn attached to a large annular gearbox fitted to the rear of the propeller hub. The annular gearbox housed epicyclic, or planetary, pitch-change gears. Pinion gears engaged the outlet drive of the gearbox and drove worm shafts entering the hub blade sockets meshed with the worm gears integral with the blade adaptors. The blades were capable of rotating 360 degrees, but cutout switches prevented that. Since the power came to the gearbox via the flexible shaft, other sources, including hydraulic, pneumatic, mechanical, or manual could be used if required.[54]

The use of a VDM propeller offered an alternative to other designs in the aeronautical marketplace, especially for military air services. It required no modifications to the engine, which avoided costly conversion programs. The decentralized location of the pitch actuation mechanism away from the propeller hub meant that the actual shaft of the propeller was hollow. This hollow shaft construction meant ready adaptation of cannon armament into the design of military, primarily fighter, aircraft. A blade pitch indicator in the cockpit gave the pilot instant information.[55]

In extreme environments, electric actuation offered another performance advantage over the much more common hydraulic propellers.

[52] Hans Ebert, "Bericht über die VDM-Verstelluftschraubenentwicklung (Report on the Development of the VDM Controllable Propeller)," January 6, 1943, German/Japanese Captured Air Technical Documents (hereafter cited as CATD), NASM. Passage translated by Sarah Richards, July 9, 2012; A. H. Metcalfe to J. S. Buchanan, January 17, 1937, AIR 2/2406,TNA; John D. Waugh, "Investigation of the VDM Propeller Works," August 15, 1945, CATD, NASM, 1; "The German VDM Electric Propeller 1940 Model Used on the Heinkel 115 Twin-engined, Mine-Laying Seaplane," n.d., B5-900010-01, PTF, NASM; *25 Jahre Vereinigte Deutsche Metallwerke*, 1–2.

[53] American test pilot and engineer Robert Stanley has been credited with the original design of the VDM propeller in 1933, but there is no documentation explaining how his design made it to Germany or if the similarity between the two is merely coincidence. See Robert M. Stanley, "Controllable-Pitch Propeller," US Patent No. 1,986,229, January 1, 1935; and National Aviation Hall of Fame, "Robert Stanley," 2011, www.nationalaviation.org/stanley-robert/ (Accessed January 6, 2015).

[54] John D. Waugh, "Details of the German VDM Electric Propeller, Part I," *Industrial Aviation* 1 (July 1944): 33.

[55] A. H. Hall to the Air Ministry, "Relative Merits of Curtiss and VDM Variable-Pitch Airscrews," January 3, 1938; and H. B. Howard, "VDM Airscrew Designed for Battle-Merlin–General Report," November 14, 1937, AIR 2/2406, TNA.

Hydraulically controlled propellers suffered from sluggish performance at high altitudes and low temperatures. The oil in the hubs simply froze. The availability of the VDM led to the decision of the Reichsluftfahrtministerium (RLM, or Reich Aviation Ministry) to no longer accept hydraulic propellers or to severely limit their use on front-line combat aircraft.[56] Overall, the German aeronautical community favored electric actuation over hydraulic, which was a strong indication of one culture's reaction to a technology that worked very well for another. With the exception of the Curtiss Electric propeller, all of the other major variable-pitch propeller designs were hydraulic.

The first VDM propeller was controllable, which depended on pilot selection to operate it at all times. VDM offered it in two diameters, ten feet six inches and eleven feet six inches, rated for 750 and 1,000 horsepower aircraft engines, and weighing 258 and 295 pounds respectively. After full approval by the DVL, volume production began during the spring of 1936. By the spring of 1937, over 1,000 propellers had been delivered and fitted chiefly to BMW and Daimler Benz engines installed in the Luftwaffe's Heinkel He 111 and Junkers Ju 86 bombers. Several individual propellers had been in the air over 300 hours, which amounted to 50,000 hours total.[57]

VDM constant-speed propellers appeared during the spring of 1937. Full-scale production was to be up and running a year later. The basic propeller was the same. The conversion consisted of a simple attachment of a small generator mounted on and driven by the engine. The pilot used a rheostat control to set engine rpm. There were no special engine fittings, but a source of current was needed for the 140 watt pitch changing motor.[58] Inside the cockpit, the pilot used a series of conventional electrical switches to choose fixed-pitch, selective-pitch, constant-speed, or feathering operation in flight.[59]

Due to VDM's origins, duralumin was the main blade material. A marked difference between the VDM blade and others was the aerodynamic shape of the blade all the way to the root, which meant that every square inch of the structure generated thrust as it whirled through the air. Blade retention consisted of screwing the blade hub into an adapter

[56] A. H. Metcalfe to J. S. Buchanan, January 17, 1937; and H. S. Royce, "The VDM Airscrew," n.d. [April 1937], AIR 2/2406, TNA.

[57] Ibid.

[58] H. S. Royce, "The VDM Airscrew," n.d. [April 1937], AIR 2/2406, TNA.

[59] Waugh, "Details of the German VDM Electric Propeller, Part I," 33.

and locking it in with a wedge ring. The mounting hardware for just one blade consisted of sixteen separate pieces that were easy to maintain and service, but to one observer represented the "German flair for intricate multiple-part mechanisms" and an overall longer maintenance process longer due to the large number of small parts. Nevertheless, Luftwaffe maintenance personnel found the blade assembly easy-to-maintain in the field.[60]

VDM divided its propeller manufacturing program between two facilities. At Frankfurt, small quantity production took place with approximately 200 highly skilled workers. These workers excelled at fabricating new prototypes and producing custom batches of propellers on order. The larger factory at Hamburg employed approximately 700 semiskilled, workers in a dedicated section alongside areas used for aircraft undercarriage and shock absorber manufacturing. By January 1937, the VDM factories produced approximately 200 propellers a month.[61]

The Nazis recognized the public relations value of introducing to the world aeronautical community their new aerial weapons at the ten-day International Air Meeting in Zurich, Switzerland, in July 1937. The Swiss Air Force sponsored the event at its Dübendorf Airfield every five years beginning in 1922. It was Europe's principal aviation meet, equivalent to the annual National Air Races in the United States. Over 300 participants from fourteen European nations, including Germany, France, Great Britain, Italy, Poland, and Czechoslovakia, challenged each other in various competitions and precision demonstrations of flying skill. On Friday, July 23, the first day of the meet, 80,000 people were in attendance.[62]

Militarism was in the air at Zurich. Among the many participants were personnel and aircraft representing seven national air forces. *New York Times* correspondent Clarence K. Streit reported by wireless that to the crowds at Dübendorf, European military aircraft, including the demonstration teams of France, the *Patrouille d'Étampes*, and Italy, the *Squadrone Italiano*, "overshadowed" the events featuring civilian aircraft, which included a spectacular glider demonstration by Nazi pilot

[60] Waugh, "Details of the German VDM Electric Propeller, Part I," 33; John D. Waugh, "Details of the German VDM Electric Propeller, Part II," *Industrial Aviation* 1 (August 1944): 37; John D. Waugh, "Particulars of the German VDM Electric Propeller," *SAE Journal (Transactions)* 52 (August 1944): 348.

[61] H. S. Royce, "The VDM Airscrew," n.d. [April 1937], AIR 2/2406, TNA.

[62] Clarence K. Streit, "Touring Airplanes Open Zurich Meet," *New York Times*, July 24, 1937, p. 4; "Zurich International–II," *Flight* 32 (August 5, 1937): 142.

Hanna Reitsch.[63] The Swiss-based international aviation news bulletin, *Interavia*, asserted that the air meet would provide an opportunity to compare them in a peaceful environment of bloodless aerial competition.[64] Perhaps more aware of the final use of the technology, the editors of *Time* magazine emphasized that the air meet "offered a fine chance to the nations of Europe to show how they were getting along with human and mechanical preparations" for going to war against each other in the air.[65]

To Reichsmarschall Hermann Göring, Lt. Gen. Erhard Milch, state secretary for aviation, and Maj. Gen. Ernst Udet, chief of the Luftwaffe's Technical Office, Zurich was the opportunity "to lift the veil" of obscurity shrouding the Luftwaffe to reveal a state-of-the-art and "formidable weapon" comparable to any other in the world.[66] That weapon included two of Germany's first modern airplanes that were central to overall Luftwaffe strategy, the fast and nimble Messerschmitt Bf 109 fighter and the sleek twin-engine Dornier Do 17 bomber. Designed by Willy Messerschmitt, the Bf 109 was a short-range interceptor capable of a high rate of climb and maneuverability. It represented a major technological jump over the Arado and Heinkel biplane fighters that it replaced. Nicknamed the "Flying Pencil" due to its long cylindrical shape, the Do 17 was faster than most fighters. Both aircraft were prototypes of the production versions that featured streamline design with enclosed cabins, all-metal stress-skin construction, and retractable landing gear that enabled speeds over 200 mph. Luftwaffe personnel installed high performance three-blade, constant-speed VDM propellers and supercharged liquid-cooled inverted V-12 inline engines specifically for the Air Meet.[67]

All eyes at the event were fixed on the Luftwaffe monoplanes and the pilots flying them as they competed in the marquee speed, climb-and-dive, and alpine circuit races. The speed race went to the pilot and airframe that flew over a closed course in the shortest amount of time. Test pilot Carl Francke in a light gray Bf 109 completed four circuits of the 31.36 mile course to win in 29 minutes 35 seconds with an average speed of 254 mph. Charles Gardner of Great Britain never came close

[63] Clarence K. Streit, "Speed Race is Won by German Flier," *New York Times*, July 26, 1937, p. 21.

[64] "Outlook on the Fourth International Air Meeting, Zurich," *Interavia* 451 (July 17, 1937): n.p.

[65] "Transport: Zurich Meet," *Time* 30 (August 9, 1937), www.time.com/time/magazine/article/0,9171,758084,00.html (Accessed May 24, 2011).

[66] "Zurich, 1937," *Interavia* 456 (August 4, 1937): n.p.

[67] "New German Planes in Zurich Contests," *New York Times*, July 23, 1937, p. 9.

in his Percival Mew Gull powered by a 205-horsepower de Havilland Gipsy Six engine and two-pitch propeller at an average speed of 217 mph in over 34 minutes. Gardner's privately entered Mew Gull was a highly successful long-distance and closed course racer in the British Empire, but it was no match for the Luftwaffe's Bf 109 and its 640-horsepower Daimler-Benz DB 600 V-12 engine and constant-speed VDM propeller.[68]

Fresh from his win in the speed race, Francke won the climb and dive competition on the 27th in the same airplane. The contest consisted of a climb to 3,000 meters (9,840 feet) and a "hell dive" to 352 meters (1,066 feet). Francke made the flight in two minutes five seconds beating his competitors by twenty seconds. The up-and-down average of 105 mph indicated the potency of the Luftwaffe's premier fighter to rise, meet, and destroy Nazi Germany's enemies in the air. The international aviation press made much that the performance was only possible with a VDM propeller.[69]

The Alpenrundflug, or circuit race over the Alps, was another important evaluation of military aircraft. The circuit involved flying over the mountain range twice, which was an indicator of both high altitude and extreme temperature performance. Major Hans Seidemann in a Bf 109 and Milch in the Do 17 won the Alpine Circuit Race for single and multiseat aircraft. Each aircraft completed the circuit at 56 and 58 minutes respectively, which was approximately fifteen minutes under the time elapsed during the previous 1932 competition.[70] Part of the aerial demonstration of the Do 17 included flying low over Dübendorf with the right propeller feathered.[71]

Germany sent the largest contingent to Zurich, thirty-five airplanes and ninety men. They won six events, which outpaced Czechoslovakia's three and France's two victories. Interviewed German fliers expressed their regret that their competition was not better. Either the countries present did not "send their best entries" in reference to the French, or did not participate in the case of the United States. The Luftwaffe, its aircraft, and the VDM propellers used on all of them had arrived on the world

[68] Karl Seyboth, "Internationales Flugmeeting Zürich," *Der Deutsche Sportflieger* 4 (September 1937): 7.

[69] Clarence K. Streit, "Reich Fliers Win 3 Races at Zurich," *New York Times*, July 27, 1937, p. 6; "An Airscrew Achievement," *The Aeroplane* 54 (May 11, 1938): 577.

[70] "Outlook on the Fourth International Air Meeting, Zurich," *Interavia*; Clarence K. Streit, "Reich Fliers Win 3 Races at Zurich," *New York Times*, July 27, 1937, p. 6.

[71] Seyboth, "Internationales Flugmeeting Zürich," 5.

stage. Company advertising boasted, "Germany wins with VDM," as it listed the major competitions at Zurich.[72]

According to one unidentified German pilot, the Zurich meet was a turning point in international military aviation. Before 1937, American military aircraft "probably would have won every event" if they had been entered. With the resurgence of the Luftwaffe, however, he implied that Germany was the clear leader in military aviation technology as shown in the bloodless competitions at Zurich. If the United States had bothered to show up, they would have lost.[73] The Nazi pilots boasted that Germany had the fastest fighter and bomber aircraft in the world, which from the results at Dübendorf appeared to be true.

Between the Zurich air meet and the invasion of Poland in September 1939, the Luftwaffe used aircraft equipped with VDM propellers to place Nazi Germany in the international news and at the forefront of international aeronautical competition. Messerschmitt's engineering test pilot Dr. Hermann Wurster achieved a monoplane world landplane speed record of 379 mph in a Bf 109 with the propeller absorbing 1,660 horsepower on November 11, 1937. Shortly after, a Dornier Do 18 flying boat powered by diesel motors set the world seaplane distance record of 5,242 miles while remaining in the air forty-three hours.[74] Heinkel and Messerschmitt fighters went on to achieve world speed records of 392.5, 463.9, and 469.2 mph equipped with VDM propellers. The latter record, made on April 26, 1939, by test pilot Fritz Wendel in a Me 209 V-1 powered by a 1,800-horsepower DB 601 V-12 engine, stood until 1969.[75]

Zurich and the subsequent record attempts were the formal public introduction of the Luftwaffe's new aircraft in a bloodless environment. They took full advantage of the spectacle of military aviation to illustrate the Luftwaffe's potential to the world. At the same time, German volunteers under the guise of the Legion Condor fought for dictator General Francisco Franco during the Spanish Civil War (1936–1939). The bitter campaigns between the German- and Italian-supported Nationalist forces and the Republicans, receiving aid from the Soviet Union, included a significant use of air power. The war proved Luftwaffe technology in combat and developed the core tactics and technology it would use in future conflicts.

[72] VDM, "Deutschland Siegt Mit VDM-Luftschrauben," *Der Deutsche Sportflieger* 4 (September 1937): n.p.
[73] "Czech Flier Beats German at Air Meet," *New York Times*, August 3, 1937, p. 9.
[74] "An Airscrew Achievement," 577.
[75] Waugh, "Details of the German VDM Electric Propeller, Part I," 33.

The aerial battleground over Spain witnessed the emergence of the Luftwaffe as a modern air force. By late 1936, Soviet-built Polikarpov I-15 Chato biplane and I-16 Mosca monoplane fighters and Tupolev SB-2 Katiuska bombers operated by Republican forces outmatched the three squadrons of Heinkel He 51 biplanes flown by Jagdgruppe (JG) 88 of Legion Condor. At the urging of the legion's commander, Generalmajor Hugo Sperrle, the Luftwaffe quickly diverted the first sixteen Bf 109Bs, a squadron's worth with four spares, to Tablada aerodrome in Seville in March 1937. Just two weeks before the opening of the Zurich Meet, the squadron went into combat over Brunete on July 6. Immediately, Legion Condor pilots possessed a fighter evenly matched to their Republican opponents, which led to innovations in tactics capitalizing on the Bf 109's superior high altitude and diving performance. By the following year, all three squadrons of JG 88 flew Bf 109s, which contributed to virtually uninterrupted Nationalist air superiority over the peninsula until the end of the conflict on March 30, 1939.[76]

Even though Messerschmitt designed the Bf 109 for the VDM propeller from the outset, the first sixteen went into combat with Schwarz two-blade fixed-pitch propellers. A Schwarz propeller was unique due to its hybrid construction. The laminated and compressed core consisted of soft wood blades, usually pine or fir, joined to a hub made up of Bakelite and hardwood layers. Special machinery pressed a hard coating of metal mesh and cellulose sheet over the entire propeller. For added strength, the leading edge received more mesh and cellulose and a final covering of brass.[77] The blades were durable and propeller manufacturers in Europe, Asia, and North America scrambled for the license, but the fixed configuration did not deliver the full performance offered by the VDM. Initially, JG 88 flew a mixture of Schwarz and VDM propellers on their Bf 109Bs. VDM caught up with the demand, which included retrofitting the Schwarz-equipped aircraft in Spain. All successive Bf 109 variants incorporated VDM constant-speed propellers. The Bf 109E, which appeared during the Catalonian campaign in early 1939, offered an astonishing

[76] Jesus Salas Larrazabal, *Air War Over Spain*, trans. Margaret A. Kelley (London: Ian Allan, 1974), 122; William Green, *The Augsburg Eagle, A Documentary History: Messerschmitt Bf 109* (London: Jane's, 1980), 24–28; Gerald Howson, *Aircraft of the Spanish Civil War, 1936–1939* (Washington, DC: Smithsonian Institution Press, 1990), 195, 196, 198, 200, 233.

[77] *Schwarz Holzpropeller und Metallpropeller* (Berlin: Propellerwerk Gustav Schwarz, March 1934); and H. A. Guthrie, "Wooden Airscrews Manufactured by Schwarz Firm of Berlin," US Naval Attaché Report 404, July 31, 1935, File B5-782000-01, PTF, NASM.

348 mph top speed and could climb to 20,000 feet in eight minutes, which in part was due to its three-blade VDM propeller.[78]

The rise of Nazi Germany facilitated the creation of the VDM propeller, which offered an alternative to what the Americans produced and the British had been struggling to innovate. Surprisingly, the company accomplished the achievement within an astonishingly short period of time. It was just four years between when Ebert started the project and the earliest constant-speed models made their debut at Zurich. More importantly for the image-conscious Nazi regime, the VDM propeller superseded the Junkers-licensed and American-designed Hamilton Standard mechanisms as the primary high performance propellers in Germany. By the spring of 1938, VDM had produced approximately 3,000 propellers with current production at an estimated 800 a month. The exact output was a secret of the RLM. VDM went on to introduce volume production of its constant-speed propeller later in November with all orders going to the Luftwaffe. VDM claimed to be largest producer of controllable-pitch airscrews in the world. Overall, it was estimated that 95 percent of the Luftwaffe's aircraft used VDM propellers.[79]

Within six months of the end of the war in Spain, the Luftwaffe's modern and all German Messerschmitt, Junkers, Heinkel, and Dornier airplanes would be crossing into Poland on September 1 to usher in World War II. Each aircraft relied upon the constant-speed propeller for maximum power and efficiency during climbing, diving, and high altitude cruise. The Luftwaffe faced the air forces of Poland, France, and Great Britain with an obvious and threatening performance advantage.

Great Britain: A Prolonged Debate and Mixed Acceptance

As American and German aircraft took to the air on a regular basis with constant-speed propellers installed during the mid-1930s, the British aeronautical community still struggled over whether the fundamental innovation was even appropriate for its aircraft. There was some movement, however, toward acceptance of variable pitch in the British aeronautical community in the mid-1930s. The pace of innovation, support, and acceptance revealed much about how the propeller fit within the

[78] Howson, *Aircraft of the Spanish Civil War*, 235.

[79] A. H. Hall to the Air Ministry, "Relative Merits of Curtiss and VDM Variable-Pitch Airscrews," January 3, 1938; A. R. Collins to the Air Ministry, "VDM Airscrews," November 2, 1938, AIR 2/2406, TNA; "An Airscrew Achievement," 577.

overall design of British airplanes and if the "time was right" for the new device.

The de Havilland Aircraft Company rose to prominence as a propeller manufacturer in Great Britain using Hamilton Standard licensed products. The trouble the company experienced in sourcing a variable-pitch propeller for the Comet racer in the pivotal MacRobertson Race led to the crucial licensing agreement in June 1934. They were the same controllable-counterweight propellers that appeared during the crucial spring of 1933 in the United States. The British called the counterweights "brackets" due to the way they appeared to enclose the blade roots at the hub. To many in the British aeronautical community, a two-pitch bracket-type propeller was an ideal compromise. It was more efficient than a fixed-pitch propeller yet lighter and less complicated to maintain and operate than a constant-speed propeller like those being developed by Gloster, Bristol, and Rolls-Royce. Equipped with the manufacturing and operational information provided by Hamilton Standard, de Havilland was able to get production up and running very quickly.

There was interest in constant-speed propellers on the part of the Air Ministry. Air Marshal Sir Hugh Dowding, Air Member for Supply and Research at the time, awarded new contracts to the Bristol Aeroplane Company and Rolls-Royce in 1934 to augment and perhaps stimulate Gloster's ongoing but glacially slow work on the Hele-Shaw Beacham design. The Gloster team achieved encouraging results in flight tests, but the problem was the design of the hub. During operation, it easily distorted and allowed oil leakage, which reduced the pressure available for pitch actuation.[80] It was time for other engineering groups to tackle the problem. Despite the awarding of the contracts, the Air Ministry's movement toward encouraging the development of a variable-pitch propeller in mid-1930s Great Britain was, according to official observers, "not undertaken in any spirit of indecent haste."[81]

The ongoing work on perfecting the Hele-Shaw Beacham design and de Havilland's introduction of Hamilton Standard propellers in Great Britain generated an opportunity for community-wide discussion on

[80] Airworthiness and Engine Departments, Royal Aircraft Establishment, "Performance and Flight Endurance Tests of Gloster Hele-Shaw Variable-Pitch Airscrew for Kestrel II.S Engine," Report A.S. 34, August 1937, AVIA 6/6963, TNA; Bill Gunston, *Fedden: The Life of Sir Roy Fedden*, Rolls-Royce Heritage Trust Historical Series No. 26 (Derby, England: Rolls-Royce Heritage Trust, 1998), 196.

[81] Ministry of Supply, "Propellers: Development and Production, 1934–1946," n.d. [1950], AVIA 46/211, TNA, 7, 16.

whether or not to adopt variable-pitch propellers. During the spring of 1935, *Flight* published a two-part series by the journal's editor, Carl M. Poulsen, which detailed the latest designs. The articles explained the operation and benefits of the technology, with a particular focus on the Hamilton Standard/de Havilland and Hele-Shaw Beacham propellers. Poulsen proclaimed that the advantages gained from the use of a variable-pitch propeller were "great enough" to justify the increased weight, cost, and complexity. His proviso, reflecting the conservative nature of his readers, was that those benefits only helped "certain types of aircraft" designed for "certain classes of work."[82]

The newly emerging generation of modern fighters fit those conditions. A biplane fighter with a relatively low top speed such as the 200 mph Hawker Fury meant that the ideal propeller pitch for takeoff, climbing, and diving was not very far from the best setting for level flight. In that situation, a fixed-pitch propeller was satisfactory. The gap between ideal pitches for those conditions, which were all crucial for a successful fighter interception, widened with the introduction of faster monoplanes with supercharged engines approaching 300 mph. Moreover, the higher wing loadings of monoplanes required better performance from the propeller. From the perspective of operational maintenance, an additional benefit was less strain on the engine and better fuel economy overall.[83]

Poulsen's articles set forth a convincing argument for the variable-pitch propeller. His most telling evidence was the fact that operators increasingly adopted them in Great Britain and abroad. There was, however, considerable debate regarding whether other propulsion innovations provided better performance, especially during the crucial periods of takeoff and climb. Two-speed reduction gear installed on an engine generated higher rpm for takeoff and climb and lower rpm for cruise. Fuel rated at 87-octane offered more power and better fuel consumption. Taking into account all three alternatives, Poulsen solicited the views of *Flight*'s readers to see which one they thought was best for the airplane.[84]

A distinguished group of airframe and propulsion specialists responded to *Flight*'s inquiry. In contrast to the contentious presentation by Hele-Shaw and Beacham to the RAeS in 1929, the discussion revealed that the majority of the respondents recognized the value of the variable-pitch

[82] C. M. Poulsen, "Controllable-Pitch Airscrews, Part I," *Flight* 27 (May 2, 1935): 465–468; and "Controllable-Pitch Airscrews, Part II," *Flight* 27 (May 9, 1935): 499–502.

[83] Poulsen, "Controllable-Pitch Airscrews, Part I," 468; Ministry of Supply, "Propellers: Development and Production, 1934–1946," 7, 16.

[84] "Controllable-Pitch or –?," *Flight* 27 (May 16, 1935): 1.

propeller. According to Charles C. Walker, chief engineer of de Havilland, two-speed gearing and 87-octane fuel did increase power, but most of that energy was lost if used with a fixed-pitch propeller. A de Havilland two-pitch bracket propeller represented a "gain of both power and efficiency." Roy Chadwick of A. V. Roe, Arthur Davenport of Westland Aircraft, and Arthur Gouge from Short Brothers echoed the consensus view that variable-pitch propellers were best for high speed monoplane aircraft providing their cost and weight could be kept down.[85]

Surprisingly, the most enthusiastic response came from England's leading designer of fixed-pitch propellers, Dr. Henry C. Watts of the Airscrew Company Ltd. In 1929, he and E. J. H. Lynam led the "anti-VP school" against Hele-Shaw and Beacham. They argued then that the "weight, cost, and complication" was not justifiable at a time when slow biplanes were the standard. Six years later, however, his views had clearly evolved. In early June 1935, Watts asserted that the variable-pitch propeller was the "logical solution" to improving aircraft performance. The only way to prevent "blade stall" on a propeller was to alter the pitch of the blades to prevent the loss of thrust at takeoff. To Watts, the dramatic increases in aircraft speeds meant the time had finally come for a variable-pitch propeller.[86]

The decision to use a variable-pitch propeller also depended on whether it was capable of two-pitch or constant-speed operation. Henry Knowles and Rex K. Pierson, the chief designers of Saunders-Roe and Vickers respectively, argued that the latter, what they called "infinitely variable," coupled with a supercharger was the best. To them, the de Havilland two-pitch propeller was too expensive and heavy to justify for takeoff performance alone. Only a constant-speed propeller could provide a sufficient range of pitch actuation to provide better performance under all conditions. Ultimately, their preference exposed the fact that no constant-speed propeller was available commercially in England. And until it was, they were not in the market for a variable-pitch propeller.

For Stanley H. Evans, an assistant designer at the Heston Aircraft Company in Middlesex, the "best argument" in favor of introducing variable-pitch propellers in England was the innovation's demonstrated performance in the United States. As a matter of fact, the American aeronautical community no longer argued the legitimacy of the device since it was a "practical reality" that delivered the "goods – or, rather,

85 "Justifying the C.P. Airscrew," *Flight* 27 (June 6, 1935): 608–610.
86 Ibid., 609–610.

horses" for increased performance. He congratulated de Havilland on its "vision" to make an innovation like the Hamilton Standard propeller available to the British aeronautical community. He asserted that British aviation, as well as the country in general, sadly needed a "little more vision" overall.[87]

With the only commercially viable product in its possession, de Havilland became the leading variable-pitch airscrew manufacturer in Great Britain in the late 1930s. The company completed its first order, for two-pitch bracket propellers, for the Japanese military in early 1935. The first major Air Ministry order for variable-pitch propellers in Great Britain went to de Havilland. The company's Stag Lane factory at Edgware was to produce 825 three-blade two-pitch bracket propellers for radial engines that powered Bristol Blenheim, Handley-Page Harrow, Armstrong-Whitworth Whitley, and Vickers Wellesley bombers beginning in September 1935.[88]

De Havilland also found a niche in commercial aviation. Imperial Airways placed the first order. The airline's massive and modern four engine Short Empire flying boats and Armstrong-Whitworth Ensign airliners required over 200 propellers for regular operations on routes that connected Great Britain to the far reaches of its Empire during the late 1930s. Orders came from aircraft manufacturers in Belgium, the Netherlands, Poland, Sweden, Turkey, and Yugoslavia. Even the diminutive Comet racer received its own pair of de Havilland two-pitch propellers in 1936.[89]

The Hamilton Standard licensing agreement did make available the latest designs to de Havilland. De Havilland constructed its first constant-speed bracket propellers in 1937.[90] The RAF's different branches ordered them as an upgrade for their multiengine aircraft. Bomber Command selected them for the Armstrong-Whitworth Whitley, Vickers Wellington, Handley-Page Hampden, Short Stirling, and Avro Manchester. Bristol Beauforts operated by Coastal Command included them as standard equipment. Fighter Command saw the need for their use on the twin-engine Bristol Beaufighter and Westland Whirlwind fighters.[91] Production

[87] Ibid., 610.
[88] Ministry of Supply, "Propellers: Development and Production, 1934–1946," 17, 26.
[89] C. Martin Sharp, *D.H.: A History of de Havilland* (Shrewsbury, England: Airlife, 1982), 152–153, 155, 272.
[90] Ministry of Supply, "Propellers: Development and Production, 1934–1946," 26.
[91] "Pitch Panic: How Hurried Changes from Two-Pitch to Constant-Speed Airscrews Were Made in Time for the Battle of Britain," *Flight* 44 (December 9, 1943): 648.

of the Hydromatic propeller began in mid-1938.[92] De Havilland aggressively marketed the "quick-feathering airscrews" for its Flamingo airliner and other commercial aircraft.[93]

The factory at Stag Lane Aerodrome served as de Havilland's sole propeller facility from 1935 where both experimental design and production took place. As part of the larger scheme to increase aircraft manufacturing capability in time of war, the British government initiated construction of a "shadow factory" for de Havilland in 1937. Built at the cost of £943,000, the new factory at Lostock, west of Bolton, Lancashire, served as a nucleus for production methods and staff in the event of rapid wartime industrial mobilization. The Air Ministry anticipated monthly production of 200 bracket-type hubs and 400 blades at Lostock after it opened in early 1938.[94]

With the Air Ministry's blessing, Bristol started work on its own propeller based on the Hele-Shaw Beacham design in 1934. The hub design featured an internal cylinder that slid on a fixed piston enclosed within the propeller hub that provided a twenty degree pitch range.[95] The Bristol team machined a stiffer hub casing from a steel forging that proved superior to the earlier Gloster design. Flight tests began on the new propeller installed on a Bristol Mercury-powered Gloster Gauntlet fighter the following year.[96]

Bristol's three-blade Hele-Shaw Beacham propeller was the first constant-speed propeller to receive approval for general use in the United Kingdom. It came in two sizes, a twelve-foot diameter, 340 pound propeller designated for the 900-horsepower Pegasus radial, and an eleven-foot diameter 300 pound unit for the 800-horsepower Mercury radial. Extensive flight tests with the latter on a Hawker Hart beginning in October 1936 yielded glowing reports by the test pilots. They commented on the "clean running" and how the governor held all engine speeds steadily with no noticeable "hunting," or continuous blade changing

[92] Sharp, *D.H.*, 155.

[93] De Havilland Propellers, "Higher Performance for Air Transport," *Flight* 37 (April 5, 1940): 469.

[94] "An Airscrew 'Shadow' Factory," *Flight* 30 (October 29, 1936): 452; Howard Kingsley Wood, "Supply and Production: Fourth Report by the Air Ministry," December 19, 1939, War Cabinet Memoranda, Records of the Cabinet Office, CAB/68/3/40, TNA, 314–315, 340, 343.

[95] Ministry of Supply, "Propellers: Development and Production, 1934–1946," 11; Bruce Stait, *Rotol: The History of an Airscrew Company, 1937–1960* (Stroud, Gloster, England: Alan Sutton Publishing, 1990), 23.

[96] Gunston, *Fedden*, 135, 195–196.

above and below the ideal angle. By June 1937, the propeller had been in the air for over 100 hours.[97]

The separate Rolls-Royce contract took the Hele-Shaw Beacham design in another direction. The fixed inner piston was still there. Rather than mounting the cylinder that operated the pitch change mechanism inside the hub, they mounted it onto the front. Push-pull rods actuated the blades directly. The design facilitated a greater range of pitch, up to thirty-five degrees, for more efficient flying and seventy-five degrees for the all-important feathering feature. Rolls-Royce engineers called it their "external cylinder" propeller.[98]

Bristol's chief engine designer, Roy Fedden, continued to be a zealous advocate of a British-designed and -manufactured variable-pitch propeller. He could not change the opinion of the Bristol board of directors that the company was not the one to take the lead. In an unprecedented move, he proposed a meeting with his contemporaries at Rolls-Royce, Arthur F. Sidgreaves and Ernest W. Hives, at the Royal Thames Yacht Club in London. They discussed the establishment of a dedicated propeller design and manufacturing firm for the benefit of both companies and the nation as a whole. With the encouragement of the Air Ministry's head of research and development, Air Marshal Wilfrid R. Freeman, Rolls-Royce and Bristol formed Rotol Airscrews Ltd. on May 13, 1937, as a joint venture. The new organization's name, a combination of "Rolls-Royce" and "Bristol," was suggested by a wife of one of the Bristol employees about to be transferred to the new enterprise.[99]

The creation of Rotol finally provided the necessary financial and engineering resources that made the original Hele-Shaw Beacham design a practical constant-speed propeller. The Air Ministry moved quickly in its support for Rotol. It placed an order for 275 internal cylinder propellers for Bristol Blenheim light bombers in October 1937. This was the second order the Air Ministry had placed for variable-pitch propellers.[100] The company's design staff, led by new chief project engineer, Pop Milner, concentrated on Rolls-Royce's variation because it facilitated

97 "The 'Bristol' Constant-Speed Controllable-Pitch Airscrew," *The "Bristol" Review* (June 1937): 40, 42.

98 Ministry of Supply, "Propellers: Development and Production, 1934–1946," 11; Stait, *Rotol*, 23; Gunston, *Fedden*, 198.

99 Derek N. James, *Gloster Aircraft since 1917* (London: Putnam, 1971), 18; Stait, *Rotol*, 2, 4; Gunston, *Fedden*, 198; Anthony Furse, *Wilfrid Freeman: The Genius behind Allied Survival and Air Supremacy, 1939 to 1945* (Staplehurst, UK: Spelmount, 2000), 74.

100 Ministry of Supply, "Propellers: Development and Production, 1934–1946," 18.

feathering.[101] The advent of the improved designs from Bristol and Rolls-Royce that solved the technical ills of the design meant that the original Gloster program ceased by late summer 1937.[102]

By year's end, Rotol was a new company that reflected the near-equal presence of its two parent organizations. Bristol and Rolls-Royce chose the location for the brand new factory at Staverton in the heart of the picturesque Cotswolds in Gloucestershire for three important reasons. It sat on the important Gloucester-Cheltenham road that provided an easy commute for workers living in either city. Across that same road, the local airport provided a venue for the flight test and evaluation of new propeller designs. On a larger scale, Staverton was almost equidistant between the cities of Bristol and Derby; the latter the home of Rolls-Royce, so each of the parent companies would have equal access as they needed to be involved in everyday operations. There was equal representation at the highest levels of Rotol. R. Stammers, a longtime employee of Fedden at Bristol, served as general manager while E. O. Cameron, an accountant at Rolls-Royce, became company secretary.[103]

A major impetus in getting Rotol up and running was the nontechnical matter of national pride. De Havilland manufactured a licensed product of American origin. Members of the British aeronautical community, as represented by Fedden, and the Air Ministry, including Freeman, met during the late summer and fall of 1938 to discuss the question of paying "large sums to America" for propeller technology. They agreed that continuing to do so would mean that American companies, primarily Hamilton Standard, were able to reinvest the license fees into further research and development, which further reinforced their dominance in propellers. The Air Ministry needed to break the American hold by supporting propeller innovation at Rotol.[104]

A way to gain a larger share of the propeller market was to use Rotol propellers on recordbreaking British aircraft. Great Britain, just like the governments of United States and Germany, used flights of spectacle to publicize its national technology, aviation organizations, and status as an aeronautical, and ultimately modern, nation. On Saturday, November 5,

[101] "Harry Lawley Milner," in *The Aeroplane Directory of British Aviation* (London: Temple Press, 1955), 503.

[102] Gunston, *Fedden*, 196.

[103] "The Rotol Airscrew," *Flight* 35 (March 23, 1939): 296; Stait, *Rotol*, 5–7.

[104] W. R. Freeman to A. H. R. Fedden, August 11, 1938; and Notes of Meeting of Air Council Committee on Supply, "Rotol Airscrews Ltd.," September 8, 1938, AVIA 10/223, TNA.

FIGURE 20 The Vickers Wellesleys of the RAF Long Range Development Unit equipped with three-blade Rotol constant-speed propellers in April 1938. Smithsonian National Air and Space Museum (NASM 2007–4084).

1938, three modified Vickers Wellesley single-engine bombers of the RAF Long-Range Development Unit (LRDU) flew south out of Ismailia, Egypt. Two of the aircraft reached Darwin, Australia, forty-eight hours later after establishing a new nonstop distance record of 7,162 miles. Building upon the platform of an operational Wellesley, which featured retractable landing gear and a unique geodetic airframe design innovated by British inventor, Barnes Wallis, the LRDU introduced several modifications. New aerodynamic features included enclosed crew compartments and a Bristol "long chord" engine cowling (reminiscent of the highly successful NACA design) to create a more streamline shape for fuel economy. Two crucial long-distance technologies were extra fuel tanks and an automatic pilot. The propulsion system consisted of an 840-horsepower supercharged nine-cylinder Bristol Pegasus XXII radial engine powered by 100-octane fuel, and a twelve foot eight inch diameter internal cylinder Rotol constant-speed propeller featuring three magnesium alloy blades.[105] The LRDU Wellesleys showcased the latest in British aeronautical technology (Figure 20).

[105] "Triple Triumph: The Long-Range Development Unit Breaks World's Distance Record," *Flight* 34 (November 10, 1938): 426–428.

Despite the desire to see Rotol succeed as the only British-designed propeller, the Air Ministry was well aware of the successful development of the VDM propeller. VDM pursued licensing opportunities outside of Germany even though its primary, and most lucrative, market was equipping the Luftwaffe. A syndicate of investors, which recruited former employees of Blackburn Aircraft Ltd. and the Saunders-Roe Company, finalized the license for design and the manufacturing in Great Britain of the VDM propeller along with a novel spinner design for £25,000.00 in December 1937. Operating under the name Constant Speed Airscrews Ltd., they acquired the defunct 45,000 square foot Avon automobile body factory on Wharf Street in Warwick for manufacturing and expected the delivery of machine tools from Germany by mid-1938.[106]

According to the newly formed company, the VDM propeller fulfilled in every possible way the longstanding need for a variable-pitch propeller in Great Britain. The plan was to offer the German propeller for civilian use initially with opportunities in the military sector to follow.[107] After receiving a controllable variant in June 1937, the RAE conducted whirl and endurance tests that led to the approval of the controllable propeller for use on civil aircraft powered by Merlin engines.[108]

Compared to American and British designs, the VDM was the lightest variable-pitch propeller available. A twelve foot diameter VDM weighed 314 pounds. The Rotol external cylinder propeller weighed 333 pounds. The Hamilton Standard-licensed de Havilland two-pitch and constant-speed bracket propellers weighed 370 and 375 pounds respectively.[109] For those in the international aeronautical community still obsessed with weight above performance, the VDM offered a clear advantage between fifteen and fifty-five pounds.

Both the Air Ministry and Constant Speed Airscrews were anxious to get examples of the constant-speed propeller, but VDM's leadership under Unholtz put them off with the reason that he felt they were not yet

[106] J. S. Buchanan, "Minute Sheet," January 5, 1937; A. H. Metcalfe to J. S. Buchanan, January 17, 1937; "Note of an Interview with Representatives of the VDM and Constant Speed Airscrews Ltd.," December 7, 1937, AIR 2/2406, TNA.

[107] A. H. Metcalfe to J. S. Buchanan, January 17, 1937; S. Davey to Constant Speed Airscrews Ltd., "VDM Constant Speed Airscrews," February 4, 1937; and Chief Superintendent, Royal Aircraft Establishment to the Air Ministry, "New Type Airscrew by Constant Speed Airscrews Ltd.," April 5, 1937, AIR 2/2406, TNA.

[108] J. S. Buchanan to Constant Speed Airscrews Ltd., "VDM Airscrew Designed for Merlin Engine," December 23, 1937, AIR 2/2406, TNA.

[109] A. H. Hall to the Air Ministry, "Relative Merits of Curtiss and VDM Variable-Pitch Airscrews," January 3, 1938; and H. B. Howard, "VDM Airscrew Designed for Battle-Merlin–General Report," November 14, 1937, AIR 2/2406, TNA.

ready for production.[110] Work on the propeller was taking place against the backdrop of the rising international tensions generated by Hitler's demand for the Sudetenland from Czechoslovakia. Avoiding war, the Munich Conference of September 1938 left Anglo-German relations, at least on the surface, on a high note. VDM promised that a constant-speed unit would be forthcoming to Great Britain as soon as it met its more pressing obligations in Germany.[111]

VDM never offered a constant-speed propeller to its licensees until it received permission from the RLM in February 1939. Even then, the Nazi leadership dictated the secrecy of the design even down to the details of shipping the drawings in a special dispatch sealed with lead.[112] As relations between Great Britain and Germany worsened, Constant Speed Airscrews found itself in an awkward position during the spring of 1939. They expended a lot of time, money, and effort into the VDM project, but the deteriorating political situation contributed to an ironic situation. Engineer Robert McGlasson wrote to the Air Ministry that testing the German propeller was a "national necessity" that would increase the efficiency of British military aircraft "from a National Defence [*sic*] point of view." If the Air Ministry saw fit, a constant-speed propeller could be in England within three weeks. After successful tests, the RAF could rely on VDM propellers to defend Great Britain and to maintain its commitments to its allies in Europe.[113]

A VDM constant-speed propeller did appear in Great Britain during the summer of 1939. Two factory mechanics escorted it, hidden away from prying eyes in a sealed container, to Derby for testing at Rolls-Royce. The engine manufacturer tested the device on a Merlin engine installed in a Fairey Battle tactical bomber. After the conclusion of the tests, the mechanics and the propeller returned to Hamburg.[114] With war clouds on the horizon, it was really too late for any credible testing and evaluation. The outbreak of hostilities in September 1939 put an end to any plans to produce the VDM propeller in Great Britain. Despite the name, Constant Speed Airscrews never produced a propeller in its Warwick factory, but

[110] C. H. Warne to A. R. Collins, July 28, 1938, AIR 2/2406, TNA.

[111] A. R. Collins to the Air Ministry, "VDM Airscrews," November 2, 1938, AIR 2/2406, TNA.

[112] VDM to Constant Speed Airscrews, "Re: Constant Speed Device for VDM-VP Airscrews," February 21, 1939, AIR 2/2406, TNA.

[113] R. McGlasson to the Air Ministry, April 19, 1939, AIR 2/2406, TNA.

[114] R. McGlasson to Mr. Collins, "Forthcoming Tests for VDM Constant-Speed Airscrew on Merlin Engine," June 5, 1939, AIR 2/2406, TNA.

did produce the lightweight and easily removable spinner for virtually all RAF aircraft during the upcoming conflict.[115]

By 1939, the British aeronautical community clearly recognized that variable-pitch propellers had their uses, but not in the universal way demonstrated by the United States and Germany. To the British, the new innovation, in the form of the two-pitch bracket propeller, was adequate for use on multiengine bombers, transports, and airliners. For single-engine fighters, however, there was continued reluctance to use anything but a constant-speed propeller. The de Havilland variable-pitch propeller of American origin was too heavy, the British designed and manufactured Rotol was not yet available, and the VDM proved to be a disappointing distraction for its British investors. For the time being, the community believed that engine power alone would overcome the performance deficiencies created by the use of a fixed-pitch propeller that resulted from the longstanding emphasis on light weight and simplicity of operation.

Differing Attitudes toward Technology

The late 1930s witnessed the realization of the international aeronautical community's two-decade quest to produce aircraft capable of flying higher, faster, and farther than ever before. To achieve that, the designers of these new all-metal streamline monoplanes incorporated a combination of synergistic advances to increase performance. The three communities of the United States, Germany, and Great Britain integrated the critical variable-pitch propeller into their aircraft in different ways and for dissimilar reasons.

In the United States, the revolutionary period of propeller innovation and acceptance was at an end and American propellers flooded the international aeronautical marketplace. A new period of refinement settled in as two manufacturers, Hamilton Standard and Curtiss-Wright, competed against each other for lucrative commercial and military contracts with improved versions of their original propellers destined for use on "modern" airliners, bombers, fighters and pursuit planes, transports. There were also murmurs of Pete Blanchard and Charles MacNeil's new design emerging from the home of all propellers, Dayton, Ohio.

Across the Atlantic, the rise of totalitarian government and intense nationalism encouraged aeronautical innovation in Nazi Germany.

[115] Waugh, "Details of the German VDM Electric Propeller, Part I," 32; Waugh, "Particulars of the German VDM Electric Propeller," 348.

Advanced aircraft appeared in the skies over civil war-torn Spain and the international competition grounds of Switzerland to reveal yet another instance of innovation in propeller technology. Rather than a long development driven by the needs of commercial aviation and incremental growth, Hans Ebert's VDM propeller was constant-speed from the start and destined for military aircraft.

The British variable-pitch propeller community adapted and persevered within a restrictive creative, technological, and financial environment. Work on the development of the Hele-Shaw Beacham constant-speed propeller continued without a single operational model taking to the sky while foreign propeller technology became increasingly available. The highly successful airframe and engine manufacturer, de Havilland, built and sold licensed versions of Hamilton Standard propellers that proved to be the backbone of the British aviation industry. European innovations in variable-pitch mechanisms and blade designs also permeated the environment. Nevertheless, there was a push for an all-British constant-speed propeller for certain aircraft.

Those optimistic for the future believed that the modern airplane and the worldwide aerial networks made possible by the new technology would usher in a new era for humankind, the Air Age, based on international harmony and cooperation. Instead, these aeronautical nations fought a world war in the air with those revolutionary modern airplanes. The United States and Germany fully recognized the importance of installing constant-speed propellers on fighter, bomber, and transport aircraft. Great Britain was still reluctant to do the same. RAF Fighter Command's experience during the first year of World War II would make that unwillingness a matter of life, death, and national survival.

10

"The Spitfire Now 'Is an Aeroplane' "

On August 24, 1940, Flight Lt. Gordon Olive led "A" Flight of 65 Squadron on an early afternoon patrol over the Thames Estuary. Controllers directed the four Spitfires to intercept an incoming raid of approximately sixty bombers protected by a heavy escort of forty fighters at about 20,000 feet. Seeing the fighters above them, Olive ordered "A" Flight to climb. They attacked "down sun," with the sun at their backs and in the eyes of their targets, from 28,000 feet. He damaged a twin-engine Messerschmitt Bf 110 fighter that dived away to escape destruction. Aware of the danger of other Luftwaffe aircraft in his vicinity, Olive climbed above them and pursued another Bf 110. As he was about to destroy his second target, five single-engine Bf 109s attacked him from above. Olive immediately climbed toward and above them. He attacked the rearmost fighter until running out of ammunition before returning to the squadron's forward airfield at Manston near the Channel Coast in Kent.[1] In a chaotic and deadly environment where altitude, or "the high ground," was the key to survival, Olive had climbed three times, met his enemies, and drove them away. Just two months before, his Spitfire struggled against the Luftwaffe's fighters in combat. The addition of a constant-speed propeller of American origin enhanced the performance of Olive's Spitfire and helped make it a truly world-class fighter.

During the spring, summer, and fall of 1940, Fighter Command of Great Britain's RAF faced Nazi Germany's Luftwaffe in an aerial duel to control the skies over France and England. After the fall of France, this struggle for the air, if lost by the RAF, would be a prelude to an invasion

[1] Flight Lt. G. Olive, Combat Report, August 24, 1940, AIR 50/25, TNA.

of the British Isles. The resultant Battle of Britain represented a dramatic change in warfare where the survival of a nation depended upon a small group of aviators, called the "few" by Prime Minister Winston S. Churchill, and their Spitfire and Hurricane fighter aircraft. It also demonstrated the importance of air power weaponry, which included well-known systems such as radar and the fighter-interceptor, but fundamental technologies such as the variable-pitch propeller as well.

The technical development of the variable-pitch propeller in Great Britain followed a different, albeit still circuitous, path than in North America and Continental Europe. As homegrown American and German propellers took to the air in the mid-1930s, there had yet to be a British-designed variable-pitch propeller put into production. For commercial and military operators, American Hamilton Standard variable-pitch propellers satisfied the void left unfilled by British metal and wood fixed-pitch manufacturers. The development and modification of the Spitfire and Hurricane into the legendary aircraft of the "few" in 1940 coincided with the final community-wide acceptance of the variable-pitch propeller in Great Britain. That transformation revealed the crucial role of users in the shaping of technology as Fighter Command pilots worked to incorporate constant-speed propellers into their aircraft for more performance.[2]

Rearmament, Air Defense, and the Modern Fighter

As Nazi Germany rose as a threat to peace in Europe in the mid-1930s, the British government prepared for war and rushed to mobilize its industry. Alarmed by the rise of the Luftwaffe, the Air Ministry initiated the expansion of the RAF to increase the number of Great Britain's fighters, bombers, and maritime aircraft in the name of national defense beginning in 1934.[3] Japanese imperialist expansion into China and the outbreak of the Spanish Civil War, both of which included systematic bombing of civilian targets, added extra impetus as the tumultuous decade continued to demonstrate the emerging horrors and importance of aerial warfare.

During the expansion program, Fighter Command transitioned from its World War I–inspired biplanes to Hawker Hurricane and Vickers-Supermarine Spitfire monoplanes. Both evolved from a 1934 Air Ministry

[2] Ronald Kline and Trevor Pinch, "Users as Agents of Technological Change: The Social Construction of the Automobile in the Rural United States," *Technology and Culture* 37 (October 1996): 764–765.

[3] Denis Richards, *The Fight at Odds*, vol. 1 of *Royal Air Force, 1939–1945* (London: Her Majesty's Stationery Office, 1953), 1–4.

specification for a defensive fighter monoplane with eight machine guns, an enclosed cockpit, retractable landing gear, and a Rolls-Royce liquid-cooled V-12 engine. At Hawker, Sydney Camm designed the Hurricane with a traditional engineering approach best reflected by its half-metal, half-wood fuselage and fabric covering. Supermarine's chief engineer and designer, Reginald J. Mitchell, and his colleagues incorporated all-metal stressed-skin construction and a distinctive elliptical wing into the state-of-the-art Spitfire. With the operational introduction of the Hurricane in December 1937 and the Spitfire in August 1938, RAF Fighter Command had two "modern" fighters capable of 300 mph.

Fighter Command, under the leadership of Air Chief Marshal Sir Hugh Dowding, relied upon a sophisticated network of radar stations and ground-based observers, a system of radio-based aircraft direction and control, and squadrons of fighter aircraft to defend the British Isles. Radar, short for radio detection and ranging, was a new technology that used radio waves to detect flying aircraft, which gave Fighter Command advanced notice and the location of enemy raids. An army of ground-based observers supplemented the detection abilities of radar. Controllers took information from these two sources and directed individual RAF units to their targets by radio. Theoretically, the integrated air defense system allowed a small number of fighters to attack a numerically superior enemy air force effectively and within a matter of minutes.[4]

Fighter Command's Spitfire and Hurricane squadrons were central elements in this air defense network. RAF fighter doctrine focused on the coordinated destruction of unescorted high speed enemy bombers. The RAF predicted that no enemy fighters could reach England. Even if they did, no dogfighting would occur due to the high speeds of the new modern aircraft of the late 1930s that would deter accurate interception and force the pilots to black out from excessive g-forces. The answer was to create sophisticated tactical maneuvers based on formation flying with the primary unit being a three-aircraft section flying in an arrow-like "vic." In the event of war, Fighter Command saw the tactics as simple enough in execution to easily train new pilots and to use them as the basis for improvisation as conditions warranted.[5]

[4] David Zimmerman, *Britain's Shield: Radar and the Defeat of the Luftwaffe* (Stroud: Sutton Publishing, 2001), 109–131.

[5] Air Ministry, *Air Publication 129: Royal Air Force Flying Training Manual, Part I – Landplanes*, November 1937, and Headquarters, Fighter Command, *Fighter Command Attacks-1939 (Provisional)*, February 1939, Royal Air Force Museum Archives and

By the late 1930s, the British aviation industry was strong and viable in terms of the number and basic quality of its aircraft compared to its allies and potential enemies.[6] As modern as the RAF believed the Hurricane and Spitfire to be, the two fighters did not incorporate all of the synergistic systems of the Aeronautical Revolution. A variable-pitch propeller was not available at the time of either fighter's operational introduction. De Havilland made plans to produce variable-pitch propellers for the Spitfire and Hurricane as early as 1936, but the Air Ministry and RAF Fighter Command did not want them. To these government and military leaders, British fighters needed to be as light and powerful as possible without the extra weight of a variable-pitch propeller.[7]

The first squadrons of Hurricanes and Spitfires entered RAF service with two-blade fixed-pitch wood propellers designed and manufactured by the Airscrew Company of Weybridge, Surrey. Founded in 1923 from the remnants of the defunct Lang Propeller Company, the Airscrew Company rose to be Great Britain's predominant propeller manufacturer by 1930. The major aircraft manufacturers, the RAF, the state-run Imperial Airways, and well-known long-distance pilots such as Sir Alan Cobham used the company's laminated wood propeller on their aircraft. The British aeronautical community called the Airscrew Company's products "Watts propellers" because of the well-known designer, Dr. Henry C. Watts, who had joined the firm in 1932. Air Ministry procurement officials thought they were ideal because they gave ultimate performance at designated fighting altitudes between 15,000 and 20,000 feet and weighed only eighty-three pounds.[8] An essential feature to the company's product was the process of composite construction provided under license by Schwarz in Germany, which the RAF adopted as standard in June 1935.[9]

The Airscrew Company designed its propellers to generate the best performance at the upper end of performance envelope. That meant the performance of the Spitfire ranged from a stall speed of 70 mph, the absolute

Library; Stephen Bungay, *Most Dangerous Enemy: A History of the Battle of Britain* (London: Aurum Press, 2001), 249–250.

6 Sebastian Ritchie, *Industry and Air Power: The Expansion of British Aircraft Production, 1935–1941* (London: Frank Cass, 1997), 2–3.

7 "Pitch Panic: How Hurried Changes from Two-Pitch to Constant-Speed Airscrews Were Made in Time for the Battle of Britain," *Flight* 44 (December 9, 1943): 648.

8 Tony Deeson, *The Airscrew Story, 1923–1998* (n.p.: Airscrew Howden, n.d. [1999]), 13, 19, 22.

9 H. C. Watts to Messrs. Gustav Schwarz, June 27, 1935, File B5-782000-01, PTF, NASM.

minimum to generate lift, to a top speed of 362 mph.[10] Supermarine test pilot Jeffrey Quill noted that the amount of torque created by the fixed-pitch propeller designed for maximum speed at 3,000 rpm at 18,000 feet meant that a Spitfire pilot was hard-pressed to take off in a straight line and had an easier time getting aloft from a wide open grass airfield rather than a narrow paved runway. Despite the long diagonal takeoff run, the two-blade fixed-pitch wood propeller generated what was considered the best handling qualities once the airplane was in the air. Contrary to what he had advocated in *Flight* in 1935, Watts and his designers believed that the power available at takeoff was so ample and the savings in weight so great that it did not matter that the blades actually stalled at the beginning of takeoff run.[11]

The terrible takeoff performance of the Spitfire Mark I reflected the emphasis on speed and performance in the air and the belief that the aircraft would be fighting in the specific conditions the designers, and the Air Ministry, envisioned. Reality relegated the new fighters to Fighter Command's largest airfield at Duxford and created dangerous situations at the smaller airfields. The first seventy-seven production Spitfires had wood, two-blade fixed-pitch propellers installed.[12]

The RAF's first Spitfires were received by 19 Squadron at Duxford in August 1938 (Figure 21). During the tumultuous days leading up to the Munich Agreement at the end of September, the RAF had only one squadron until Spitfires destined for 66 Squadron arrived in late October. The two squadrons immediately began to evaluate their new aircraft. Besides a revised canopy shape for better visibility, higher quality oil seals for the engine, and stronger gearing for the starter motor for increased durability, an emerging consensus among the Duxford squadrons was the new Spitfire needed a variable-pitch propeller for better takeoff performance and safety.[13] According to Squadron Leader Henry Illife Cozens

[10] Aeroplane and Armament Experimental Establishment, Martlesham Heath, "Spitfire K.9787–Merlin II–Performance Trials," January 6, 1939, Collection of Mike Williams, www.spitfireperformance.com/k9787.html (Accessed December 28, 2011).

[11] H. F. King, "With the Squadrons," *Flight* 33 (May 26, 1938): 517; Eric B. Morgan and Edward Shacklady, *Spitfire: The History*, rev. ed. (Stamford, CT: Key Books, 2000), 28, 30.

[12] Deeson, *The Airscrew Story, 1923–1998*, 22; Douglas P. Tidy, "Myths of the Battle of Britain," *Military History Journal* 5 (June 1980), http://samilitaryhistory.org/vol051dt.html (Accessed February 18, 2010); Alfred Price, *Spitfire in Combat* (Stroud, England: Sutton Publishing, 2003), 21.

[13] "Pitch Panic," 648.

FIGURE 21 RAF Fighter Command's 19 Squadron received its new Spitfire fighters equipped with two-blade Airscrew Company fixed-pitch propellers. Courtesy of the Trustees of the Royal Air Force Museum.

of 19 Squadron, a stalled propeller at lower speeds during takeoff was a serious performance disadvantage.[14]

Squadron Leader Donald S. Brookes, an engineering specialist assigned as the RAF's liaison officer to the Supermarine factory at Southampton, was aware of the problem. With the RAF since 1919, Brookes still flew and was in touch with his fellow fighter pilots. He personally delivered the first Spitfire assigned to his previous command, 74 Squadron, at Hornchurch fighter station in February 1939.[15] On his own initiative, Brookes contacted de Havilland at Stag Lane to see if they could supply a variable-pitch propeller. De Havilland provided a three blade two-pitch bracket design with lightweight blades as used on the Fairey Battle. With the new propeller, RAF pilots enjoyed a takeoff run decreased from 420 to 320 yards, an increase in rate of climb from 175 mph to 192 mph, and

[14] Henry Iliffe Cozens, "Into Service," in Alfred Price, *Spitfire at War* (London: Ian Allan Ltd., 1974), 18.
[15] Richard C. Smith, *Hornchurch Scramble: The Definitive Account of the RAF Fighter Airfield, Its Pilots, Groundcrew and Staff*, vol. 1 (London: Grub Street, 2000), 41.

an increased maximum speed by five mph, and service ceiling raised by 3,000 feet to 34,000 feet. The tradeoff was an increase of 193 pounds to the overall weight of the aircraft.[16]

The paradigms of weight and simplicity guided Fighter Command's original decision to equip its fighters with fixed-pitch propellers. The initial squadron trials at Duxford and Hornchurch proved that approach to be impractical. The RAF responded by designating the de Havilland controllable two-pitch bracket propeller as standard equipment for all new Spitfire and Hurricane eight-gun fighters and the Boulton-Paul Defiant four-gun turret fighter leaving the factory.

The two-pitch bracket propeller offered an increase in performance, but there was a problem. Young and eager fighter pilots simply forgot to operate them correctly. In March 1939, Al Deere, a twenty-one year old pilot officer from New Zealand, and his fellow pilots in 54 Squadron became the third RAF unit to convert from Gloster Gladiator biplanes to Spitfires. Excited over his new fighter, he forgot to change his pitch setting to cruise after takeoff, which would decrease engine rpm in flight, after taking delivery of it at Eastleigh airfield near the Supermarine factory at Southampton. His Merlin engine over-revved for the entire flight across southern England.

Arriving over Hornchurch, Deere could not resist an unauthorized "beat-up" of the station before landing. Flying low and fast and performing aerobatics over the fighter station was not an acceptable activity. The howling noise of the maxed-out engine and the propeller racing with the tips going supersonic was an indication of Deere's oversight to everyone on the ground. Upon landing, the sight of the front half of his brand-new Spitfire smothered in engine oil was the confirmation. A red-faced Deere faced the ire of his flight leader more for his inability to operate the two-pitch propeller correctly and nearly destroying his engine rather than for his breach of regulations over Hornchurch.[17]

[16] The weight of the de Havilland two-pitch bracket propeller was 345 pounds, which also required 66 pounds added to the rear of the airplane to correct the center of gravity. Subtracting the eighty-three pound weight of the Airscrew Company propeller and 135 pounds of lead from the engine bearers needed to balance the Spitfire, the total increase in weight was 193 pounds. Aeroplane and Armament Experimental Establishment, Martlesham Heath, "Spitfire K.9793–Merlin II–Short Performance with 2 Pitch Metal Airscrew," July 12, 1939, AVIA 18/636, TNA; "The Spitfire," *The Aeroplane* 58 (April 12, 1940): 518; Bill Sweetman, *Spitfire* (New York: Crown Publishers, 1980), 8; Price, *Spitfire in Combat*, 21.

[17] Alan C. Deere, *Nine Lives* (London: Hodder and Stoughton, 1959), 36–37.

Careless pilots would also land with their propellers set to cruise and forget to set them to takeoff for their next flight. That setting was dangerous because the engine and propeller could not produce maximum power to accelerate for takeoff leaving it in a near-stall condition. It was possible when flying from one of the RAF's long and wide grass runways such as Duxford to get off the ground, but in the heat of a squadron scramble, however, where every second counted, that would mean a Spitfire or Hurricane would take much longer to get off the ground, if at all. Squadron Leader Cozens of 19 Squadron remembered his dislike for the new propeller because "there was no half-way house" in terms of performance.[18]

For a Fighter Command pilot, a two-pitch propeller was not ideal for the dynamic conditions of aerial combat, or general flight for that matter. It offered the performance of two fixed-pitch propellers, one set for takeoff and the other for cruise, but with a wide gap in between. Moreover, it was one more flight system that required attention in the cockpit that took away from their central mission of flying and fighting.

By the spring of 1939, the mainstream British aeronautical community was discussing the use of the two-pitch bracket propeller in comparison to the performance and safety benefits of the constant-speed propeller. Air Ministry officials discussed the reluctance of Imperial Airways to adopt new technologies in light of the imminent establishment of the new British Overseas Airways Corporation the following fall. Under Secretary of State for Air, Harold Balfour, stressed that "nearly every country in the world" had abandoned the two-pitch airscrew in favor of constant-speed for its airliners. Pilots of the four-engine Short Empire flying boats complained of difficult takeoffs when they operated in restricted expanses of water. Constant-speed propellers alleviated that problem. The transition would be costly, but Balfour stressed that "frequent and heavy expenditure for a comparatively minor technical advance" was part of being in the "aviation business." Before the new airline and the Government that controlled it came under legitimate international scrutiny, Balfour believed the airliners had to be brought up to what had become "standard practice" in other nations.[19]

[18] Cozens, "Into Service," 18.

[19] "Imperial Airways Ltd.: Employment of De-Icer Equipment and Constant-Speed Airscrews," April 3, 1939, AVIA 2/1493, TNA.

Learning the Lessons of War

As the British aeronautical community continued to evaluate its position on variable-pitch propellers, the threat of conflict gained momentum across the English Channel. On September 1, 1939, the Nazi Wermacht, with the Luftwaffe flying above, stormed into Poland and started World War II. The next day, the twelve bomber and fighter squadrons of the RAF's Advanced Air Striking Force flew to France. Great Britain officially declared war on Germany on September 3. The British Expeditionary Force (BEF), accompanied by its own tactical Air Component of thirteen fighter, reconnaissance, and liaison squadrons, followed in October. With the outbreak of war, the Air Ministry and the RAF worked at a desperate pace to get as many fighters as possible. Secretary of State for Air Sir Howard Kingsley Wood reported to the War Cabinet in December 1939 that Spitfire and Hurricane production increased by 50 percent since the previous July.[20]

The British war machine faced a significant challenge when it came to providing variable-pitch propellers for the RAF's fighters. Planners faced severe shortages in material, skilled labor, tooling, and other areas as they tried to forecast how to provide a complicated piece of technology in the numbers needed. The fact that only two variable-pitch manufacturers – de Havilland and Rotol – existed to establish a foundation was sobering.[21] The bitter reality was that neither company was in a position to produce their high performance Hydromatic and external cylinder constant-speed propellers.

The Air Ministry financed the expansion of the industry to maintain the momentum of supply. The burden rested squarely with de Havilland. The Air Ministry decided to increase production of the technically less superior bracket airscrew at Lostock rather than risk the time for retooling the equipment and retraining the work force for making Hydromatic propellers. A £350,000 grant doubled the hub manufacturing area. As the bitter winter of 1939–1940 settled in, the workers at Lostock worked double shifts to produce approximately 3,000 bracket propellers. At one point, heavy snows prevented any outside access to the factory except by animal-drawn sledge. Soon after, five "dispersal factories," formerly defunct cotton mills, opened in the Lancashire countryside to augment

[20] Howard Kingsley Wood, "Supply and Production: Fourth Report by the Air Ministry," December 19, 1939, CAB/68/3/40, TNA, 314–315, 340, 343.

[21] Ministry of Supply, "Propellers: Development and Production, 1934–1946," n.d. [1950], AVIA 46/211, TNA, 14.

Lostock's production with machining of hubs, shaping of blades, and assembly.[22]

The need for more propellers spread out to the greater British industry in the following months. The British automobile industry proved ideal for the type of work needed to manufacture variable-pitch hub components for de Havilland. Alvis and the Standard Motor Car Company of Coventry and ABC Motors of Walton-on-Thames in Surrey became subcontractors to de Havilland under Air Ministry contract. Standard received £13,000 in April 1940 for machine tools that it would use to fabricate component parts such as hubs, blade adapters, and the ever-important constant-speed governors designated at the time for Bomber Command aircraft.[23]

The fact that Rotol was only beginning to produce small batches of constant-speed propellers exacerbated the situation. Several factors contributed to the problem. First of all, Rotol was working toward delivery of 1,936 propellers for installation on Armstrong-Whitworth Whitley medium bombers, Fairey Albacore torpedo planes, Bristol Blenheim light bombers, and Miles Master trainers, contracted in 1938 with delivery completed in the spring of 1940.[24] The community-wide belief that multiengine aircraft benefitted most from variable-pitch propellers placed the Air Ministry's emphasis in that direction.

Second, Rotol was committing most of its engineering resources to the adaptation of the Curtiss Electric propeller for the British market. Earlier in September 1936, Fedden at Bristol had arranged for the licensing rights to manufacture and distribute the propeller. Fedden echoed the belief that a hydraulically actuated propeller was less than ideal due to the likelihood of the device leaking oil. An electrically actuated propeller would alleviate this problem. Rotol took over the project after its formation. Despite the success of Curtiss in the United States, the long-suffering Rotol program centered on correcting a pitch actuation mechanism found to be too delicate for service use, especially after the first production batches were completed during the spring of 1940. From the manufacturing perspective, the design required new production techniques, which was unacceptable to the RAF. The entire electric distraction wasted

[22] Ibid., 20–21, 25.

[23] Kingsley Wood, "Supply and Production: Fourth Report by the Air Ministry," 320; "Airscrew Components," *The Aeroplane* 58 (April 5, 1940): 496; Ministry of Supply, "Propellers: Development and Production, 1934–1946," 17, 21.

[24] W. R. Freeman to A. H. R. Fedden, August 11, 1938, AVIA 10/223, TNA.

effort in terms of both engineering and worker resources and adversely affected production of hydraulic propellers.[25]

Finally, unlike the case of de Havilland, Rotol was not ready to manufacture propellers in large quantities. It was not until after the invasion of Poland in September 1939 that the Air Ministry and Rotol worked together to create a network of contractors and subcontractors to facilitate volume production.[26] In the meantime, Rotol received £300,000 for expansion at the Staverton factory.[27]

The installation of Rotol propellers on RAF fighters had not been overlooked entirely. Ernest W. Hives, representing Rolls-Royce and Rotol, and the Air Ministry's Committee of Supply discussed propellers for Spitfire and Hurricane fighters during the summer of 1938. The Air Ministry wanted an experimental order produced for the purposes of training personnel and working out manufacturing methods. Rotol asked in return that the Air Ministry provide detailed information pertaining to how many actual production propellers would be ordered and in what time frame. The Air Ministry offered to place a substantial order for 1,500 provided a satisfactory prototype was available in December, featured wood blades, and a production unit was comparable in cost to a de Havilland propeller.[28] It is not clear if the large order ever materialized or if Rotol simply could not produce the propellers in quantity fast enough after it delivered the prototype.

Regardless of those problems, as soon as Rotol propellers became available, the RAF introduced them to its operational units on a limited trial basis. One of the first was in early March 1939 when Squadron Leader Desmond Cooke of 65 Squadron conducted flight tests of a Rotol variable-pitch propeller with his Gladiator biplane fighter. He noted that the new propeller was "now running well" after twelve hours in the air. The tests would be short-lived since 65 Squadron became the fourth RAF unit to convert to the Supermarine Spitfire Mark I fighter

[25] T. A. Sims to Chief, Materiel Division, Liaison Section, "Controllable-Pitch Propellers," September 16, 1936, Folder "Curtiss Electric Propeller, 1934–1938," Box 6302, RD 3330, RG 342, NARA; Ministry of Supply, "Propellers: Development and Production, 1934–1946," 23–26.

[26] Ministry of Supply, "Propellers: Development and Production, 1934–1946," 19.

[27] Kingsley Wood, "Supply and Production: Fourth Report by the Air Ministry," 314–315, 340, 343.

[28] W. R. Freeman to A. H. R. Fedden, August 11, 1938; R. Stammer to the Secretary of the Air Council Committee on Supply, August 17, 1938; and Notes of Meeting of Air Council Committee on Supply, "Rotol Airscrews Ltd.," September 8, 1938, AVIA 10/223, TNA.

with its standard two-pitch bracket propeller three days later.[29] Later on November 1, Squadron Leader Cozens collected the first "Rotol Spitfire" for 19 Squadron to conduct extensive flying and reliability trials at Duxford.[30] 54 Squadron at Hornchurch received a unit's worth of Rotol Spitfires later that December.[31] 41 Squadron received a handful of similarly equipped Spitfires in January 1940.[32]

These fighter propellers were the external cylinder type, which Rotol designers believed was best due to the wider range of movement at thirty-five degrees.[33] Observers keeping up with the new technology installed on the Mark I variants of the Hurricane and Spitfire could identify the new Rotol propellers by their wide blade roots and blunt, rounded spinners covering the hub. De Havilland propellers as found on 65 Squadron's Spitfires featured longer pointed spinners and narrow blade roots.

In theory, the Air Ministry, the RAF, and Fighter Command had committed to the idea of constant-speed airscrews for its Hurricane and Spitfire fighters by the spring of 1940, but only to the Rotol design. The reasons were clear. Comparison flight tests of Spitfires equipped with Watts fixed-pitch, de Havilland two-pitch, and Rotol constant-speed airscrews at the Aeroplane and Armament Experimental Establishment at Boscombe Down in March 1940 revealed the latter generated the highest performance at takeoff, climb, and in level flight.[34] In addition, the Rotol was of British origin from a company with a proven engineering foundation through its connections to Rolls-Royce and Bristol.

The de Havilland bracket propeller was an unattractive alternative for two main reasons. For many, the design left much to be desired. Philip Lucas, one of the Hawker test pilots involved in the development of the Hurricane, remembered how the de Havilland propeller leaked and slung oil into the airstream, which would blow onto the windscreen obscuring the pilot's view. He also complained of the slow throttle response of the constant-speed version, especially during combat maneuvers. The Rotol

[29] Pilot's Flying Log Book for D. Cooke, January 1935-July 1940, AIR 4/18, TNA, 81–86.

[30] Royal Air Force Operations Record Book: History of 19 Squadron, November 1, 1939, AIR 27/252, TNA, 33.

[31] Royal Air Force Operations Record Book: History of 54 Squadron, December 10, 12–14, 1939, AIR 27/252, TNA, 61.

[32] Royal Air Force Operations Record Book: History of 41 Squadron, January 20–21, 1940, AIR 27/424, TNA, 68.

[33] "The Latest Rotol Airscrew," *Flight* 37 (May 23, 1940): 472.

[34] Aeroplane and Armament Experimental Establishment, Boscombe Down, "Spitfire N.3171–Merlin III/Rotol Constant Speed Airscrew–Comparative Performance Trials," March 19, 1940, Collection of Mike Williams, www.spitfireperformance.com/n3171.html (Accessed December 28, 2011).

neither leaked nor was it slow to respond.[35] Second, Rotol engineered its design to be the lightest variable-pitch propeller available at 333 pounds. The de Havilland two-pitch bracket propeller weighed 370 pounds. The addition of the components to make the propeller capable of constant-speed operation added five more pounds.[36]

The combination of those factors ensured that the de Havilland two-pitch bracket propeller was already too heavy so its constant-speed version was probably not even considered as a practical long-term solution to improving fighter performance. The British aeronautical community was still an engineering culture that believed weight had to be kept down at all costs.

The major problem with the Rotol propeller was availability. By the spring of 1940, Hurricane Mark I fighters were just emerging from the factory with Rotol propellers, which increased their top speed from 318 to 328 miles per hour at 16,200 feet.[37] In the meantime, the de Havilland two-pitch bracket propeller would remain the standard for Spitfire Mark I aircraft. Fortunately for a service committed to one design and only able to get the other, Rolls-Royce incorporated a universal propeller shaft capable of accepting either the Rotol or the de Havilland propeller into the engine chosen for both aircraft, the 1,030 horsepower Merlin III.[38]

Beginning in February 1940, the introduction of another propulsion innovation, 100 octane fuel, expanded further the power of the Merlin engine. The more efficient fuel burned cleaner and facilitated a higher manifold pressure, twelve pounds per square inch, which increased output by approximately 200 horsepower. Pushing the boost cut-out control, or "pulling the plug," gave the fighter even more "emergency power," which was a performance-enhancing option pilots did not hesitate to use in combat.[39] The increase in power was wasted on a fixed-pitch propeller because it exacerbated the poor takeoff characteristics of that configuration.

[35] Philip G. Lucas, "A Hurricane History," in *Sydney Camm and the Hurricane: Perspectives on the Master Fighter Designer and His Finest Achievement*, ed. John W. Fozard (Washington, DC: Smithsonian, 1991), 164–165.

[36] A. H. Hall to the Air Ministry, "Relative Merits of Curtiss and VDM Variable-Pitch Airscrews," January 3, 1938, AIR 2/2406, TNA.

[37] "The Latest Hurricane," *The Aeroplane* 58 (May 3, 1940): 611; Francis K. Mason, *The Hawker Hurricane* (Garden City, NY: Doubleday, 1962), 36.

[38] "The Rolls-Royce 'Merlin'," *Jane's All the World's Aircraft, 1939* (London: Sampson, Low, Marston and Company, 1939), 37d.

[39] "Hundred Octane," *Flight* 37 (March 28, 1940): 277; Mike Williams, "Spitfire Mk. I versus Me 109E: A Performance Comparison," 2011, www.spitfireperformance.com/spit1vrs109e.html (Accessed January 3, 2012).

As the Phoney War gave way to the Nazi invasion of the Low Countries and France on May 10, the RAF was operating Hurricanes equipped with a mix of Airscrew Company fixed, de Havilland two-pitch bracket, and Rotol constant-speed propellers. 1 Squadron provided fighter escort for the Fairey Battle and Bristol Blenheim bombers of the RAF's Advanced Air Striking Force. The squadron received a single Hurricane with a Rotol constant-speed propeller at its airfield at Berry-au-Bac, northwest of Paris on April 18, 1940. The delivery pilot instructed 1 Squadron's pilots in the use of the propeller and collected the input of the unit's more senior officers on what they would require of future fighters. The commanding officer, Squadron Leader P. J. H. "the Bull" Halahan, and his pilots were "greatly pleased" with its performance. The squadron collected four additional constant-speed fighters, which came to be better known as "Rotol Hurricanes," on Wednesday, May 1.[40]

The following day, 1 Squadron conducted a preliminary assessment of the performance and fighting qualities of the new Hurricane and a captured Messerschmitt Bf 109 fighter at Orleans. The German fighter clearly displayed better takeoff and rate-of-climb. Once at altitude, Halahan and his pilots believed the Rotol Hurricane was an even match in the air during combat simulations at 15,000 feet, which included a head-on approach followed by a traditional dogfight and a mock surprise line astern attack by the Bf 109. The problem was that Luftwaffe offensive tactics favored a diving attack from above with the element of surprise followed by climbing away to repeat the attack or find other prey. The most common defensive maneuver was a half-roll into a vertical dive. The success of both tactics relied upon rapid changes in engine power, which, in turn, was dependent on an efficient constant-speed propeller to turn that energy into thrust. For the Luftwaffe, its fighter aircraft had VDM constant-speed propellers. Halahan stressed that "until all Hurricane aircraft" had Rotol airscrews, RAF pilots would have little chance to engage Bf 109s effectively.[41]

Unfortunately, not all of 1 Squadron's pilots were lucky enough to fly a Rotol Hurricane during the Battle of France. Flying Officer Paul Richey was one of those pilots whose older Hurricane was known as a

[40] Royal Air Force Operations Record Book: History of 1 Squadron, AIR 27/1, TNA, 128, 137.

[41] P. J. H. Halahan to the Under Secretary of State, Air Ministry, "Report of Comparative Trials of Hurricane versus Messerschmitt 109," May 7, 1940, Collection of Mike Williams, www.wwiiaircraftperformance.org/hurricane/hurricane-109.pdf (Accessed December 29, 2011).

"wooden-blader" due to its Watts propeller. Having just shot down two twin-engine Messerschmitt Bf 110 fighters over Sedan on May 11, Richey found himself surrounded by five more at 6,000 feet. Knowing he did not have the speed to "dive away and beat it," he engaged the Messerschmitts since his Hurricane and its fixed-pitch propeller were designed for excellent maneuverability at that altitude. The encounter was a long suffering fifteen minutes of thrust and parry until one of his opponents got the upper hand. Richey jumped from his crippled Hurricane and parachuted to safety, receiving a severe concussion in the process. He was back in the air a few days later in a new fighter. His replacement aircraft was a "nice new Rotol" Hurricane.[42]

The experience of Flight Lt. Ian Gleed during the Battle of France in May 1940 further illustrated the level of integration of variable-pitch propellers for fighters in Great Britain. Gleed commanded "A" flight of 87 Squadron. Attached to the Air Component of the BEF, Gleed and his fellow pilots had the job of protecting the army on the ground and reconnaissance aircraft in the air. His Hawker Hurricane Mark I was a "brand new one, with the latest variable-pitch Rotol airscrew" that were just being delivered to RAF squadrons. Gleed's description of a scramble from his airfield near Lille to intercept Luftwaffe bombers and fighters highlighted why military aircraft, especially fighters, needed constant-speed propellers:

> From the end of the aerodrome a red Verey [flare] soared up; we sprinted to our 'planes – it was the emergency signal, meaning every 'plane off the ground. We slammed the throttles open; 'planes were taking off in every direction, the Rotol airscrew 'planes getting off first, scraping over the [two-pitch] Variable-pitch models, who in turn soared over the wooden prop versions.[43]

During the brief terror of rising to meet the Luftwaffe in his "Hurrybox," Gleed unknowingly recognized in the hostile skies over France the far from orderly integration of the three important stages of propeller development by the RAF.

By the late spring of 1940, the Luftwaffe had destroyed the Polish Lotnictwo Wojskowe, the French l'Armee de l'Air, and the RAF units sent to fight on the continent. The RAF's Hurricane pilots fought valiantly against the Luftwaffe and suffered catastrophic losses. Between the Air Component and the Advanced Air Striking Force, the RAF lost 261 Hurricanes in France both in the air and on the ground, which amounted

[42] Paul Richey, *Fighter Pilot* (London: B. T. Batsford Ltd., 1941), 57–58, 74.
[43] Ian Gleed, *Arise To Conquer* (London: Victor Gollancz Ltd., 1942), 28, 36.

to a quarter of Fighter Command's operational strength.[44] Virtually unopposed, the Wehrmacht rolled forward pushing columns of refuges and the shattered Allied armies west.

A Matter of Life and Death

Fighter Command kept its Spitfire squadrons in Great Britain in strategic reserve as the Luftwaffe mauled the Hurricane squadrons in France. With the war on the Continent lost, Spitfire units began to clash with the Luftwaffe. One of those units was 65 Squadron stationed at Hornchurch and under the command of Squadron Leader Desmond Cooke.[45] At the time, he made a brief flight in one of the Rotol Spitfires assigned to 54 Squadron, also stationed at Hornchurch, on May 19, 1940.[46] Interested in any performance advantage, Cooke had the previous frame of reference from flying his Rotol-equipped Gladiator to compare the two-pitch bracket de Havilland propeller installed on his Spitfire to 54 Squadron's Rotol fighter.

As pilots from other squadrons came in to try out 54 Squadron's Spitfires, the performance advantages of the constant-speed propeller over the two-pitch bracket propeller became part of their daily conversations. Located northeast of London, Hornchurch was a central fighter station for the RAF and ideally situated to allow Fighter Command squadrons to protect London and British shipping in the English Channel from attack. Overall, the RAF was a small organization, but the units operating just outside of London were especially close-knit. Up to four

44 Mason, *Hurricane*, 50; Zimmerman, *Britain's Shield*, 195.

45 Cooke was born on July 28, 1907 to Harry de Lancey and Dorothy Cooke in George Town, Penang, which was part of the British Straits Settlements colony on the Malayan Peninsula. He entered the RAF College at Cranwell in 1925 and received a permanent commission in July 1927 upon his graduation. After flying assignments in England, India, and the Middle East, he took over 65 Squadron in October 1937 and rose to the rank of Squadron Leader the following April. As the leader of one of the RAF's main fighter squadrons, "Cookie" led his pilots flying Gloster Gladiator biplane fighters. "Birth," *Singapore Free Press and Mercantile Advertiser*, July 2, 1907, p. 4; Royal Air Force Operations Record Book: History of 65 Squadron, October 13, 1937 and March 21, 1939, AIR 27/592, TNA, n.p.; Christopher Shores, *Those Other Eagles: A Tribute to the British, Commonwealth, and Free European Fighter Pilots Who Claimed between Two and Four Victories in Aerial Combat, 1939–1982* (London: Grub Street, 2004), 119; Gordon Olive and Dennis Newton, *The Devil at 6 O'Clock: An Australian Ace in the Battle of Britain* (Loftus, NSW: Australian Military History Publications, 2001), 54.

46 Pilot's Flying Log Book for D. Cooke, 115–116; Royal Air Force Operations Record Book: History of 54 Squadron, December 10, 1939, AIR 27/511, TNA, 61.

Spitfire squadrons operated there at any given time, including 54 and 65 squadrons, which constituted the "Hornchurch Wing." While not on alert or on patrol, the pilots of the respective units intermingled with each other in the mess, checked on each other for the latest details of missions in the dispersal areas, and traveled together on leave. An ongoing topic of conversation was tactics and technology, especially in the dynamic environment of engaging the Luftwaffe over southeastern England, the Channel, and France.[47]

On May 23, Pilot Officer Al Deere and his wingman, Pilot Officer Johnny Allen, met twelve Bf 109s over Calais in the first significant combat action for 54 Squadron. During the violent encounter, Deere met the superior force and shot down two Bf 109s. He might have destroyed a third had he not run out of ammunition. Deere returned to Hornchurch convinced that the Rotol propeller gave his Spitfire superior climb performance, which accentuated its already unmatched maneuverability. He "found no difficulty" climbing in a spiral turn as he zoomed through the sky to out fly his opponents. The two-pitch bracket propellers that equipped the majority of Fighter Command's Spitfires and Hurricanes could not provide adequate performance during a sustained climb, which was crucial to the rapidly evolving fighter interception tactics of the RAF. Deere believed his view was in the minority, but he felt that would be short-lived as he shared his exploits with his fellow pilots.[48]

During the frenzied week of May 26 to June 3, Fighter Command's Spitfire squadrons fought off the onslaught of Luftwaffe fighters and bombers trying to destroy Operation Dynamo, the evacuation of British and French ground forces at Dunkirk, located just six miles from Belgium on the coast of northern France. Squadron Leader Cooke, 65 Squadron, and other Spitfire units from 11 Group flew multiple sorties against large numbers of Luftwaffe Bf 109s and BF 110s as they simultaneously evaluated the performance of their aircraft. They relished the fact that the Spitfire was more maneuverable than the Bf 109 in a turning dogfight due to the latter's higher wing loading, which meant the smaller area of the Messerschmitt wing handled almost double the aerodynamic load in flight.[49]

These pilots realized how the limitations of the two-pitch propeller could be the difference between life and death.[50] Fighter Command losses

[47] Deere, *Nine Lives*, 48, 78.

[48] Ibid., 49–52, 90.

[49] "Pitch Panic," 648; Olive and Newton, *The Devil at 6 O'Clock*, 61.

[50] Two other Spitfire modifications that increased pilot survivability was the installation of armor plate and bullet-proof windscreens surrounding the cockpit. Work began

during the week was fifty-two pilots killed, eight captured, and eleven wounded.[51] The Spitfire squadrons were operating at a serious disadvantage. The Bf 109 had a higher operational ceiling in excess of 30,000 feet and a clear advantage in acceleration, rate of climb, and power diving due its VDM constant-speed propeller. Changing from takeoff to cruise, or fine to coarse pitch, as the Spitfire climbed was equivalent to changing from bottom to top gear in a small four speed car, which meant the engine was under strain and underpowered during a crucial and vulnerable moment in the air. If a Luftwaffe pilot wanted to dive away to escape, all he had to do was zoom away. A Spitfire pilot had to throttle his engine back in a dive to avoid over-revving his engine if he pursued his quarry. The consensus gathered by de Havilland engineers and test pilots in contact with Fighter Command was that a constant-speed propeller was needed.[52]

The service trials conducted by 54 Squadron and the criticisms of the two-pitch bracket propeller confirmed Fighter Command's belief in the superiority of the Rotol propeller. On June 16, 1940, Fighter Command informed the Air Ministry that Rotol-equipped Spitfires exhibited the following advantages over aircraft equipped with the de Havilland two-pitch propeller: better takeoff, climb, ceiling, maneuverability, diving speed, and endurance. In the opinion of the pilots conducting the evaluations at Hornchurch, a Rotol Spitfire was "superior to any enemy aircraft yet encountered" over Great Britain and France. Fighter Command requested that Rotol constant-speed propellers be incorporated into all new aircraft leaving Supermarine factories as a "matter of supreme urgency."[53]

The acknowledgment that the Rotol propeller was a key element to Spitfire and Hurricane performance was important. Rotol would be ready for volume production of hubs and blades for the next variant of the Spitfire, the Mark II. These improved Spitfires would emerge from the giant shadow factory at Castle Bromwich with an anticipated operational introduction of August 1940, which was two months away.[54] New Hurricanes delivered by Hawker and other manufacturers continued to

in November 1939, but intensified and came to a conclusion in May 1940. Williams, "Spitfire Mk. I versus Me 109E."

[51] Those numbers include pilots of Hurricane, Defiant, and other Fighter Command aircraft, but the majority of units engaged flew Spitfires. Norman Franks, *The Air Battle of Dunkirk* (London: William Kimber, 1983), 194.

[52] "Pitch Panic," 648; Olive and Newton, *The Devil at 6 O'Clock*, 61.

[53] H. R. Nicholl to the Air Ministry, "Rotol Airscrews for Spitfire," June 16, 1940, AIR 2/2824, TNA.

[54] Morgan and Shacklady, *Spitfire: The History*, 99.

be equipped with Rotol constant-speed propellers when they were available.[55] Unfortunately, Rotol's production problems collided with Fighter Command's dire need for constant-speed propellers during the early summer of 1940. For the immediate future, the majority of operational Spitfire and Hurricane units had to make do with the de Havilland two-pitch bracket propeller.

Despite the unavailability of the Rotol propeller, there was a push to improve the performance of the majority of Fighter Command's operational units. This simultaneous effort was from the bottom-up and centered on Hornchurch. It was a well-known fact that the addition of a constant-speed governor made the de Havilland two-pitch bracket propeller into a constant-speed propeller. An unknown engineering officer from one of the Hornchurch squadrons contacted de Havilland directly by telephone on Sunday, June 9. He asked if the company could perform a sample conversion on a Spitfire "without a lot of paperwork and fuss." De Havilland agreed and immediately went to work gathering the personnel and components needed for the trial. Four days later on June 13, a crew of select de Havilland specialists, led by engineer Arthur W.F. Metz, worked thirty-six hours nonstop to convert a Spitfire for evaluation by Fighter Command.[56]

The improved Spitfire, with a de Havilland constant-speed governor, arrived at Hornchurch on June 15. Squadron Leader Cooke and the pilots of 65 Squadron conducted the flying evaluations with de Havilland test pilot, Eric Lane-Burslem. The Spitfire displayed a reduced takeoff run from 320 to 225 yards, an increased rate of climb to 20,000 feet from 11 minutes and 8 seconds to 7 minutes 42 seconds, an increase in the operational ceiling by 7,000 feet from 32,000 feet to 39,000 feet, and improved maneuverability. The end result was that the constant-speed propeller maximized the design of the fighter to its fullest performance potential at all altitudes and conditions.[57] The modified Spitfire transitioned from being a performance-specific airplane designed to perform under predetermined conditions with a two-pitch propeller to one that

[55] Director General, Research and Development, Ministry of Air Production to Air Officer, Commanding-in-Chief, RAF Fighter Command, "Conversion of Hurricane Aircraft D.H. Two-Pitch Airscrews to Constant Speed," June 29, 1940, AIR 2/2822, TNA.

[56] "Pitch Panic," 648; Arthur W. F. Metz, "'Constant-Speed' Recollections," *Flight* 69 (May 11, 1956): 598.

[57] Pilot's Flying Log Book for D. Cooke, 119–120; "Pitch Panic," 648; Ministry of Supply, "Propellers: Development and Production, 1934–1946," 22; Price, *Spitfire in Combat*, 21.

exhibited performance-general characteristics capable of constant-speed operation in all conditions.

The adaptation of the constant-speed propeller into the technical system of the Spitfire produced an aircraft evenly matched with the Messerschmitt Bf 109. Each fighter had their strengths and weaknesses. The Spitfire was faster in the climb while the Bf 109 was quicker in the dive. The Spitfire was more maneuverable in a traditional dogfight where it could easily turn a tighter circle. If a Bf 109 pilot needed to escape, a rapid nose-over into a vertical dive easily separated him from his pursuer. The Spitfire could not follow because the negative-G maneuver made the Merlin engine cut out due to fuel starvation in its gravity-fed carburetor, which was a serious tactical disadvantage. The Messerschmitt's 1,200 horsepower Daimler-Benz DB 601 inverted V-12 engine, which featured a direct fuel injection system, was immune from this problem.[58] The overall result in converting the Spitfire was that the key to victory and survival did not lay in simply having the best aircraft, but in how a pilot used the technology at his disposal.[59] Rapidly evolving doctrinal adjustments at the squadron level focused on changing the rigid prewar Fighting Area Attacks tactics to looser and more easily handled formations based on two aircraft, but not all units recognized or implemented those ideas.[60]

Flight Lt. Gordon Olive was one of the pilots to fly the converted Spitfire. Having flown with 65 Squadron since before the war, the Australian immediately recognized the performance difference between two-pitch and constant-speed propellers. With the two-pitch bracket propeller, a Spitfire struggled to reach 25,000 feet and to him was "just flyable" at that altitude. With the addition of the constant-speed governor, the transformed fighter would "climb beautifully." Olive believed the effect the improved propeller had on the performance of the Spitfire was "truly magical."[61]

[58] This was one problem that was not fixed before the end of the Battle of Britain. The immediate solution was the adoption of a half-roll followed by pulling back on the stick to dive before rolling back to correct the position of the Spitfire. The addition of a fuel restrictor, called "Miss Shilling's orifice" after its developer at the Royal Aircraft Establishment, alleviated the problem temporarily in early 1941 until better designed pressure carburetors not dependent on gravity for fuel feed became available. Alec Lumsden, *British Piston Aero-Engines and Their Aircraft* (Shrewsbury: Airlife Publishing Ltd., 1994), 31–32.

[59] Williams, "Spitfire Mk. I versus Me 109E"; Price, *Spitfire at War*, 28–29.

[60] A pilot's individual decision to alter his fighter's gun harmonization range, the distance at which the bullets from all eight machine guns converged at one point, from 450 to a more deadly 250 yards, was another evolutionary battlefield innovation.

[61] Olive and Newton, *The Devil at 6 O'Clock*, 131.

From Hornchurch, Cooke went through official channels to recommend a service-wide conversion program. He used his firsthand experience with the converted Spitfire and the data compiled by his engineering officer, Flight Lt. McGrath – who may have been that mysterious officer that initially contacted de Havilland – to argue for the improvement of the combat capability of his and other Fighter Command units. Enthusiasm swelled in Fighter Command, especially among the pilots who had already met the Luftwaffe in combat over France and Dunkirk.[62]

On June 17, Fighter Command forwarded to the Air Ministry the results of the Hornchurch comparative tests between Rotol and de Havilland two-pitch and constant-speed-equipped Spitfires, along with Cooke's recommendations. The tests decisively exposed the inferiority of the two-pitch propeller installed on the overwhelming majority of Fighter Command's Spitfires. They revealed how the converted de Havilland constant-speed fighter flown by 65 Squadron generated a performance improvement equal to those demonstrated by 54 Squadron's Rotol Spitfires. More importantly, the change could be made at the operational squadron level. Learning from that first Spitfire, de Havilland estimated that full conversion would take an estimated twenty hours. Dowding at Fighter Command requested the Air Ministry give "immediate consideration" to approving a high priority retroactive conversion program for all Spitfire fighters.[63] Not wanting to waste any time, Cooke had the de Havilland technicians convert his personal Spitfire on Thursday, June 20. The following day, he led his squadron of two-pitch Spitfires on a combat patrol over France against the Luftwaffe.[64]

Recognizing the urgency of the situation, on Saturday, June 22, the same day France signed an armistice agreement with Nazi Germany, the Air Ministry instructed de Havilland via telephone to convert all Spitfires, Hurricanes, and Defiants from two-pitch to constant-speed. De Havilland agreed to start the conversion program at twelve Spitfire stations on Tuesday, June 25. The work was to take precedence over all other existing contracts with the immediate focus placed on Fighter Command's Spitfires. Both de Havilland and the Air Ministry optimistically believed that by July 20 all Spitfire squadrons would have constant-speed propellers. The Ministry of Aircraft Production anticipated it would not be

[62] "Pitch Panic," 648.

[63] Air Officer Commanding-in-Chief, Fighter Command to the Air Ministry, "Spitfire Conversion of Two-Pitch de Havilland Airscrews to Constant Speed," June 17, 1940, AIR 2/2824, TNA.

[64] Pilot's Flying Log Book for D. Cooke, 119–120.

until mid-September before operational two-pitch Hurricanes could be converted.[65]

Having obligated itself to the program, de Havilland faced two simultaneous challenges. One was making sure the conversion teams had the necessary components. The company had a small reserve of constant-speed governors produced for the recently collapsed French l'Armee de l'Air, which would fill the gap until production began. A special Royal Navy team recovered the units in Nazi-occupied territory under the cover of night and brought them back to England. By the time that initial supply dried up, de Havilland's Lostock factory would be well into delivering 500 conversion sets. Specifically, twenty sets needed to go to the teams in the field and another twenty to the Supermarine factory at Southampton on a daily basis until the work was completed.[66]

A conversion kit consisted of five main components. The constant-speed governor, simply called the "unit," made the two-pitch propeller capable of automatic actuation. A long flexible hardened steel rod with splines at each end, called a quill shaft after the first writing pens, connected the unit to the engine. Four external engine oil pipes ensured consistent hydraulic pressure and oil delivery to the propeller hub. In the cockpit, the pilot operated the propeller with a lever and a conduit relayed his inputs, when needed, back to the unit. Finally, there was a myriad assortment of brackets and assorted hardware to ensure everything fit correctly.[67]

Some of the more crucial components in the conversion kits were not produced by de Havilland and Ivor Jones of the Progress Department coordinated the schedule with various companies. Rolls-Royce manufactured the quill shafts that connected the constant-speed unit to the engine and the crucial engine oil pipes. Unable to collaborate due to the unprecedented production demand for Merlin engines, Rolls-Royce provided the design data to de Havilland, which in turn, used its own resources to produce those parts. A factory previously used to manufacture the civilian Gipsy engine, a power plant not necessary for wartime production, provided over 1,000 sets of those components. Other outside

[65] Director General, Research and Development, Ministry of Air Production to Air Officer, Commanding-in-Chief, RAF Fighter Command, "Conversion of Hurricane Aircraft D.H. Two-Pitch Airscrews to Constant Speed," June 29, 1940, AIR 2/2822, TNA; "Pitch Panic," 648.

[66] "Pitch Panic," 648; Ministry of Supply, "Propellers: Development and Production, 1934–1946," 21.

[67] "Pitch Panic," 648.

suppliers, such as MRC Ltd. of Chadwell Heath, Essex, produced the propeller cockpit control using its highly successful Teleflex cable. Tasked with producing cable control systems for most British aircraft, MRC adapted to the feverish pace to get the job done.[68]

The actual conversion of an already fully assembled Spitfire was the other major challenge. This work involved the installation of the five main components, not a completely new propeller. The propellers did not have to be completely removed and replaced since the basic Hamilton Standard design allowed modification for constant-speed operation. Nevertheless, each airscrew had to be dismantled in situ to access internal pins that set the range of blade movement. A simple move of the so-called index pins expanded the blade angles from two set positions for takeoff and cruise to a broader range that generated higher and more efficient performance.[69]

De Havilland and Fighter Command improvised special-purpose tools to make the job easier since the work could take place anywhere from a fully equipped maintenance hangar to a dispersal revetment at the edge of a combat-ready fighter station. Newly machined arbor stands facilitated airscrew disassembly by allowing complete subassemblies to be set aside while the teams focused on the work at hand. Custom-made offset spanners, or wrenches, enabled removal without resorting to the further complication of removing crucial engine components, such as the glycol coolant tanks mounted on the forward top of the Merlin.[70]

Due to the severity and uniqueness of the situation, the conversion program started without the traditional bureaucratic paperwork documenting a government production contract. The Air Ministry desperately needed constant-speed propellers for its fighter aircraft. Personnel at de Havilland were divided over the unconventional circumstances. A fastidious clerk in the contracts department sighed, "We shall probably never get paid for this," in response to the undocumented verbal obligations the company had made. An engineer involved in the conversion planning replied matter-of-factly, "If it isn't done, we may never live to be paid for anything."[71]

[68] "Pitch Panic," 648; "For Remote Control," *Flight* 36 (July 6, 1939): 19.

[69] "Pitch Panic," 648.

[70] Ibid., 649.

[71] De Havilland was still tying up the loose ends of the contracts, primarily documenting which aircraft received constant-speed conversions and if the company was adequately compensated, as late as 1943. "Pitch Panic," 649.

With France out of the war, the Luftwaffe shifted its focus to England. De Havilland frantically worked to get the necessary parts for the constant-speed kits and the modification process solidified. The conversion program officially started Tuesday, June 25, when thirteen de Havilland engineers, including Arthur Metz, ventured out by car to twelve Spitfire stations. Each car had at least six conversion kits crammed inside. Upon arrival, a young propeller engineer would convert the first aircraft while simultaneously instructing the RAF personnel, primarily much older engine mechanics, or "fitters," and senior flight sergeants, how to do the job. The engineer and the RAF crew converted the second airplane together. The crew, under the supervision of the engineer, completed the third airplane by itself. If the de Havilland engineer was satisfied with the third airplane, he moved on to the next fighter station. The RAF personnel remained and would expand the process by training additional crews the same way. Fighter Command and de Havilland felt it would be ideal if a fighter station could have as many as three aircraft being converted at a given time with full conversion taking approximately ten days.[72]

Training the mechanics to convert a Spitfire was one thing. Instructing the Spitfire pilots on how to properly operate them was another. With the two-pitch mechanism, the pilot had a choice of two blade angles or, more importantly, two engine rpm settings. With the addition of the constant-speed governor, the pilot used the propeller control lever to set an rpm level appropriate to a situation. Once the rpm was set, pitch changed automatically through the full range of blade angles. Advance instructions from the Air Ministry gave specific details on setting the propeller for take-off, cruise, aerobatics, combat, and landing to get the most power out of the Merlin engine. The process was not entirely foolproof. After shutting the Merlin down, a pilot needed to reset his pitch control to takeoff rpm so that he would be ready for the next scramble. Additionally, Eric Lane-Burslem flight tested the majority of the first converted aircraft at each station and instructed the pilots on how to fly with the new technology. Thrust in the middle of a chaotic situation, Lane-Burslem provided the voice of flying experience while simultaneously encouraging the joint industry-military team to get their jobs done as quickly and efficiently as possible.[73]

[72] "Pitch Panic," 648–649; Arthur W. F. Metz, "'Constant-Speed' Recollections," *Flight* 69 (May 11, 1956): 598.

[73] C. M. Banter, "Spitfire I fitted with de Havilland Constant Speed Airscrew," June 22, 1940, AIR 2/2824, TNA; "Pitch Panic," 648–649; Metz, "'Constant-Speed' Recollections," 598.

The pace was hectic and the results were dramatic. 609 Squadron, based at Northolt west of London, was enjoying a lull in operations on Wednesday, June 26. Out of nowhere and without prior notice, a de Havilland conversion team arrived at the fighter station. The team finished one Spitfire that evening and it took to the air. The core group of pilots in 609 Squadron consisted mainly of civilians from the Auxiliary Air Force. Weary from the Dunkirk campaign where they had lost one-third of their comrades over the course of just three days, the squadron was elated and amazed by the performance of the converted Spitfire. Their general consensus was that "the Spitfire now 'is an aeroplane.'" Squadron Leader H. S. "George" Darley anticipated the full conversion would be completed eight days later on July 4 once Northolt's fitters took over. The Luftwaffe would not wait and four days later, the squadron was back to operational readiness defending England on Sunday, June 30, as their fitters continued working.[74]

During the course of July, the urgency to convert Fighter Command's Spitfires to constant-speed reached a crescendo. The working times of the de Havilland and RAF engineers and mechanics averaged about fifteen to sixteen hours a day with specific instances of twenty hour days. Many of the squadrons recuperating from the Dunkirk campaign were in southern Wales and other areas away from the immediate combat zones in south-eastern England. They sent their Spitfires, in groups of one or two, to the larger fighter stations, maintenance units, and even to the de Havilland factory at Stag Lane for conversion on a piecemeal basis. For squadrons temporarily out of action for rest, teams converted and test flew as many as four and five Spitfires in a day.[75] At RAF Lyneham in Wiltshire, Metz and Flight Lt. Maynard, the engineering officer for 33 Maintenance Unit, oversaw the conversion of nine Spitfires over the course of three days alone.[76] The pace was much slower for operational units. 611 Squadron at Digby in Lincolnshire had its aircraft converted between combat missions at the rate of one Spitfire a day beginning June 28.[77]

Hornchurch continued as a center of the program. A. G. Watts and six of his fellow fitters from 92 Squadron arrived there to work alongside the

[74] Royal Air Force Operations Record Book: History of 609 Squadron, June 26, 1940, AIR 27/2102, TNA, n.p.; Leslie Hunt, *Twenty-One Squadrons: The History of the Royal Auxiliary Air Force, 1925–1957* (London: Garnstone Press, 1972), 197.

[75] "Pitch Panic," 649.

[76] Metz, "'Constant-Speed' Recollections," 598.

[77] Royal Air Force Operations Record Book: History of 611 Squadron, June 28, 1940, AIR 27/2109, TNA, 118.

maintenance personnel of 65 and 74 Squadrons. As the pilots flew their Spitfires in from their rest field at Pembrey in Wales near Swansea, Watts and his fellow fitters completed twenty-one aircraft over the course of twenty-three hectic days.[78]

As the conversion program was gaining full momentum, Cooke and the pilots of 65 Squadron took their modified Spitfires into the air against the Luftwaffe. Blue and Green sections of "B" flight, under the command of Flight Lt. Gerald A. W. Saunders, left Hornchurch at just after 8:00 P.M. on July 7, 1940, to intercept enemy raiders over the English Channel. Information from the Observer Corps provided to Fighter Command estimated that there were several formations of bombers flying at approximately 8,000 feet with protecting fighters flying above at 12,000 feet. On approaching the coast between Dover and Folkestone, the flight sighted enemy aircraft through a cloud formation. As they positioned themselves to attack, half a dozen unseen Bf 109s dove on Blue Section from the rear. The formation broke up and each of the three aircraft engaged in individual combat with their pursuers. Flight Sergeant William H. Franklin claimed two Bf 109s and Flight Lt. Saunders one. Blue Section survived the attack, engaged their enemy, and escaped repeated attacks before returning to Hornchurch. The squadron's intelligence officer noted in his report for the day that "our pilots report that due to the new DH constant-speed airscrews, they were able to out-maneuver enemy aircraft in every instance."[79] The new propeller was making a difference for Fighter Command.

The next day, July 8, Cooke himself led eight Spitfires from 65 Squadron to intercept Luftwaffe raiders over Dover around 3:30 pm in the afternoon. They spotted several Bf 109s flying in pairs and threes and the Squadron Leader ordered the group to break up into three sections in line astern formation in preparation to attack. The formation reflected standard RAF fighter tactical guidelines where each aircraft directly followed the one in front of it to facilitate efficient movement as a section and allowed consecutive individual attacks on a specific target. During the ensuing melee where 65 Squadron pilots claimed two Bf 109s, Cooke led the two aircraft of his section into a large cloud. When it re-emerged into the open the Squadron Leader and his Spitfire were gone, never to be seen or heard from again.[80] The unofficial talk among 65 Squadron was

[78] A. G. Watts, "'Constant-Speed' Flashback," *Flight* 69 (April 6, 1956): 400.

[79] A. Hardy, "Intelligence Patrol Report: Combat of 65 Squadron at 20.40 hours on 7th July 1940," Royal Air Force Operations Record Book Appendices, AIR 27/596, TNA.

[80] Royal Air Force Operations Record Book: History of 65 Squadron, AIR 27/592, TNA.

that between seventy and eighty undetected Luftwaffe aircraft attacked them from above. Fighter Command intended the line astern attack to be made against unescorted bombers, but combat against the Luftwaffe's fighters revealed that it left RAF pilots vulnerable to attack from behind. The two pilots in Cooke's section that survived, who just barely got away with their lives and landed in totally wrecked Spitfires, last saw their squadron leader with six Bf 109s on his tail.[81]

For Gordon Olive and the pilots of 65 Squadron, the loss of Cooke was a shock. Their squadron commander acted as their "guide, philosopher, and friend," especially for Olive who had been with the squadron since 1937. The tragic irony was that Cooke postponed a safe desk-bound promotion to Wing Commander to stay and fight with them during the crucial propeller conversion phase. Even more saddening was the fact that Cooke planned to marry his long-time fiancé just three days after the July 8 mission.[82] Shaken by the loss of their leader, 65 Squadron continued on in their aerial struggle against the Luftwaffe. Two days later, the Battle of Britain began on July 10.

Throughout the rest of the month and into August, Fighter Command and the Luftwaffe dueled high over the English coast. They fought over the vital shipping convoys that brought supplies and material to Britain. The Luftwaffe concentrated its forces and severely tested Fighter Command. At that critical moment, the conversion program reached an important milestone. On Friday, August 2, just forty-four days after the conversion program started, all of Fighter Command's Spitfires, numbering approximately 300 serviceable aircraft, were capable of constant-speed operation. The gypsy de Havilland and RAF conversion teams expanded their focus to the remaining Hurricane and Defiant fighters while the factories switched to direct production for the Supermarine and Hurricane assembly lines.[83]

The Luftwaffe shifted its focus to achieving air superiority over England during Unternehmen Adlerangriff, or Operation Eagle Attack, on Tuesday, August 13. The successful neutralization of the RAF's integrated air defense system, primarily the radar stations and the coastal fighter airfields, would be the prelude to the Wehrmacht's invasion of the British Isles. The Luftwaffe called the first day of this battle for air superiority over England, Adlertag, or Eagle Day.

[81] Olive and Newton, *The Devil at 6 O'Clock*, 98.

[82] Olive and Newton, *The Devil at 6 O'Clock*, 98; Dean Sumner, "The First Nine Days of a Summer Month," *Dover Mercury*, June 30, 2011, p. 14.

[83] "Pitch Panic," 649; Bungay, *Most Dangerous Enemy*, 419–426.

FIGURE 22 Constant-speed-equipped Spitfires of 65 Squadron taking off from Hornchurch in August 1940. Imperial War Museum.

It was 65 Squadron that rose to meet the Luftwaffe on Adlertag (Figure 22). During the early afternoon, Gordon Olive, in command of "A" Flight, joined with "B" Flight over the English Channel near Dover. The six aircraft met approximately twenty Bf 109s at about 19,000 feet. With his pilots engaged, Olive saw four more Bf 109s at 23,000 feet. He climbed to meet them and promptly shot down the last Messerschmitt before the other three dove for the safety of France. On his way back to the main engagement, Olive saw an additional four Messerschmitts at 26,000 feet. He climbed once again, positioned himself above them and down sun to gain the advantage of surprise, and destroyed another Bf 109. As Olive started to once again rejoin "A" Flight, thirty Bf 109s attacked him in level flight. Olive climbed above them into the sun and boldly attacked the nearest fighter. Sent into disarray, the Messerschmitt formation broke up. Olive pursued a lone Bf 109 headed toward France, expending the rest of the ammunition from his eight Browning machine guns before it disappeared into the protection of a cloud. He successfully took on superior numbers of fighters all by himself in three consecutive encounters, out-climbed them, and shot down two Bf 109s and perhaps a third. Olive returned to Manston and prepared for his next mission.[84]

Three days later, on Friday, August 16, the frantic conversion program reached its climax. Over the course of fifty-six days, 1,051 Spitfires and Hurricanes had been converted to constant-speed operation at an average of approximately eighteen aircraft per day.[85] Of those aircraft,

[84] Flight Lt. G. Olive, Combat Report, August 13, 1940, AIR 20/25, TNA.

[85] "Pitch Panic," 649.

approximately 600 were operational in the field at the RAF's fourteen fighter stations and six airfields ready to face the Luftwaffe.[86]

The addition of constant-speed and the continued evolution of tactics enhanced the performance of Fighter Command. The primary mission of the Spitfire squadrons became the disruption and destruction of the high-flying Luftwaffe fighter escorts so that Hurricane units could attack the bombers menacing England. That meant the Spitfire pilots had to climb higher than ever before to get to an advantageous position above their targets as quickly as possible. Ideally, they would then dive down sun toward their quarry. Once engaged, they needed all of the power of the Merlin engine and speed of the Spitfire overall to maneuver against and destroy their enemy. The first to "bounce" in that manner was often the victor.

A mission flown by Flight Lt. David M. Crook of 609 Squadron on Monday, September 30 symbolized the merging of the manipulation of advanced technology with the art and ritual of air combat. At 7:30 A.M., he began his day by checking over his aircraft, which included setting his airscrew to fine pitch for takeoff, ensuring he would be ready for a scramble at a moment's notice. Three hours later, the squadron's orderly alerted the pilots of their first mission for the day and within minutes all twelve aircraft of 609 Squadron rose from Middle Wallop fighter station in Hampshire. A controller informed them of their target, more than 100 enemy aircraft approaching Swanage in Dorset in southwestern England at 20,000 feet. The Spitfires continued to climb from 10,000 feet on up to 27,000 feet when a large formation of Luftwaffe fighters appeared on their left. The Luftwaffe sent them to lure the RAF up to destroy its pilots and aircraft and 609 Squadron positioned itself above the Bf 109s with the sun at its back. By this phase of the Battle of Britain, Crook knew very well that "superior height" was the "whole secret of success in air fighting."[87] In preparation for the attack, Crook set his trigger to "Fire," increased the engine rpms to 3,000 by moving the constant-speed propeller control all the way forward, and "pulled the plug" for full emergency power.

The German fighters caught a glimpse of the Spitfires and quickly dove away, and 609 Squadron followed them down to 1,000 feet at speeds approaching 600 mph as they lined up their individual targets. Crook gave his a "terrific burst of fire at very close range" and watched the Bf

[86] Bungay, *Most Dangerous Enemy*, 419–426.

[87] David M. Crook, *Spitfire Pilot* (London: Faber and Faber, 1942), 56.

109 "burst into flames" before diving straight down into the sea. Seeing that a "very nice little massacre was in progress," Crook chased another Bf 109 headed back to France across the Channel and destroyed it just off the coast near Cherbourg. Pleased with himself, Crook flew at low level over the Channel and back to Middle Wallop. Five Bf 109s were destroyed by 609 Squadron during the mission.[88] Crook and his fellow 609 Squadron pilots flew their Spitfires to their limits, which included the operation of their de Havilland constant-speed propellers for maximum efficiency. Toward the end of the Battle of Britain, they became the first Fighter Command unit to destroy 100 Luftwaffe fighters and bombers.

The Fighter Command pilots that experienced and benefitted from the transition to constant-speed, including Al Deere, Paul Richey, Ian Gleed, Gordon Olive, and David Crook, used the new technology in combat as the Battle of Britain raged on into October. In the end, Fighter Command destroyed the majority of the 1,014 bombers and 873 fighters the Luftwaffe lost over England. The cost was approximately 770 de Havilland- and Rotol-equipped Hurricanes and Spitfires.[89]

Users as Innovators

The performance of the Spitfire and Hurricane over the course of the Battle of Britain illustrated the vital importance of the constant-speed propeller to combat aircraft. The technology enhanced the ability to climb, dive, and fight and increased survivability. In the process, Fighter Command's use of the two fighters has become legendary. It took more than the constant-speed propellers installed in the Spitfire and Hurricane to make it possible for Fighter Command to win the Battle of Britain, but the RAF may not have prevailed without them.

The story of how the Spitfire and Hurricane got their constant-speed propellers is a lesson in understanding how aircraft evolve. Widely recognized as the first "modern" British fighters, they did not spring forth from their factories in the late 1930s incorporating all of the latest innovations generated by the Aeronautical Revolution. The absence of constant-speed propellers on the majority of these fighters during the spring and early summer of 1940 reveals how the adoption of fundamental technologies differed from community to community. Primarily, the

[88] Ibid., 78–82.

[89] Fighter Command lost a total 1,023 fighters of different types. Bungay, *Most Dangerous Enemy*, 368, 372.

long development period of the constant-speed propeller in Great Britain, shaped by its aeronautical community's resistance to the variable-pitch propeller, ensured that the new innovation would appear at a later period than other countries. Once open to the idea, the RAF favored the Rotol airscrew for the reasons of design practicality, economy, and good old-fashioned technological nationalism, which outweighed the dire fact that it would not be available in larger numbers until months later. The realities of air combat over France and England overrode those choices and resulted in the last-minute conversion program coordinated by Fighter Command and de Havilland during the summer of 1940. Lacking what was considered to be neither a refined lightweight design nor a British lineage, the constant-speed bracket propeller generated near equal performance and, most importantly, it was available.

The crucial element in the final acceptance of the variable-pitch propeller in Great Britain was the pilots of Fighter Command. They realized their need for the constant-speed propeller to improve the performance of their Supermarine Spitfire and Hawker Hurricane fighters. The process began as soon as they received the new aircraft beginning in late 1937 and it intensified as they clashed with the Luftwaffe in early 1940. These technologists shaped the already sophisticated aeronautical systems they operated by questioning the status quo. In the grand scheme of World War II, the contributions of engineering- and performance-minded pilots such as Donald Brookes and especially Desmond Cooke were not combat victories.[90] Theirs was ensuring that their fellow Fighter Command pilots had the best equipment available in the hope that Great Britain would prevail in the rapidly unfolding and great struggle ahead. Institutionally, they made the process of integrating variable-pitch propellers into the system of the British fighter airplane go much faster than it otherwise would have. For them, the lives of their fellow pilots and the survival of their nation depended on it.

[90] Brookes continued as an engineering officer over the course of the war and retired in 1946. Cooke's official aerial combat tally was two aircraft destroyed and one shared. Since he disappeared two days before the RAF's recognized official start of the Battle of Britain on July 10, 1940, Cooke is not listed as one of "the few" on the Roll of Honor. He is listed among the over 20,000 RAF pilots and aircrew lost over the United Kingdom and Europe that have no known graves and recognized at the Air Forces Memorial at Runnymede in Surrey.

11

A Propeller for the Air Age

During the early winter of 1955, well-known travel writer William W. Yates of the *Chicago Daily Tribune* paused to reflect on the state of commercial aviation in the ten years since the end of World War II. Every five seconds saw the takeoff of a propeller-driven airliner on a long-distance flight that would take passengers and cargo across continents and oceans. Flying through all types of weather to one of the 3,500 international airports around the world was as routine as driving a car along Lake Shore Drive in Chicago. More than 324 million passengers had taken flight since 1945. Propeller-driven airliners had turned the "ocean of the air" into a "highway for the traffic of all nations." Yates acknowledged that aircraft capable of flying "higher, faster, and farther" made that worldwide travel revolution possible.[1] The technology at that crucial intersection of altitude, speed, and range found in modern high performance aircraft since the 1930s was the propeller.

As one of the most important innovations of the Aeronautical Revolution, the variable-pitch propeller was a transformative technology

[1] Yates based his article on *Every Five Seconds ...*, a booklet distributed jointly by the International Civil Aviation Organization and International Air Transport Association in recognition of the tenth anniversaries of postwar commercial aviation and their respective organizations. They attributed the increased technological sophistication of postwar commercial aircraft to the "military necessity" of World War II and perhaps the early Cold War while overlooking that trend's existence since the early days of flight. William W. Yates, "Civil Aviation Makes Great Strides in Last Ten Years: Flights to Every Point in World Become Routine," *Chicago Daily Tribune*, November 27, 1955, section G, p. 1; International Civil Aviation Organization and International Air Transport Association, *Every Five Seconds ...: Ten Years of Postwar International Civil Aviation* (Montreal: ICAO/IATA, 1955).

that made the modern airplane a weapon of war and instrument of global travel through the rapid onset of the Air Age during the 1940s and 1950s. That triumph was the result of the perseverance of propeller specialists over the preceding two decades. Ironically, within twenty years after World War II, a significant portion of the aeronautical community and society in general came to consider the propeller an obsolete relic of aviation's past. The variable-pitch propeller was no longer a technology capable of making commercial and military airplanes fly even higher, faster, and farther. With the introduction of the jet, a new cultural resistance against the continued use and further refinement of the propeller emerged.[2]

Modern Propellers at War

During the 1940s, nations in Asia, Europe, and North America fought each other around the world with airplanes that were, at their technical foundations, the product of the interwar period. The Aeronautical Revolution of the 1920s and 1930s created the modern airplane and World War II solidified its position as a global technology. It was clear that the revolution in propeller design and construction was well over by the spring of 1941. A propeller designed for use on a high-performance military or commercial airplane bore little resemblance to a laminated piece of wood designed during World War I for takeoff, climb, or cruise. It was a product of a modern industrial age made from the latest alloys, steel, and composite materials and incorporated sophisticated design. In other words, it was "a highly complicated mechanism" that required "the most delicate kind of craftsmanship."[3] Like the other segments of aeronautics, variable-pitch propellers and the communities that created them became important components of national war machines and the globalized aerial peace that followed.

As the road to world war intensified in the late 1930s, the American propeller community found itself caught up in the frenzy of European

[2] To echo Eric Schatzberg's assertions about the nature of the transition from wood to metal in airframe materials, the American public after 1960 drew upon existing symbolism to link the jet engine with progress and modernity, which influenced the interpretation of the propeller as a backward and obsolete technology and stymied its continued refinement in the United States. Eric M. Schatzberg, *Wings of Wood, Wings of Metal: Culture and Technical Choice in American Airplane Materials, 1914–1945* (Princeton, NJ: Princeton University Press, 1998), 3.

[3] "Propeller Producers Widening Bottleneck despite Unusual Dependence on Hand Labor," *Wall Street Journal*, May 16, 1941, p. 28.

rearmament. President Franklin D. Roosevelt's recognition of the United States as an "arsenal of democracy" in December 1940 reflected overall American mobilization toward equipping Great Britain and France in their struggle against Nazi Germany. Large numbers of aircraft and other equipment for the British and French crossed the Atlantic. The British and French air commissions were a consistent presence through their negotiations to get Hamilton Standard and Curtiss propellers from the United States and to get license-built versions manufactured in Canada. The fall of France saw the complete shift of all contracts to Great Britain.[4]

Soon, the quantity production of aircraft propellers became a major element of the rapidly mobilizing American war machine. Manufacturing involved a significant amount of handwork and training. It took six months for a worker to become fully versed in one operation such as blade making. Regardless of whether the material was solid duralumin or hollow-steel, shaping them into a blade was a time-consuming, labor-intensive process. The *Wall Street Journal* noted that "behind all of the production records being hung up in propeller production stand years of painstaking and unceasing research and development work."[5]

For a nation "interested primarily in results," that was an impressive achievement. Hamilton Standard was the most prepared manufacturer in terms of the transition to wartime production. During the period 1935 to 1939, the East Hartford factory produced approximately 380 propellers a month. The company committed itself to delivering 60,000 propellers to the US Government by June 1942. After the expansion of the shop floor to 310,000 square feet, the workers produced 1,900 propellers per month by the spring of 1941. The conversion of an old textile mill into a 200,000 square foot satellite factory at Pawcatuck, Connecticut, added another 1,000 propellers a month by the following September. Besides this initial expansion, Hamilton Standard's production strategy included subcontracting individual components and assemblies out to seventy-five different manufacturers and licensing its products to makers of non–war related products.[6]

Hamilton Standard was the primary propeller manufacturer for the Allies during World War II. Virtually the entire Army Air Forces and BuAer front-line inventory, from multiengine bombers to fighter and transport aircraft, as well as a significant majority of RAF aircraft employed

[4] British Air Commission, "Propellers," January 1940–December 1941, AVIA 38/662, TNA.

[5] "Propeller Producers Widening Bottleneck," 28.

[6] "Propeller Producers Widening Bottleneck," 28; "Propellers from Pawcatuck," *The Bee-Hive* 16 (May–June 1941): 1.

Hydromatic propellers. Hamilton Standard and its three licensees – refrigerator manufacturers Frigidaire and Nash-Kelvinator, and office equipment maker Remington-Rand – produced 530,135 Hydromatic propeller assemblies during the war. Most of those were the ubiquitous three-blade Model 23E50 Hydromatic.[7] Hamilton Standard's Norwich, Connecticut, factory and the Canadian Pratt & Whitney Aircraft Company-operated subsidiary in Montreal, Canadian Propellers Ltd., manufactured an additional 38,000 and 12,497 controllable counterweight propellers respectively for light transports and training aircraft.[8] All told, approximately 76 percent of all Allied aircraft featured a Hamilton Standard propeller.

For the United States, the Hydromatic propeller was a central technology in the Army Air Forces' strategic bombing campaigns over Europe and Japan. Young pilots and aircrew flew heavy bombers like the Boeing B-17 Flying Fortress and the Consolidated B-24 Liberator that incorporated the innovations developed during the Aeronautical Revolution – streamlined design, all-metal construction, cantilever monoplane wings, four high-output radial engines – in a long-distance aerial war against the Axis. During combat operations, the successful operation of Hydromatic feathering systems was commonplace (Figure 23). The Army Air Forces exhibited much more concern over isolated failures. Pilots and flight engineers in both the European and Pacific theaters of operation identified a major problem with the engine and the propeller sharing the same oil. Damage inflicted on an engine or oil tank from antiaircraft artillery, or flak, could result in the loss of the oil supply, which deprived the propeller of the needed hydraulic pressure for feathering. The resultant recommendation was for a separate system to ensure the return of more bombers from Axis territory.[9]

The deadliest bomber of them all was the Boeing B-29 Superfortress. The addition of tricycle landing gear, pressurized cabins, and advanced

[7] Aircraft equipped with the Model 23E50 included the Avro Lancaster, Boeing B-17 Flying Fortress, de Havilland Mosquito, Douglas SBD Dauntless, Grumman TBF Avenger, and North American B-25 Mitchell bombers, Douglas C-47 Skytrain and C-54 Skymaster transports, and Grumman F6F Hellcat and Vought F4U Corsair fighters.

[8] "Hartford Fosters World Air Group," *Montreal Gazette*, May 5, 1943, p. 7; Hamilton Standard Propellers, *Wherever Man Flies* (Hartford, CT: United Aircraft Corporation, 1946); Irving Brinton Holley, Jr., *Buying Aircraft: Matériel Procurement for the Army Air Forces* (Washington, DC: Government Printing Office, 1964), 563.

[9] Joseph V. LeBarbera, "Feathering Props on B-17," May 30, 1944; Allan R. Willis, "Auxiliary Feathering System," October 30, 1944; Donald C. Burrows, "Feathering Fuel," November 9, 1944, Collection A1276, Roll 116002, HRA.

FIGURE 23 The pilots of this US Fifteenth Air Force Boeing B-17 Flying Fortress feathered the propeller on their number four engine after it caught fire during a mission over northern Italy in June 1944. National Archives and Records Administration.

radar and fire control systems to the synergy of innovation also made the B-29 the world's most advanced propeller-driven airplane in 1945. The Twentieth Air Force used them to attack Imperial Japan from the Marianas Islands 1,500 miles away to the southeast in the Pacific. Combat formations of B-29s carried 20,000 pounds of bombs or aerial mines each to Japan at a cruising speed of 220 mph and at an altitude of 30,000 feet.[10] The Twentieth began its campaign following the core doctrine of daylight high-altitude precision bombing. High-altitude winds, known today as the "jet stream," pushed bombers over their targets quickly or bogged them down at the mercy of Japanese defenders, consistently hazy

10 John M. Campbell, *Boeing B-29 Superfortress* (Atglen, PA: Schiffer Military/Aviation History, 1997), 27.

weather obscured targets, and the widely dispersed industry in fragile wood buildings deterred effective results. The strategic bomber force's new commander, Gen. Curtis E. LeMay, introduced low-level night-time area attacks with incendiary bombs. Over the course of a single twenty-four hour period on March 9–10, 1945, the Twentieth's fire bombers killed upwards of 100,000 people and destroyed one-fourth of Tokyo.

A B-29 had four Hydromatic propellers. Each had four blades spanning a diameter of sixteen feet six inches and weighed 870 pounds, making them the largest ever installed on an airplane up to that time. They gave the necessary performance that ensured that the heavily laden lumbering aircraft operated in and out of airfields safely and cruised at high speeds with their deadly cargoes. The feathering mechanism prevented the destruction of the plane and crew in the event of engine damage.[11]

The Hydromatic was crucial to the safety of the Twentieth's crews. One crew from the 60th Squadron, 39th Bomb Group of the Twentieth Air Force participated in a raid on the Otake oil refinery on the main Japanese island of Honshu on the morning of May 10, 1945. The bomber withdrew from the target over the Inland Sea of Japan and received antiaircraft damage on its right inboard engine, which began to lose oil pressure and caused the aircraft to lose altitude. Facing a ten-hour flight over the vast Pacific Ocean to its home base on Guam, the airplane commander, Capt. Chester G. Juvenal, ordered the malfunctioning engine shut down and the propeller feathered. The last thing Juvenal and the crew wanted to experience was bailing out and spending hours, maybe days or weeks, in a life raft with no guarantee of rescue. The precautionary measure allowed the bomber to land safely on three engines at the emergency airfield on Iwo Jima. Flight engineer Lt. Elmo F. Huston recalled, "Had we not been able to feather the prop, we would have been in real trouble."[12] The B-29's Hamilton Standard Hydromatic constant-speed propeller ensured that the crew of twelve young men, like so many bomber crews during World War II, lived to fight another day.

The careers of the two men most responsible for the Hydromatic, Frank Caldwell and Erle Martin, rose during the war years. Caldwell moved up the corporate and professional ladder. In 1940, he became the

[11] "Target–Tokio," *The Bee-Hive* 20 (June 1944): 12–13; "Biggest Props Drive B-29," *Atlanta Constitution*, July 5, 1944, p. 18.

[12] Elmo F. Huston, e-mail to author, September 4, 2001; and February 17, 2003; Edward T. Reilly, e-mail to author, February 18, 2003; Edward T. Reilly, "Account of Mission Seven," *39th Bomb Group (VH), Crew 7*, 2001, http://39th.org/39th/aerial/60th/crew07a.html (Accessed February 22, 2003).

director of the UAC Research Division at Hartford, Connecticut, where he supervised the design and construction of one of the world's leading industrial propeller and wind tunnel testing facilities.[13] Caldwell also served as the president of the important IAS during 1941 as the aviation industry made the transition to direct American involvement in World War II.[14] In Caldwell's absence, Erle Martin became engineering manager at Hamilton Standard in 1940 and guided the company's development of propellers through the war.[15]

While Hamilton Standard was the dominant propeller manufacturer for the Allies, its longstanding rival, Curtiss, was starting production from a much smaller scale. In mid-1938, the company employed 111 workers in a 17,000 square foot corner of the Curtiss Aeroplane Division plant in Buffalo, New York. The creation of the Propeller Division led to the move to a dedicated 64,000 square feet factory in northern New Jersey in July 1938. By the end of the year, 300 workers produced an average of twenty-five propellers a month. To meet its commitments to the American military, Curtiss expanded its workforce to 3,400 to produce 650 propellers a month by May 1941. Curtiss was producing on a large scale on the eve of American entry into World War II. Unlike Hamilton Standard, which subcontracted out to other firms, Curtiss-Wright expanded its own manufacturing base to meet its production contracts. The ultimate goal was 1,000 propellers a month from 15,000 employees working in 1.3 million square feet of manufacturing space divided between five factories by the spring of 1942.[16]

The Propeller Division never achieved the high volume production it anticipated at its primary factories in Caldwell, New Jersey, and Neville Island and Beaver, Pennsylvania (Figure 24). All told, it accounted for only 21 percent of total American propeller output. Despite the inability to compete with Hamilton Standard, the Curtiss Electric propeller provided a clear alternative to the Hydromatic for high-performance aircraft and was a contribution to the war effort. Curtiss manufactured

[13] Walter H. Caldwell, telephone interview by author, July 23, 1999; "Frank Walker Caldwell," *Collier's Encyclopedia*, 322–323; Eugene E. Wilson, *Slipstream: The Autobiography of an Air Craftsman* (Palm Beach, FL: Literary Investment Guild, 1967), 256.

[14] For his work, the IAS made Caldwell an honorary fellow in 1946. *Technology Review* 43 (February 1940–1941): 1; *Technology Review* 49 (April 1946–1947): 3.

[15] Russell Trotman, "Turning Thirty Years," *The Bee-Hive* 24 (January 1949): 27, 28; "Erle Martin, Inventor, Dead; Improved Aircraft in the 30s," *New York Times*, December 14, 1981, File CM-133000-01, BF, NASM.

[16] "Propeller Producers Widening Bottleneck," 28.

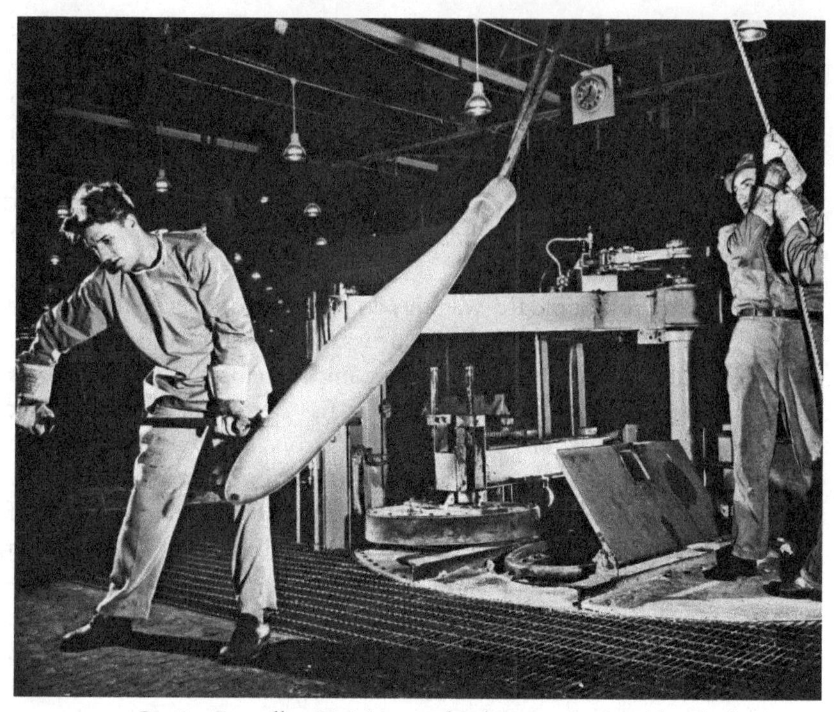

FIGURE 24 Curtiss Propeller Division workers removing a red hot hollow-steel blade from a rotary pit furnace in 1943. The next step was to place it into a hydraulic press for die quench hardening to retain the final blade contour. Smithsonian National Air and Space Museum (NASM 9A12115).

144,863 propellers for Allied frontline aircraft, including the Curtiss P-40 Warhawk, Lockheed P-38 Lightning, Republic P-47 Thunderbolt, and Grumman F4F Wildcat fighters, the Martin B-26 Marauder bomber, the Consolidated PB2Y Coronado flying boat, and the Curtiss C-46 Commando transport.[17]

The inventor of the Curtiss Electric propeller fared very well during the war. For his intense and expensive work beginning in 1909, Wallace R. Turnbull did not receive any substantial technical or financial awards during the decade of the 1930s. During World War II, however, his efforts reaped unprecedented financial awards. Turnbull received a generous royalty for each Curtiss Electric propeller during the war. In one year alone, he earned $10 million in royalties. The impressive payments led

[17] "Curtiss Wright: When the Difference Depends on a Split Second or Less!," *LIFE* 7 (August 9, 1943): 93; "The Third Part of a Plane: Curtiss Electric Propellers," n.d. [1942], File B5-260140-01, PTF, NASM; Holley, *Buying Aircraft*, 563.

the cash-strapped US government to reevaluate its contract agreements as it pertained to paying royalties to inventors.[18]

The Curtiss Electric propeller played an important role in a defining moment of World War II. Col. Paul Tibbets, commander of the world's first atomic bomber force, the 509th Composite Group, ordered the installation of reversible propellers as part of a series of modifications to the unit's B-29s, which included the removal of armor and defensive turrets to lighten the airframe for more speed and the addition of pneumatic bomb bay door actuators for faster bomb delivery and target egress. The ability to reverse pitch facilitated two things. The reversible mechanism also made ground handling easier when taxiing the airplane and positioning it over dedicated bomb pits for loading their sensitive cargo. Unlike the Twentieth Air Force's B-29 crews, a 509th crew could not simply drop their payload into the sea in the event of an aborted mission. The Electric propellers would slow down an overloaded bomber carrying a five-ton atomic weapon after landing on one of the 8,500 foot runways at North Field on Tinian. Combined with a refined and fuel-injected Wright R-3350 radial engine, the Curtiss Electric propeller provided the added performance and flexibility to operate the 509th's B-29s. The atomic bombers, *Enola Gay* and *Bock's Car*, ushered in another era in human history, the atomic age, when they attacked the Japanese cities of Hiroshima and Nagasaki in August 1945.[19]

Besides the Hamilton Standard Hydromatic and Curtiss Electric Propeller, the American war machine produced a third constant-speed propeller for the war effort. Pete Blanchard and Charles MacNeil's Unimatic propeller became a reality when GM acquired Engineering Projects and formed the Aeroproducts Division in June 1940. Blanchard and MacNeil continued on as general manager and chief engineer respectively while their original employees had the option to work for the new

[18] J. H. Parkin, "Wallace Rupert Turnbull, 1870–1954: Canadian Pioneer of Scientific Aviation," *Canadian Aeronautical Journal* 2 (January–February 1956): 41–42; Mac Trueman, "On a Wing and a Prayer: Wallace Rupert Turnbull," *The New Brunswick Reader* 4 (December 7, 1996): 10. Turnbull passed away at his home in Saint John, New Brunswick at the age of eighty-four on November 26, 1954. His obituary made a point to mention that he had never flown in an airplane. "R. Turnbull, 84, Aviation Pioneer," *New York Times*, November 27, 1954, p. 13.

[19] Propeller Division, Curtiss-Wright Corporation, "Curtiss Propellers: Thirty Years of Development," September 4, 1945, File B5-260000-01, PTF, NASM; Richard H. Campbell, *The Silverplate Bombers: A History and Registry of the Enola Gay and Other B-29s Configured to Carry Atomic Bombs* (Jefferson, NC: McFarland, 2005), 14.

company. One of Blanchard's new hires was Glenn Peterson, the first African-American engineer to work at GM.[20]

GM provided for a new factory in what was originally a cornfield in Vandalia, Ohio, just north of Dayton and across the road from the city's municipal airport. The factory became operational in May 1941 after months of chaotic work. Blanchard and MacNeil initiated a five year plan to develop the Unimatic as a practical propeller, but the Japanese attack on Pearl Harbor in December dramatically accelerated the program. The Vandalia factory, full of machinery specifically designed to fabricate Unimatic components, produced seventy-three propellers that same month. By the following year, the Unimatic, better known as the "Aeroprop," saw service on fighter aircraft such as the Bell P-39 Airacobra and P-63 Kingcobra, and the North American P-51 Mustang. Workers at Vandalia and a second factory produced 12,500 propellers by February 1944. Overall, Aeroproducts' 2,500 employees, 99 percent of whom had never worked on a propeller before, produced 20,773 three- and four-blade constant-speed propellers, which was approximately 3 percent of total American production during World War II.[21]

The stress of starting Aeroproducts from scratch took its toll. MacNeil, just thirty-four years old, suffered a heart attack on November 18, 1944, while flying home from an aeronautical convention in Kansas City. He landed his Cessna monoplane near Brazil, Indiana, before dying in a farmhouse close by. Blanchard continued on despite the loss of his engineering partner.[22]

The American military entered World War II with what amounted to a readily available "modern" propeller. As part of their contribution to the war effort, NACA researchers worked to refine the aerodynamic properties of propeller blades. In preparation for war, the NACA hosted a

[20] "Better Idea for Aircraft Propeller Design Led to Formation of Aeroproducts Division," n.d. [1950], GM; Aeroproducts Division, "Propellers," n.d. [1950], Charles Kettering Collection (hereafter cited as CKC), Kettering University Archives (hereafter cited as KUA), Flint, Michigan; "Werner J. Blanchard" and "Charles Seward Jadis MacNeil," *Who's Who in Aviation*, 1942–1943 (Chicago, IL: Ziff-Davis Publishing Company, 1942), 44, 269; Richard A. Tunney, "The Aeroproducts Division of General Motors Corporation," n.d. [1950], in Juliet Blanchard, *A Man Wants Wings: Werner J. Blanchard, Adventures in Aviation*, 1986, www.margaretpoethig.com/family_friends/pete/wings_bio/index.html (Accessed May 24, 2012), 81–82, 85.

[21] Aeroproducts Division, General Motors Corporation, *Blades for Victory: The Story of the Aeroproducts Propeller and the Men and Women Who Build It* (Dayton, OH: Aeroproducts Division, General Motors Corporation, 1944), 22, 34; Aeroproducts Division, "Propellers," n.d. [1950], CKC, KUA; Holley, *Buying Aircraft*, 563.

[22] "Stricken During Flight: C.S.J. MacNeil, General Motors Engineer, Dies in Indiana," *New York Times*, November 19, 1944, p. 51.

conference on propeller research attended by the Army, Navy, and the manufacturers in April 1939. They agreed to move forward with what amounted to an emergency program that would increase the overall performance of high-speed and high-power propellers.[23]

Blade design reflected a compromise where most of the blade was an airfoil, but the portion where it attached to the hub, or root, was round for structural strength. Beginning in the spring of 1939, researchers in the Langley Full-Scale Tunnel fabricated streamlined cuffs in the profile of an airfoil for each blade root that covered over 45 percent of the blade. They calculated that the increased blade area enabled a prototypical fighter to reach a speed of 400 mph at 20,000 feet.[24] The use of cuffs also facilitated increased cooling for radial engines and the installation of deicing equipment for cold weather operation. Langley carried on with extensive cuff research that allowed for the modification of existing blade designs. To increase thrust further, NACA researchers investigated wider, or paddle, blades that featured larger chord length from the leading to the trailing edge of the blade. Langley discovered that blades one and a half times wider than conventional blades led to increased efficiency at higher altitudes and speeds.[25]

Propeller manufacturers introduced paddle blades and blade cuffs as part of their contribution to the refinement of military aircraft. Following what they believed to be the best practice, Hamilton Standard and Aeroproducts favored only paddle blades while Curtiss used both cuffs and paddle blades on their designs. Regardless of the manufacturer, high performance aircraft like the North American P-51 Mustang escort fighter and the Boeing B-29 Superfortress strategic bomber benefitted from their application.

Those two innovations transformed the performance of one of America's front-line fighters in the strategic air war against Nazi Germany, the Republic P-47 Thunderbolt, in late 1943. Combat units like the famed

[23] NACA Headquarters, "Conference on Propeller Research," April 12, 1939, Folder "AP356-1 (676)-Propeller Investigations, 1939," Box 242, NACA Research Authorization Collection, NASAHO.

[24] David Biermann, Edwin P. Hartman, and Edward Pepper, "Full-Scale Tests of Several Propellers Equipped with Spinners, Cuffs, Airfoil and Round Shanks, and NACA 16-Series Sections," NACA Special Report No. 168 (October 1940), 1-12; NASA Cultural Resources, "Langley Facility 643 (30 X 60 Full Scale Tunnel): Test 128-XP-47 Stability, Cooling (Part 1)," 1941, http://crgis.ndc.nasa.gov/historic/643_Test_128_-_XP-47_Stability,_Cooling_(Part_1) (Accessed April 28, 2015).

[25] Louis H. Enos, "Recent Developments in Propeller Blade Design," *Aero Digest* 37 (August 1940): 50–51; George W. Gray, *Frontiers of Flight: The Story of NACA Research* (New York: Alfred A. Knopf, 1948), 212–213, 217–218.

56th Fighter Group of the Eighth Air Force were experts at maximizing the P-47's superior diving capability, rugged construction, and unrivalled firepower of eight .50-caliber machine guns through aggressive tactics. Army Air Forces evaluations and combat operations, however, revealed that the Thunderbolt exhibited poor climb and maneuverability against the Luftwaffe's Messerschmitt and Focke-Wulf fighters below 25,000 feet.[26] The installation of a Curtiss Electric propeller with cuffed paddle blades thirteen feet in diameter and water injection equipment for the 2,000 horsepower Pratt & Whitney R-2800 radial engine increased top speed by 10 mph and the rate of climb at low altitudes to 600 feet per minute. The improved Thunderbolt could climb to 30,000 feet in approximately thirteen minutes rather than twenty.[27]

The Eighth Air Force converted its Thunderbolts and all new P-47s subsequently incorporated the refinements. First Lt. Robert S. Johnson of the 56th flew his modified Thunderbolt on New Year's Day 1944. He was skeptical initially about the "fuss" being made over his new propeller by the engineering officers, but quickly realized the overwhelming boost in performance, which in his words was "worth 1,000 horsepower, and then some," as he was now able to outclimb and outdive both the Messerschmitt Bf 109 and Focke-Wulf Fw 190.[28] All of the 56th's Thunderbolts were converted by January 4 as the group flew a bomber escort mission to Münster in northwestern Germany. The group went on to be one of the highest scoring American fighter units of World War II.[29]

Long-time competitors, Hamilton Standard and Curtiss, and the upstart Aeroproducts Division of GM, produced 695,771 constant-speed propellers on a massive scale during World War II. Their contribution was not unnoticed. Each company received the US Government's Army-Navy "E" award for excellence in the production of material for the overall war effort at least once during the course of the conflict. Prolific African-American aviation author and veteran pilot of the Spanish Civil War, James H. L. Peck, wrote in *Popular Science* that the "amazingly ingenious" propellers pulling American fighters and bombers through the air in four major theaters of operations around the world were "more than

[26] Army Air Forces Proving Ground Command, "Final Report on the Tactical Suitability of the P-47C-1 Type Aircraft," December 18, 1942, www.wwiiaircraftperformance.org/p-47/p-47c-tactical-trials.html (Accessed April 28, 2015).

[27] Roger A. Freeman, *Zemke's Wolf Pack: The Story of Hub Zemke and the 56th Fighter Group in the Skies over Europe* (New York: Orion, 1989), 138.

[28] Robert S. Johnson, *Thunderbolt!* (New York: Rinehart, 1958), 239–241.

[29] David R. McLaren, *Beware the Thunderbolt!: The 56th Fighter Group in World War II* (Atglen, PA: Schiffer Military/Aviation History, 1994), 52.

a little responsible" for growing Allied air superiority over the Axis.[30] At the war's end, the leadership of all three manufacturers looked forward to a bright future through a world connected by propeller-driven airplanes in the postwar period.

Having survived the tense days of 1940, the British war machine committed itself to the volume production of constant-speed propellers. The aircraft production program was gaining momentum, which allowed Great Britain to take the war to the enemy. Operational fighter and bomber units ventured out to far-flung combat zones such as North Africa confident they could get the requisite spare propellers since those areas lacked adequate repair facilities. Unfortunately, there was, at one crucial moment or another, a shortage in the supply of propellers for almost every type of aircraft flown by the RAF and Fleet Air Arm until production stabilized in 1943. The situation became so extreme at certain points that once an airplane arrived from the factory to an operational unit, fitters removed the propellers and shipped them back to the manufacturer for installation on the next airplane to be delivered.[31]

The root causes for the production problems were logistical and cultural. Great Britain had only two propeller manufacturers. One of those, Rotol, was not even close to volume manufacturing with a practical product until the fall of 1940. The other, de Havilland, was not producing its most up-to-date Hydromatic until 1941 due to the focus on the outdated constant-speed bracket propeller as an expedient to fulfill the need for better fighter performance. The small size of the industry and the immediate need for propellers impaired a quick expansion program. There was also competition with the overall war industry for raw materials, skilled labor, and machine tools. The Air Ministry and the Ministry of Aircraft Production focused on enlarging the capacity of both de Havilland and Rotol.[32] Moreover, there was the initial performance- and nationalist-based desire to have the Rotol over the de Havilland airscrew. All of those problems were rooted in the cultural resistance to the variable-pitch propeller during the 1920s and 1930s. The overarching emphasis on light weight, simplicity, and national pride put the British in a deadly and precarious position in the early 1940s.

[30] James H. L. Peck, "Propellers for Our Fighting Planes," *Popular Science* 143 (November 1943): 122.

[31] Ministry of Supply, "Propellers: Development and Production, 1934–1946," n.d. [1950], AVIA 46/211, TNA, 27, 32, 40–41, 51.

[32] Ibid., 14, 41.

The two British propeller manufacturers reacted to their situations differently. De Havilland initially faced the difficult transition from manufacturing the three-blade bracket airscrew to the much more complicated Hydromatic in both three- and four-blade versions. Nevertheless, the company produced 97,773 bracket and Hydromatic propellers at its Stag Lane and Lostock factories, which accounted for approximately half of all British propellers during the war. Its workers assembled an additional 37,801 Hydromatics from imported American-made components. The RAF also imported complete Hamilton Standard propellers for use on its primary strategic bomber, the four-engine Avro Lancaster. Overall, forty distinct aircraft-engine combinations used three- and four-blade de Havilland propellers including the Lancaster, Bristol Blenheim, Beaufort, and Beaufighter, de Havilland Mosquito, Handley Page Hampden and Halifax, Hawker Typhoon and Tempest, Short Sunderland and Stirling, and the Vickers Wellington. All of the RAF and Fleet Air Arm aircraft equipped with de Havilland airscrews featured duralumin blades.[33]

With the help of the Ministry of Aircraft Production, Rotol had a number of subcontractors representing different areas of British industry to rely upon in 1941. The foundries and machine shops of the Armstrong-Whitworth, David Brown, and Laystall Engineering companies produced hubs, gears, and components. As the demands increased in 1942, vacuum cleaner manufacturer Hoover Ltd. of Tottenham, Vickers-Armstrong, and the No. 1 Shadow Group consisting of automobile manufacturers that included Austin, Daimler, and Rover initiated production of complete propellers. By the end of the war, workers produced approximately 98,034 airscrews in three-, four-, and five-blade versions for more than sixty types of production aircraft. Of that number, Rotol provided 30,000 airscrews toward the production of Great Britain's most famous fighters, the Spitfire and Hurricane (Table 3).[34]

Unlike de Havilland and in contrast to American practice, Rotol propellers featured composite wood blades due to duralumin shortages. Without a market for its two-blade fixed-pitch propellers, Henry C. Watts and the Airscrew Company manufactured Schwarz composite wood detachable blades for Rotol.[35] In 1938, the Airscrew Company's

[33] Ministry of Supply, "Propellers: Development and Production, 1934–1946," 12–13, 17, 28, 32; C. Martin Sharp, *D.H.: A History of de Havilland* (Shrewsbury, England: Airlife, 1982), 156, 208.

[34] Ministry of Supply, "Propellers: Development and Production, 1934–1946," 28, 30, 63.

[35] Watts resigned as director of the Airscrew Company in July 1946. "News in Brief," *Flight* 50 (July 11, 1946): 35.

TABLE 3 *United States and British constant-speed propeller production during World War II*

Manufacturer	Output
Hamilton Standard	530,135
Curtiss Propeller Division	144,863
Rotol Ltd.	98,034
de Havilland Aircraft Company Ltd.	97,773
Aeroproducts Division of General Motors	20,773
Total	891,578

new division, Jicwood Ltd., pioneered the use of joining compressed wood roots to birch laminates that resulted in blades as strong and light as duralumin components. By war's end, the company produced an estimated 200,000 individual blades for British military aircraft, including the Fairey Barracuda, Halifax, Hurricane, and Wellington. Two other much smaller wood blade companies, Horden Richmond and Jablo Propeller, specialized in Lancaster and Spitfire blades, respectively.[36] While seen as an alternate pathway in comparison to aluminum alloy and hollow-steel blades, the composite wood blade did not survive into the postwar period once the need for more propellers and the resultant material shortages disappeared.

The other Allies used a mixture of licensed American and British technology and innovated some of their own designs. Ratier in France built upon its pneumatic propeller to introduce its own hydraulic and electric propellers before the fall of France in June 1940. The American-built North American B-25 Mitchell, Curtiss P-40 Warhawk, and Bell P-39 Airacobra aircraft flown to the Soviet Union through the Lend-Lease program featured Hamilton Standard, Curtiss, and Aeroproducts propellers. The Soviet Union also produced licensed versions of Hamilton Standard and Rotol Винт Изменяемого Шага (Vint Izmeniaemogo Shaga), or "screws with variable-pitch," for its fighter, ground attack, and bomber aircraft.

The communal attitudes toward the variable-pitch propeller during the 1920s and 1930s had a direct effect on the production programs of World War II. The United States led the way in terms of production and

[36] The Airscrew Company Ltd., "Wooden Airscrews," *Flight* 37 (March 14, 1940): 65; Tony Deeson, *The Airscrew Story, 1923–1998* (n.p.: Airscrew Howden, n.d. [1999]), 25, 30–32; Ministry of Supply, "Propellers: Development and Production, 1934–1946," 58.

technology. Controllable and constant-speed propellers were a proven technology in early 1930s and clearly seen as a part of high-performance aircraft. The United States' arsenal of democracy also had the benefit of initializing rearmament without a war to get in the way while having enough orders from its allies to justify expansion.

Great Britain suffered from its communal response to the variable-pitch propeller. Rotol produced an all-British and elegant design, but the first Hele-Shaw and Beacham-based propellers were just taking to the air as American propellers were readily available shelf items for any airline or air force to purchase. The longstanding paradigms of light weight and simplicity were still dominant in significant quarters of British aviation as the Luftwaffe crossed into Poland in September 1939. As seen with the experience of RAF Fighter Command, Great Britain went about building a propeller production program while waging war against Germany.

The Axis powers equipped their aircraft, which were central elements of their planned conquests all over the world, with propellers of German and American origin. The aerial arsenal of Nazi Germany's Luftwaffe – from the Messerschmitt Bf 109 fighter to the Heinkel He 219 night fighter – fought World War II almost exclusively with VDM propellers.[37] As a result, VDM consolidated its aviation operations into a separate corporation from the other metalworking companies. Bernard Unholtz found himself replaced by a "Party man" better suited to lead the company while maintaining a stronger connection with the government. Propeller design occurred at Frankfurt under the leadership of Hans Ebert and manufacturing took place in Hamburg.[38]

Unlike the United States, Nazi Germany's industrial landscape was a battleground. As the US Eighth Air Force and RAF Bomber Command consistently targeted German industry, planners dispersed the production efforts of critical manufacturers like VDM beginning in 1943. Planners located plants responsible for the production of component parts such as

[37] There were, however, other German propeller manufacturers. Both Argus Motorenwerke and Junkers Flugzeug und Motorenwerke produced variable-pitch propellers, but in much smaller numbers and primarily for trainers and transports. Ferenc A. Vajda and Peter Dancey, *German Aircraft Industry and Production, 1933–1945* (Warrendale, PA: SAE International, 1998), 312.

[38] L. H. G. Sterne, G. A. Luck, and H. G. Ewing, "Propeller Development at VDM, Frankfurt," August 1945, CATD, NASM. Besides propellers, VDM produced landing gear, engine radiators, and hydraulic components in factories at Hildesheim, Köln, Gross-Auheim, and Metz. United States Strategic Bombing Survey (hereafter cited as USSBS), *Aircraft Division Industry Report*, 2nd ed. (Washington, DC: United States Strategic Bombing Survey, 1947): 110–111; *25 Jahre Vereinigte Deutsche Metallwerke A.G.* (Frankfurt-on-the-Main: Vereinigte Deutsche Metallwerke, 1955), 1–2.

blades and hubs in safer and less concentrated areas. Blade production actually increased in 1944. The high point was the production of 30,000 individual blades a month from August to October 1944.[39] When aluminum became scarce, the Luftwaffe encouraged manufacturers to develop magnesium blades in addition to increasing the demand for Schwarz compressed wood blades for VDM propellers.[40]

In August 1944, the bombers of the American Eighth Air Force and RAF Bomber Command attacked Frankfurt and Hamburg and destroyed the strategic VDM factories. The main manufacturing effort moved south to Marburg. The detail design and experimental group led by Hans Ebert moved to the safety of a railway tunnel manufacturing complex near Hasselborn northwest of Frankfurt and in the shadows of the Taunus Mountain range. Inside the tunnel, Ebert and his technical team investigated new ideas and solutions while approximately 1,500 forced laborers produced propellers for the Nazi war effort. The manufacturing equipment included a 30-ton hydraulic press that turned out aluminum blades in a single operation.[41]

The war ended for VDM on March 30, 1945, the day after Frankfurt fell to advancing US Army infantry and armored units. Even though exact records are not available, the manufacturing record of VDM was impressive. The estimation that 95 percent of the Luftwaffe's approximately 140,000 aircraft used VDM propellers between 1934 and 1945 suggests production to be between 260,000 and 400,000.[42]

In the wake of the Nazi collapse during the spring of 1945, Allied technical intelligence personnel scrambled to harvest the Luftwaffe's aeronautical legacy. They collected VDM propellers and any reports and information available and shipped them west for evaluation. US Army intelligence officers found Hans Ebert wandering among the ruins of the Hasselborn railway tunnel. Ebert's interview revealed that development work on propellers substantially stopped as the German aviation industry moved toward rocket- and gas turbine-powered craft, which was in stark contrast to the American emphasis on winning the war with established technology like the constant-speed propeller. Hitler's orders to restrict the

[39] USSBS, *Aircraft Division Industry Report*, 107, 110–111.

[40] John D. Waugh, "Details of the German VDM Electric Propeller, Part I," *Industrial Aviation* 1 (July 1944): 33.

[41] George C. McDonald, "Report on the Manufacture of Aircraft Propellers by Vereinigte Deutsches Metal Werke, Heddernheim, Frankfurt am Main," May 23, 1945, B5-90000-01, PTF, NASM.

[42] Vajda and Dancey, *German Aircraft Industry and Production, 1933–1945*, 133.

distribution of technical data further curtailed innovation in propeller design. Plans to develop high-speed scimitar-shaped blades and a hydraulic propeller capable of instantaneous pitch change never materialized.[43]

Despite those grandiose plans, American intelligence and engineering personnel found Nazi propeller technology inferior. John D. Waugh, an aeronautical engineer working for Lockheed, acknowledged the VDM to be "versatile, well-designed, and well-constructed," but he felt American and British designs were superior. Specifically, a pitch change rate of one degree per second was not adequate for high performance aircraft.[44] Adam Dickey, as the civilian head of the Army's propeller program in the 1930s and 1940s, traveled to Germany with other Wright Field engineers to inspect and evaluate German aeronautical developments. Dickey confirmed that while the Germans were ahead of the United States in many areas, primarily high-speed aerodynamics and rockets, they were dramatically behind the United States in propeller development. He specifically believed that the "early vision" of the propeller unit at McCook Field, under the leadership of Frank Caldwell, was the reason.[45]

The Axis power in the Pacific, Imperial Japan, adapted and modified foreign variable-pitch designs for its military and naval aircraft. The two primary manufacturers were the Propeller Division of Sumitomo Metal Industries and the musical instrument manufacturer Nippon Gakki Company (the modern day Yamaha Corporation). Sumitomo acquired the license for both the Hamilton Standard constant-speed counterweight propeller and the VDM in the late 1930s as Japan continued its military expansion into China. The company had two plants in the Osaka area producing 672 propellers a month by January 1941. The majority were the Hamilton Standard-type and for use on the Imperial Navy's aircraft, including the Mitsubishi A6M2 Reisen fighters, Aichi D3A1 dive bombers, and the Nakajima B5N torpedo bombers used in the December 7, 1941, attack on Pearl Harbor. Production of the VDM began in 1942 for aircraft such as the Mitsubishi Ki-67 Hiryu medium bomber and the Kawanishi N1K2 Shiden Kai fighter. Seen as more advanced, the VDM required 70 percent more work than the Hamilton Standard to

[43] McDonald, "Report on the Manufacture of Aircraft Propellers by Vereinigte Deutsches Metal Werke, Heddernheim, Frankfurt am Main."

[44] John D. Waugh, "Particulars of the German VDM Electric Propeller," n.d. [1944], CATD, NASM.

[45] D. A. Dickey, "Messerschmitt Propeller Activity," July 1945, CATD, NASM; Daniel Adam Dickey, Interview by Lois E. Walker, September 2, 16, 30, 1983, interview K239.0512-1712, transcript, United States Air Force Oral History Program, HRA, x.

manufacture. By July 1944, Sumitomo operated four manufacturing plants producing 5,247 propellers a month under the direction of long-time general manager Osamu Sugimoto, an aeronautical engineer educated at MIT.

Using its expertise in making pianos, Nippon Gakki began manufacturing wood two-blade fixed-pitch propellers in 1921. As Imperial Japan's need for propellers increased after the invasion of Manchuria in September 1931, the company began the production of ground-adjustable-pitch propellers with aluminum alloy blades at its main factory in Hamamatsu. Nippon Gakki sent a delegation of engineers to the United States and Europe to survey manufacturing processes. After the Japanese Army took over the company in 1937, Nippon Gakki produced more propellers than pianos and organs. During the war, three factories employing 10,000 people fabricated constant-speed counterweight type, wood fixed-pitch, and hybrid propellers with metal hubs and wood blades. With the help of automotive pioneer Honda Sōchirō, Nippon Gakki developed enhanced manufacturing processes that would prove instrumental to the success of Japanese industry after the war.[46]

Material and labor shortages, design changes, and poor planning led to a dramatic reduction in output during the fall of 1944. Continued bombings by the US Twentieth Air Force led to the dispersal of the overall Japanese propeller industry to new areas, which included the basement of the Sogo Department Store in downtown Osaka. B-29 attacks during June and July 1945 effectively destroyed the Japanese propeller industry. Sumitomo produced 66 percent, or 89,885, of all propellers for Japan's aerial war effort while Nippon Gakki produced 46,304, or 34 percent to account for a total of 136,189 propellers (Table 4). The intelligence personnel of the United States Strategic Bombing Survey estimated that the state of Japanese propeller technology was five or more years behind developments in the United States.[47]

The propeller communities of Nazi Germany and Imperial Japan did not survive World War II. Their destruction from aerial bombardment resulted from the folly of their national governments' expansionist desires. Unlike other German technologies and the engineers and scientists that

[46] Jeffrey W. Alexander, *Japan's Motorcycle Wars: An Industry History* (Vancouver: University of British Columbia Press, 2008), 146–148.

[47] USSBS, *Sumitomo Metal Industries, Propeller Division*, Corporation Report No. VI (Washington, DC: United States Strategic Bombing Survey, 1946): 1, 3, 10, 12; and *The Japanese Aircraft Industry* (Washington, DC: United States Strategic Bombing Survey, 1947): 100–102.

TABLE 4 *German and Japanese propeller production, 1934–1945*

Manufacturer	Output
VDM	Estimated 260,000 to 400,000
Propeller Division of Sumitomo Metal Industries Ltd.	89,885
Nippon Gakki Company Ltd.	46,304
Total	396,189 to 536,189

developed them, specifically rocket and jet propulsion, and transonic and supersonic aerodynamics, there was no legacy.[48] At least to the victors, their propeller technology was not worth saving anyway. The triumphant United States and Great Britain stood poised to dominate world aviation in the postwar period.

The Triumph of the Modern Propeller

World War II opened up a global world connected by the airplane. In the wake of the war's end, air forces and airlines maintained large fleets of piston-engine bombers, transports, patrol aircraft, and airliners. Hamilton Standard, the Propeller Division of Curtiss-Wright, and Aeroproducts in the United States and Rotol and de Havilland in Great Britain continued to supply variable-pitch propellers for existing and new aircraft and engine designs. The refined modern airplanes of the late 1940s and 1950s such as the gigantic Consolidated B-36 Peacemaker and the sleek Lockheed Constellation featured tricycle landing gear, high-output radial piston engines, and pressurized cabins, and were capable of high-altitude flight at transcontinental and transoceanic ranges. The Air Age that had arrived was not the utopia the airminded culture of the 1920s and 1930s envisioned, but it was an era where the airplane was a central element of everyday life.

These newly refined modern airplanes and the ones from the previous generation played vital roles during the potentially volatile early days of the Cold War. In Europe, the victorious Allies divided Germany and its capital, Berlin, into two occupation zones after World War II. The Soviet Union held the eastern half while the United States, Great Britain, and France administered the western half. As tensions rose over the future

[48] Michael J. Neufeld, "The Nazi Aerospace Exodus: Towards a Global, Transnational History," *History and Technology* 28 (March 2012): 57.

of Germany, Joseph Stalin ordered a ground blockade that severed all road, rail, and water connections to Berlin, located in the Soviet zone, in preparation for occupying the entire city. The United States and Great Britain staged an around-the-clock airlift of Berlin from June 1948 to May 1949. Military pilots and aircrew, flying transports like the four-engine Douglas Skymaster, kept 2.5 million people fed and warm with 2.3 million tons of supplies during the Berlin Airlift. During the Korean conflict, the Air Force's Military Air Transport Service operated Boeing C-97 Stratofreighters converted into flying hospitals. For a wounded soldier or Marine, nonstop flights from Tokyo to Hawaii and then to the continental United States meant it was a one-week journey from the battlefield to a hospital near his home to ensure the best recovery. These operations gave the world a valuable lesson in long-range aerial supply and transport.

There was a new segment of aviation that embraced variable-pitch propellers. The disparate community of general aviation, which included any flying that was not covered by scheduled commercial airlines and military aviation, entered the postwar period enthusiastic for new aircraft and performance. At the first annual National Aircraft Show in Cleveland in November 1946, seven of the thirteen manufacturers displaying propellers offered variable-pitch designs. Aeromatic of Baltimore, Maryland, and Iso-Rev of Long Island, New York offered constant-speed models while Beech Aircraft of Wichita, Kansas, Freedman Aircraft Engineering of Cincinnati, Ohio, and Hartzell Propellers exhibited their controllable-pitch designs. These manufacturers offered propellers for a new generation of air-cooled opposed engines that ranged between 65 and 330 horsepower. They made them from materials including steel and magnesium for the hub mechanisms and aluminum, composite, and laminated wood for the blades. Eyeing a new market segment, both Aeroproducts and Hamilton Standard offered their constant-speed propellers for 450-horsepower engines.[49]

The general aviation community incorporated these propellers into its aircraft for the same reasons that guided the commercial and military aviation communities: higher performance. Whether a pilot was flying an aerial demonstration at an air show, crop dusting an agricultural field, transporting cargo to a remote village, travelling for business, fighting a forest fire, flight training, competing in an air race, or just plain having

[49] "Personal *Flying* Review of the National Aircraft Show: Propellers," *Flying* 39 (December 1946): 98.

fun for recreation, they wanted maximum power, safety, and reliability.[50] New aircraft like the high-performance low-wing monoplanes with retractable landing gear like the Beech Bonanza and Ryan Navion; utility aircraft like the Republic Seabee amphibian and de Havilland Beaver, and light planes like the Stinson Voyager and Cessna 190 benefitted from variable-pitch propellers.

The general aviation community also welcomed a sizable number of war surplus aircraft released from the military and the airlines that incorporated variable-pitch mechanisms. Ex-military transports like the Curtiss C-46 Commando and the Douglas C-47 Skytrain enabled entrepreneurial pilots to make a living hauling cargo. Former fighters such as the North American P-51 Mustang and Vought F4U Corsair became "warbirds" as enthusiasts collected and flew them for fun or competed in the postwar National Air Races at Cleveland.

The performance, durability, and longevity of military, commercial, and general aviation aircraft brought aviation permanently to a greater portion of the world. One thing had become clear within the overall aviation community and implied in the general populace in the 1950s. Whether they knew it or not, when someone referred to a propeller on an airplane, they were talking about one with a variable-pitch mechanism. The modern propeller had become part of the fabric of everyday life.

The Age of the Jet

In October 1958, Pan American World Airways initiated nonstop passenger flights from New York-to-London and -Paris in its new fleet of Boeing 707 jetliners. The use of the 707 and its contemporaries, the Douglas DC-8 and de Havilland Comet by international airlines, in comparison to propeller-driven piston engine airliners, transformed air travel into a shorter, more comfortable, and more affordable experience. In anticipation of those first 707 flights, the *New York Times* announced that in "the Age of the Jet" travelers would fly "higher, faster, and farther" than ever before. Capable of cruising at almost 600 mph, which cut flight times to far off destinations nearly in half, the jet airliner was a "giant step forward in the history of transportation."[51] Since World War II, a second revolution in aeronautics had been taking place in military

[50] Beechcraft Propellers, *Beechcraft Controllable Propeller* (Wichita, KS: Engineering Service Division, Beech Aircraft Corporation, 1944), 1, File B5-110060-01, PTF, NASM.

[51] "The Age of the Jet," *New York Times*, September 6, 1958, p. 16.

and commercial aviation based on a new propulsion technology, the jet engine.

The propeller and the piston engine were the main form of aeronautical propulsion for the first half century of flight. The introduction of gas turbine technology, specifically the turbojet in the 1940s, rendered the piston engine obsolete as a source of power for high performance military and commercial aviation. To increase output beyond 3,000 horsepower, aircraft engine makers introduced ever-increasing complexity and sophistication into their designs during the course of the Aeronautical Revolution and World War II. Pratt & Whitney's Wasp Major radial featured four spiraled rows of seven cylinders. The Turbo-Compound Cyclone from Wright Aeronautical incorporated three small turbines that converted the heat from the exhaust into energy that a gearing system transmitted back to the engine crankshaft. Such byzantine engines were finicky in operation and required rigorous and expensive maintenance programs every few hundred hours of flight time. Jet engines offered longer times between overhauls, fewer moving parts, and greater distances and speeds. Military leaders embraced the turbojet for their front-line bombers and fighters during World War II and the early days of the Cold War. Airline executives soon followed with the introduction of jet-powered airliners in the 1950s. The introduction of the turbofan, which incorporated a large, enclosed multiblade fan to harness the efficiency of the propeller while developing the high thrust of the turbojet, further expanded the use of jet aircraft. By the 1960s, the Jet Age was in full swing.

For high speed aircraft, the variable-pitch propeller was obsolete as well. Propellers are the most efficient moving a large mass of air at a low velocity. As aircraft speeds increased past 500 mph, compressibility burble, the marked decrease in efficiency characterized by a loss of lift caused by increased drag, ensured that propellers could not generate the needed thrust to fly higher, faster, and farther. Frank Caldwell and his fellow researchers at McCook Field discovered the limitation just after World War I, but it had taken that long for the aerodynamic design of aircraft to catch up. As the inventors of the jet engine, Frank Whittle and Hans von Ohain envisioned, if airplanes were to go faster, they would not have propellers.[52]

The apparent primacy of the jet in the immediate years after World War II led one popular aviation writer to ask, "Has the propeller a

[52] Edward W. Constant, *The Origins of the Turbojet Revolution* (Baltimore, MD: Johns Hopkins University Press, 1980), 15.

future?"[53] The propeller community conducted numerous wind tunnel and flight research experiments in conjunction with similar investigations by industry to see if they could provide the answer. One focus was on developing supersonic propellers capable of taking long-range transport aircraft into the transonic regime between Mach 0.8 and 1.2.[54] At the Langley Aeronautical Laboratory in Virginia, NACA engineers created a "propeller research airplane" by installing an Aeroproducts propeller in the nose of a McDonnell XF-88B Voodoo jet fighter. The flight program, which ran from 1953 through 1956, revealed a propeller design that was 79 percent efficient at a speed of Mach 0.95, or approximately 700 mph.[55]

Such promising programs appeared to be futile in the early days of the Jet Age. At a 1949 NACA conference on transonic aircraft design, a Hamilton Standard engineer, having just reported on the results of wind tunnel tests on a supersonic propeller for a US Air Force contract, remarked that compared to the jet engine, "even if the propeller is good, it is not wanted."[56] The aeronautical community faced many development problems with supersonic propellers in the 1950s, which seemed unnecessary if jet technology provided equal or better performance already. A major challenge was reducing the noise that resulted from the shockwaves at the blade tips. The three-blade supersonic Aeroproducts propeller on the Air Force's experimental Republic XF-84H fighter, which offered mediocre performance overall, generated such high-intensity noise that it rendered bystanders sick.[57]

[53] William Winter, "Has the Propeller a Future?," *Popular Mechanics* 89 (February 1948): 171.

[54] Eugene C. Draley, Blake W. Corson, Jr., and John L. Crigler, "Trends in the Design and Performance of High-Speed Propellers," in *NACA Conference on Aerodynamic Problems of Transonic Airplane Design: A Compilation of Papers Presented, September 27–29, 1949*, NASA TM-X-56649, 483–498; John V. Becker, *The High Speed Frontier: Case Histories of Four NACA Programs, 1920–1950*, NASA SP-445 (Washington, DC: NASA, 1980), 135–138.

[55] Jerome B. Hammack, Max C. Kurbjun, and Thomas C. O'Bryan, "Flight Investigation of a Supersonic Propeller on a Propeller Research Vehicle at Mach Numbers to 1.01," NACA Research Memorandum No. L57E20 (1957), 5–6; Jerome B. Hammack and Thomas C. O'Bryan, "Effect of Advance Ratio on Flight Performance of a Modified Supersonic Propeller," NACA Technical Note No. 4389 (September 1958), 5.

[56] Thomas B. Rhines, "Summary of United Aircraft Wind Tunnel Tests of Supersonic Propellers," in *NACA Conference on Aerodynamic Problems of Transonic Airplane Design: A Compilation of Papers Presented, September 27–29, 1949*, NASA TM-X-56649, 448.

[57] Stephan Wilkinson, "ZWRRWWWBRZR: That's the Sound of the Prop-driven XF-84H, and It Brought Grown Men to Their Knees," *Air & Space* (July 2003), www.airspacemag.com/how-things-work/cit-wilkinson-july03.html?c=y&page=1 (Accessed September 9, 2012).

The NACA, which targeted fundamental problems facing the future American aeronautics, lost interest and disbanded the longstanding Subcommittee on Propellers for Aircraft in 1957. Propellers as a major research area disappeared in the early days of its successor, the National Aeronautics and Space Administration (NASA).[58]

Industry observers, however, like Mark Twain, said the "obituary notices" announcing propellers as obsolete were "grossly exaggerated." The variable-pitch propeller filled a particular niche in aviation: efficient flight below 500 mph at short and medium distances.[59] Piston engine-powered regional airliners and cargo aircraft, military transports, and other types of what were suddenly considered to be lower performance and less spectacular aircraft continued to serve humankind. Those aircraft needed propellers.

While there was a need, propeller makers faced difficult choices. Their work required considerable engineering and production resources. Rather than clinging to their singular specialty, they expanded into other areas. That enabled them to transfer their expertise in hydraulics, precision engineering, and electronics to new technologies, which also meant considerable effort was placed on bringing in new methods and practices. As their markets shrank, specialty propeller manufacturers evolved into diversified general equipment suppliers that ensured survival in a perilous economy.

As those changes transformed the propeller industry, the variable-pitch propeller became an artifact of the Jet Age. The combination of a gas turbine with a variable-pitch propeller resulted in the turbine propeller, or "turboprop," engine. For many, a turboprop was the best of both worlds. Compared to a piston engine, it had a better power-to-weight ratio, meaning it was lighter and more powerful, and was capable of generating more thrust. Judged against the jet engine, a turboprop offered the near-instantaneous response of the piston engine. Since a turboprop ran at a consistent speed, which was near full rpm all the time, changing the propeller pitch generated thrust quickly. Overall, turboprops generated less noise than both piston and turbojet engines and were efficient in terms of thrust and fuel economy. The combination of the best qualities of both the piston and turbojet engines – rapid response and increased power – made the turboprop a crucial propulsion technology, albeit at a much smaller scale than the heyday of World War II.

[58] Becker, *The High Speed Frontier*, 136.

[59] Charles J. McCarthy, "Air Miracles Take Time," *The Bee-Hive* 22 (Summer 1947): 15; Oliver Stewart, "Airscrews and All That," *Aeronautics* 26 (May 1952): 23.

The introduction of the turboprop led to further improvement in design. The propeller community introduced refined and more efficient pitch-changing mechanisms, stronger structures, and more precise control systems better able to handle the increased power. New rectangular and thin blade designs, which reminded many observers of windmill vanes, reflected the emphasis on getting the most out of every square inch of the surface. They absorbed more energy and turned it into thrust more efficiently while offsetting compressibility losses.[60] Designers could also choose a pair of propellers arranged in tandem and turning in opposite directions, called contra-rotating and first introduced by Rotol in 1945.[61]

Turboprop-powered airliners first began commercial operations in the early 1950s. For passengers, turboprop airliners were vibration-free, smooth in flight, and very comfortable at high-altitudes compared to piston-engine airliners. For airlines, they presented an economical alternative for short- to medium distance routes and could operate from short fields when compared to turbojet airliners that used fuel voraciously and needed long runways. In Great Britain, the Vickers Viscount was the first turboprop airliner to enter passenger service, powered by four Rolls-Royce Dart turbine engines connected to Rotol propellers. European and American airlines bought over 400 Viscounts after their introduction in July 1950. The Dart propeller became an important product for Rotol, which the aviation industry heralded as the result of "pioneering work" in guaranteeing the future of propellers.[62]

The transition to turboprops and the challenge to the continued use of propellers overall shaped the future direction of the industry. Erle Martin became president of Hamilton Standard in 1946. His primary focus was on the refinement of the company's marquee product, the hydraulically actuated constant-speed Hydromatic propeller, in the early days of the Cold War. The Super-Hydromatic incorporated three important innovations. Reversible pitch allowed heavily loaded airliners extra braking power at landing. A separate gearbox and oil system from the engine, a development that was a critical need during the war, ensured reliable operation. Hollow-steel blades, a development Hamilton Standard avoided for years, became necessary as engine power and propeller size dictated the need for lighter weight.

[60] "New Square-Tipped Blade, Similar to Windmill Vane, Developed by Hamilton Standard Propeller," *The Bee-Hive* 21 (Summer 1946): 8.

[61] Ministry of Supply, "Propellers: Development and Production, 1934–1946," 52.

[62] Stewart, "Airscrews and All That," 23; Bruce Stait, *Rotol: The History of An Airscrew Company, 1937–1960* (Stroud, Gloster, England: Alan Sutton Publishing, 1990), 129.

For gas turbine applications, Hamilton Standard marketed the propeller as the Turbo-Hydromatic and won important contracts for the four-blade 54H60 propeller designed for the Allison T56 engine in 1958.[63] The 54H60 and the T56 became the standard propulsion system for the primary propeller-driven military and naval aircraft of the United States and other nations around the world, the Lockheed C-130 Hercules transport and P-3 Orion patrol aircraft and the Grumman E-2 Hawkeye airborne early warning aircraft and the C-2 carrier onboard delivery aircraft. The orders and subsequent maintenance and spare parts contracts sustained Hamilton Standard throughout the Cold War.

There were changes taking place in Hartford. As jet engine technology supplanted the propeller-piston engine propulsion system, Frank Caldwell directed his energies to turbine research at the UAC Research Division before his retirement in 1955.[64] Despite Hamilton Standard's strong market position, Martin made the decision to diversify into environmental, flight, and fuel controls in 1948. A year later, on the twentieth anniversary of the merger of Hamilton Aero and Standard Steel, the company eliminated the word "Propellers" from its official title to be known simply as the Hamilton Standard Division, UAC. Hamilton Standard environmental control and life support systems flew with the crew of Apollo 11 to the moon in 1969 and on the Space Shuttle *Columbia* in 1981. Martin went on to serve as a UAC vice president in 1952, board member in 1958, and chief technology officer in 1960 before his retirement in 1972.[65] Reflecting overall changes in the aerospace industry, UAC renamed itself the United Technologies Corporation (UTC) in 1975 to indicate its involvement in a diverse field of industries.

The British propeller manufacturers, de Havilland and Rotol, also faced challenges after World War II. De Havilland continued to produce and market propellers licensed from Hamilton Standard for the British aviation market. The division formally became de Havilland Propellers Ltd. in April 1946. Seeing a new opportunity, the company's directors expanded the longstanding Hamilton Standard license to include aircraft

[63] "Jets Plus Propellers," *The Bee-Hive* 25 (Summer 1950): 25; "Hamilton Standard's Hollow-Steel Blades," *The Bee-Hive* 23 (Spring 1948): 18.

[64] On December 23, 1974, Caldwell died at his home in West Hartford, Connecticut. "F.W. Caldwell Dies," *Hartford Courant*, December 24, 1974, p. 4;"F.W. Caldwell Dies at 85," *West Hartford News*, January 2, 1975, p.4.

[65] Martin died on December 12, 1981. Frank J. Delear, "He Decided on Aviation," *The Bee-Hive* 30 (Spring 1955): 10; Trotman, "Turning Thirty Years," 27, 28; "Erle Martin, Inventor, Dead"; Hamilton Sundstrand Corporation, "Hamilton Standard History," 1999, www.hamiltonsundstrandcorp.com (Accessed February 26, 2002).

air-conditioning systems and started development of radar scanners and air-to-air missiles. The Firestreak infrared homing missile was the first of its kind in Great Britain. It became standard equipment for British jet fighters, including the English Electric P.1, the Gloster Javelin, and the de Havilland Sea Vixen, after its introduction in 1957. The company soon changed its name to de Havilland Dynamics to reflect the new emphasis on missiles and space technology. It was a seamless transition in the sense that the same personnel, who had been propeller specialists, became missile engineers. The Hawker Siddeley group of companies merged with the parent de Havilland company in 1959. Design and production work continued at Hatfield and Lostock under the name of Hawker Siddeley Dynamics until the creation of British Aerospace in 1977.[66]

Rotol sought a larger role as a specialty manufacturer. After its initial success with the Dart propeller, the company expanded and acquired landing gear manufacturer British Messier in 1952 and used its internal resources to develop fuel systems for turbojets, accessory power turbines, turbochargers for diesel engines in 1957. A year later, Rotol's owners, Bristol and Rolls-Royce, agreed to the sale of the company to the Dowty Group of Companies for £2,235,000. The Dart propeller and gearbox were the major selling point for the acquisition. The turboprop propellers installed on the behemoth Bristol Brabazon airliner and the contra-rotating designs for the Royal Navy's Westland Wyvern and Fairey Gannet also gave Rotol considerable notoriety. Rotol's new owners established Dowty Rotol Ltd. in April 1960 as a diversified aerospace equipment supplier, which included propellers as a core product.[67]

Optimistic for its postwar prospects, the Propeller Division of Curtiss-Wright, Hamilton Standard's main competitor, accounted for a large portion of the corporation's overall profits. Both the military and the commercial airlines used the Electric Propeller in great numbers. The Division unveiled its Turboelectric Propeller during the fall of 1950 for use on large, multiengine long-distance aircraft.[68] The failure to find a large market application for the Turboelectric, other than the US Air Force's gargantuan Douglas C-133 Cargomaster, and a change in the overall direction of Curtiss-Wright away from airframe and engine

[66] "More About British Guided Missiles ...," *Flight* 71 (May 3, 1957): 568; "Firestreak: The Design of Britain's First Air-to-Air Missile," *Flight* 77 (February 26, 1960): 277; Sharp, *D.H.*, 339–341.

[67] Stait, *Rotol*, 121, 125, 134, 145, 148–150.

[68] Curtiss-Wright Corporation Propeller Division, *Curtiss Turboelectric Propellers for Turbo-Prop Engines* (n.p., 1950), File B5-260250-01, PTF, NASM.

production led to a more severe form of diversification. The organization became known as the Curtiss Division in 1961 to reflect new work on mechanical controls and systems for aircraft, missiles, and submarines. Other than spare parts and maintenance, the propeller business consisted of the design and fabrication of the fiberglass blades for the failed X-19 vertical- and short-takeoff and landing (VTOL/STOL) aircraft project.[69] General Manager, C. E. Ehinger, announced in January 1967 that the division's primary focus was no longer propellers and related systems, but innovating component systems for military aircraft and helicopters.[70]

Aeroproducts, too, started out with much optimism during the early years of the Cold War as it turned its focus to single and contra-rotating propellers for turboprops. One of the division's propellers was on the first turboprop-powered aircraft to fly in the United States in December 1945, the Consolidated-Vultee XP-81 fighter. Engineers in Dayton started work on developing a series of propellers for turboprop engines developed by their corporate peers, Allison in Indianapolis. Tragically, as Aeroproducts moved forward, Pete Blanchard and two employees died in the crash of his private airplane in December 1948.[71] During the 1950s, Aeroproducts engineers designed propellers for several proposed and experimental commercial and military aircraft, which included the Convair Turboliner, the Douglas XA2D Skyshark ground attack airplane, and Convair R3Y Tradewind flying boat for the US Navy. Difficulties with the development of the Allison engines and devising suitable propeller controls led to the cancellation of each project. During those troubled times, Aeroproducts ceased to be an independent division within General Motors in 1953.[72]

Aeroproducts introduced a new propeller, the four-blade Model 606, for the civilian version of Allison's T56, called the 501, in 1957. Lockheed chose Allison to provide the engine for its commercial airliner, the L-188 Electra. As the first American turboprop airliner, the Electra entered commercial service with American Airlines in December 1958. Considered to be the highest development of a propeller airplane up to that time,

[69] Curtiss-Wright Corporation, *1961 Annual Report*, 12–13, Folder 6, Box1, CWC, NASM; Louis R. Eltscher and Edward M. Young, *Curtiss-Wright: Greatness and Decline* (New York: Twayne Publishers, 1998), 145–151.

[70] "Curtiss Gets Four Orders at Year End," *The Curtiss-Wrighter* 1 (January 1967): 1, Folder 7, Box1, CWC, NASM.

[71] Blanchard, *A Man Wants Wings*, 83.

[72] Aeroproducts, *Turbine Propellers* (Dayton, OH: Aeroproducts Operations, Allison Division of General Motors Corporation, n.d. [1954]), File B5-007000-02, PTF, NASM; Daron L. Gifford, "Propeller Life of Type Manufacturing" (B.S. thesis, General Motors Institute, May 1980), 9.

a four-engine Electra could carry up to ninety-eight passengers at just over 370 mph while still excelling at short to medium range flights to runways that were too short for jet aircraft.[73] A structural design flaw in the early Electras allowed the engine mounts to oscillate to the point where they would resonate with the natural frequency of the wing spar until destruction, causing catastrophic wing failure. The resultant accidents damaged permanently the public's perception of the Electra before Lockheed solved the problem. The introduction of the Boeing 707 jetliner just months before did not help the situation either.

The fortunes of Aeroproducts went along with those of Lockheed and the airlines flying the Electra. The division needed more business to survive in the ever-dwindling propeller market. The loss of important contracts for the naval version of the Electra, the P-3, and the C-130 transport to Hamilton Standard, led to downsizing in 1958. The Vandalia plant closed two years later and all operations moved west to the Allison factory at Indianapolis. A shadow of the operation that had grown during World War II, Aeroproducts fulfilled a ten-year contract to provide propellers for Convair, primarily for the 240-series piston engine airliners. Workers manufactured the last complete propeller in January 1974. Spare parts production continued with a final lifetime supply of blades being manufactured in Indianapolis in early 1977. Recognizing the ever-dwindling market for Electra propellers, Allison sold the last fifty-six complete Model 606 propellers to Hamilton Standard in March 1990.[74] The demise of Aeroproducts was one of the perils of competing to fill a niche while failing to diversify in an ever-shrinking aeronautical marketplace.

Higher, Faster, and Farther versus the Right Technology for the Job

The consistent desire to fly higher, faster, and farther was a driving force behind the development of the modern airplane of the 1930s and influenced the adoption of jet engine in the 1940s and 1950s. Quickly, aircraft delivered passengers, cargo, and bombs and opposed each other with machine guns, cannons, and missiles at higher altitudes, increased speeds,

[73] Ted Sell, "Lockheed Demonstrates Smooth New Electra," *Los Angeles Times*, October 3, 1958, p. 18; Gifford, "Propeller Life of Type Manufacturing," 9–10.

[74] "End of an Era," *Power News* (February 1977): 6; and E. Jerry Mayfield, "End of an Era," *AllisoNews* (April 1990): 5, both Rolls-Royce Heritage Trust, Allison Branch; Blanchard, *A Man Wants Wings*, 140–142.

and longer distances than ever before. It appeared, on the surface, that the continued progress of the airplane was limitless as supersonic fighters and subsonic airliners ranged over the globe. Operational high performance aircraft speeds and altitudes, which had been consistently increasing since December 1903, peaked at Mach 3.3, or 2,400 mph, and 85,000 feet around 1970. The choices of designers changed to reflect an emphasis on other parameters. Issues of aerodynamic and propulsive efficiency, lower costs, improved reliability and safety, and awareness of environmental issues related to emissions and noise augmented and often superseded the decades-old driving philosophy of aircraft design based solely on altitude, speed, and range.[75] That technical reality and changes at the highest political, economic, and cultural levels during those decades of the Jet Age illustrated the need for a niche technology like the propeller.

With oil prices low through most of the 1960s, there seemed to be no concerns over fuel efficiency in aviation. With cheap fuel and turbofan engines, which operated at about 65 percent efficiency, the airlines were making increased revenues. That perspective changed as the Organization of Petroleum Exporting Countries (OPEC), upset over Western ties with Israel, affected an oil embargo in 1973 that sent the United States and Europe reeling. No longer would energy sources be unlimited or cheap in Western societies whose technological infrastructure depended on oil. While auto makers reacted with compact, fuel efficient cars and drivers adapted by traveling less, the airlines were in a dire situation. They could not reduce service or raise ticket prices as they faced an escalation of fuel prices that were estimated to be between 20 to 50 percent.[76] Successive oil crises that stymied the West since the 1970s caused a serious reevaluation of what constituted performance in aviation. Specifically, fuel consumption became a matter of economic survival.

As the commercial aviation industry reeled from the OPEC oil embargo, the US Congress passed the Airline Deregulation Act of October 1978. Before then, the Civil Aeronautics Board regulated fares, routes, and schedules that permitted airlines to connect major cities with direct point-to-point service. Deregulation facilitated the creation of short-haul routes that on paper resembled a wagon wheel. In this new hub-and-spoke

[75] For John Anderson, "farther," or distance, is a result of design requirements for specific aircraft and is not as fundamental a performance parameter as "higher" (altitude) and "faster" (speed). John D. Anderson, Jr., *Introduction to Flight*, 7th ed. (New York: McGraw-Hill, 2012), 46–49.

[76] Roy D. Hager and Deborah Vrabel, *Advanced Turboprop Project*, NASA SP-495 (Washington, DC: NASA, 1988), v.

system, passengers started their journey at a regional airport and traveled along a "spoke" to the "hub," or major airport. From there, they made a connecting flight on a jet airliner to their final destination. Regional commuter airlines, which had been rapidly growing in number since the mid-1960s, became an important component of deregulation as they carried between fifteen and fifty passengers each flight along the short and medium length spokes.

Both the world oil crisis and deregulation stimulated the increased use of turboprop aircraft, which became known as "puddle jumpers," in commercial aviation in the 1980s. They offered several advantages for the rapidly expanding short-haul commuter market. Turboprops were cheaper to operate because they generated the highest fuel efficiency, between 80 and 90 percent, at the lower altitudes (under 25,000 feet) and shorter route distances (below 500 miles) necessary for regional operations. They were more maneuverable than larger jet airliners and capable of taking off and landing safely on short runways common to small cities and towns.[77] In the social and cultural sense, while jet airliners connected urban metropolitan areas with each other and to the rest of the world, it was the regional airlines that served as the first leg between rural and remote communities to the outside world.

New twin-turboprop aircraft designed and manufactured in North and South America and Europe became instantly recognizable at airports. The Beechcraft 1900 carried nineteen passengers in the comfort of a pressurized cabin. Brazil's leading aircraft manufacturer, Embraer, introduced the low-wing thirty-passenger EMB-120 Brasilia as a follow-on to its successful twenty-passenger Bandeirante. The ATR-42 resulted from a French-Italian collaboration to create a fifty-passenger airliner. The STOL high-wing de Havilland Canada Dash 8 initially carried forty passengers with later "stretched" models carrying up to eighty. These puddle jumpers were capable of cruising in the 300–400 mph range and flying from airfields measuring no more than a mile.

A new generation of "commuter propellers" powered these regional airliners. Hamilton Standard, Dowty Rotol, and Hartzell introduced further refinements to the variable-pitch propeller. Advanced electronic control allowed for immediate response in flight and on the ground. Echoing trends in fuselage construction, propeller makers engineered blades made

[77] Ernest Holsendolph, "Turboprops for Smaller Lines," *New York Times*, October 19, 1982, section D, p. 1; R. E. G. Davies and I. E. Quastler, *Commuter Airlines of the United States* (Washington, DC: Smithsonian Institution Press, 1995), 45, 115–121.

from composite materials, which included carbon fiber, Kevlar, fiberglass, and foam, that resulted in a 50 percent reduction in weight. These new blades also featured innovative airfoil profiles that benefited from pioneering aerodynamic research by the Aircraft Research Association in Great Britain and NASA in the United States. To handle the increased power of turboprop engines like the Pratt & Whitney Canada PW100 series without increasing noise or the individual diameter or blade width of a propeller, designers chose five- and six-blade configurations.[78]

Turboprop aircraft filled the niche as the main vehicles for regional air transportation marvelously through the 1980s and 1990s. There was, however, a problem with them. A significant portion of the flying public rejected them in comparison to large high-flying jet airliners. Smaller turboprops flew in more turbulent air and their propellers generated noise. Passengers endured rough and bumpy flights in cramped cabins unable to carry on normal conversations nor have enough time to be served by a flight attendant's well-stocked refreshment cart. A general description of a flight in a turboprop equated it to "riding on the back of a mosquito." Moreover, smaller propeller aircraft in general had a greater rate of fatal accidents, estimated to be two to three times higher in the 1990s, due to the fact that they flew at lower altitudes where they and their less-experienced crews were susceptible to the dangers of bad weather.[79] Comfort and safety were concrete reasons that deterred passengers from actually choosing to fly on turboprops. The power of personal preference was an important indicator of how individuals and groups perceived technology.

There was also a more abstract reason that influenced the less-than-favorable public opinion of propeller-driven aircraft when compared to jet-powered military and commercial aircraft. Major airline advertising in the form of a jet airliner publicity campaign beginning in the late-1950s and reaching market saturation within ten years fostered the implication that the flying public should perceive propellers as backward technology. Air carriers like Pan American World Airways endeavored to domesticate jet travel to make it more appealing to a large consumer public. They emphasized the novelty of corporate jet-powered long distance airlines

[78] Patrick Hassell, "A History of the Development of the Variable Pitch Propeller," Paper presented before the Royal Aeronautical Society Hamburg Branch, Hamburg, Germany, April 26, 2012; George Rosen, *Thrusting Forward: A History of the Propeller* (Windsor Locks, CT: United Technologies Corporation, 1984), 84–90.

[79] Adam Bryant, "On More and More Routes, Only Propeller Planes Fly," *New York Times*, November 5, 1994, p. 12.

that dissociated themselves from their rustic, romantic, and propeller-driven past. Promotional materials stressed a level of comfort comparable to any American home. The culture of the Jet Age celebrated size, speed, abundance, and access, which offered progressive and unlimited possibilities for society in the ongoing triumph of the new over the old.[80]

The idea of higher, faster, and farther reached an entirely new level with jet aircraft as society elevated them to the status of the superior aviation technology. The seemingly unlimited faith in advancing technology created a culture that welcomed and expected rapid change. The public believed propeller-driven aircraft were antiquated, uncomfortable, unsafe, technologically less-sophisticated, and culturally old-fashioned. In such miraculous times, it was difficult to recognize that the propeller's evolution into a niche was still cutting-edge and relevant to modern aviation.[81]

Turboprop aircraft had an image problem. Civic organizations and travel agents observed that many business and recreational travelers avoided propeller-driven aircraft at all costs, which often meant a long drive to and from a larger airport farther away. Cities like Amarillo, Texas, and Allentown, Pennsylvania, fearful of projecting a less than progressive image, guaranteed American Airlines and TWA profits to keep jets at their local airports. A Sioux City, Iowa, entrepreneur lamented when his city lost its jet service, "It is hard to present ourselves as a state-of-the-art, high-tech company when you have to get on a 19-seat turboprop to get here."[82] When flying on a turboprop could not be avoided, many modern air travelers complained bitterly. To them, the turboprop was a symbol of an inferior and past aerial age, an archaic technology whose time had passed.[83] Jet aircraft were just better. It did not matter that turboprops were jet aircraft, too, that filled the niche best for the specific job required and that the actual commuter aircraft were on average one to two years newer than larger jets.[84] The problem was that turboprops had propellers.

[80] Jenifer L. Van Vleck, *Empire of the Air: Aviation and the American Ascendancy* (Cambridge, MA: Harvard University Press, 2013), 248–253.

[81] David Edgerton, *The Shock of the Old: Technology and Global History since 1900* (New York: Oxford University Press, 2007), 27.

[82] Adam Bryant, "New Wrinkle in the Jet Age: Propeller Planes," *New York Times*, May 23, 1994, section A, p. 1; and Bryant, "On More and More Routes, Only Propeller Planes Fly," 12.

[83] Davies and Quastler, *Commuter Airlines of the United States*, 164, 171.

[84] Bryant, "On More and More Routes, Only Propeller Planes Fly," 12.

In contrast to the American disdain for propellers, the Soviet Union placed a considerable emphasis on turboprop aircraft for its military since the end of World War II. The result was propeller efficiencies at cruise speeds that approached those of jet aircraft. The swept-wing Tupolev Tu-95, introduced in 1956 and still in service in the early twenty-first century, served as a front-line strategic bomber and surveillance aircraft throughout the Cold War. Kuznetsov NK-12M turboprop engines powered four eight-blade contra-rotating and reversible propellers that were over eighteen feet in diameter. They enabled the Tu-95 to cruise at speeds up to Mach 0.75 at a maximum range of 7,800 miles. The fact that a Soviet propeller-driven bomber was capable of flying nonstop from Moscow to Havana, or to Washington, DC, for that matter, clearly demonstrated in no uncertain technical and political terms that the United States may have overlooked an important technology during the tense days of the Cold War.[85]

It was in this environment that a potential revolution in aircraft propulsion technology – a second reinvention of the airplane propeller – emerged. As the world oil crisis reached its zenith, the US Senate directed NASA to investigate solutions into increasing aircraft efficiency in 1974. Two concepts emerged from NASA's Advanced Turboprop Project that addressed the development of a radical turboprop configuration capable of generating the equivalent performance of a turbofan while using less fuel. Engineers at the Lewis Research Center in Cleveland (known today as Glenn Research Center) and Hamilton Standard began work on a futuristic multiblade variable-pitch propeller called a "propfan" combined with an Allison turboprop engine. GE's version, the Unducted Fan (UDF), incorporated two rows of counter-rotating scimitar-shaped blades into the rear of a turbofan engine. Both programs consisted of extensive wind tunnel evaluations and original investigations into aerodynamics, structures, component systems, and acoustics. Flight tests of the UDF installed on the rear nacelles of Boeing 727 and McDonnell Douglas MD-80 airliners in 1986–1987 and the propfan on a modified

[85] NASA Office of Aeronautics and Space Technology, *Aircraft Fuel Conservation Technology Task Force Report*, September 10, 1975, 44, 48; Mark D. Bowles, *The "Apollo" of Aeronautics: NASA's Aircraft Energy Efficiency Program, 1973–1987*, NASA SP-2009–574 (Washington, DC: NASA, 2010), 18–19, 125–126. A derivative of the Tu-95, the Tu-114 airliner, set a long-standing official world record speed of 540 mph for propeller-driven aircraft in 1960. See Fédération Aéronautique Internationale (FAI), "FAI Record Search Results," www.fai.org/records (Accessed June 24, 2014).

Gulfstream II business jet in 1987 confirmed the initial predictions of an astounding 20 to 30 percent increase in fuel savings.[86]

There was much enthusiasm for these new propeller-driven propulsion systems. The March 1985 issue of *Popular Science* ran a cover story that asked the seemingly controversial question, "So long, Jets?," as the aviation industry pondered the possibility that "propellers may be on the way back."[87] The Advanced Turboprop Project revealed the promise of the new technology in terms of aircraft efficiency and operating costs, but NASA and the manufacturers could not simply say they were new propellers. The head of GE Aircraft Engines, Brian Rowe, recognized that the public believed fans found in jet engines were "modern" and that the "old technology was propellers," which shaped how they presented the new advances. NASA and its industry partners coined "propfan." GE and a Pratt & Whitney/Allison team developing another counter-rotating pusher turboprop called their designs "ultra-high bypass" engines to emphasize their connections to turbofans.[88] McDonnell Douglas engineers, considering the installation of advanced turboprop engines on a new MD series of airliners, called them "propulsors."[89]

Despite the potential fuel savings and a marketing campaign that attempted to overcome the public's resistance, propfan-driven and UDF-powered aircraft did not appear in the 1990s. A drop in oil prices negated the need for manufacturers and airlines to reequip with advanced turboprop aircraft that was estimated to cost between $3 and $10 billion to develop. They continued to use existing turbofan-powered aircraft. The technological achievement of reinventing the variable-pitch propeller, however, did not go unnoticed. The NAA awarded NASA and its industry partners the 1987 Collier Trophy for their work in the Advanced Turboprop Project. NASA alone invested approximately $200 million in the project. Today, that knowledge and

[86] Roy D. Hager and Deborah Vrabel, *Advanced Turboprop Project*, NASA SP-495 (Washington, DC: NASA, 1988), 1–5; Mark D. Bowles and Virginia P. Dawson, "The Advanced Turboprop Project: Radical Innovation in a Conservative Environment," in *From Engineering Science to Big Science: The NACA and NASA Collier Trophy Research Project Winners*, ed. Pamela E. Mack (Washington, DC: NASA, 1998), 339–340.

[87] Jim Schefter, "New Blades Make Prop Liners as Fast as Jets," *Popular Science* 226 (March 1985): 66.

[88] Martha M. Hamilton, "Firms Give Propellers a New Spin," *Washington Post*, February 8, 1987, section H, p. H.

[89] Greg Johnson, "Something New for Airliners: Propellers," *Los Angeles Times*, June 16, 1986, p. SD-C1

technology is "on the shelf" waiting for the next fuel crisis that may justify its implementation in the future.[90]

"A Wrinkle in the Jet Age"

The failure of the flying public and the commercial aviation industry to embrace fully the turboprop, advanced or not, did not mean the end of the propeller nor the surviving companies primarily responsible for shaping its modern form. Analysts continue to argue that turboprops represent an economical and environmentally friendly alternative to comparable jet-powered aircraft for regional transportation, especially in light of high fuel prices that appear consistently on a cyclical basis.[91] In the technical and cultural sense, the continued use of the propeller revealed a "wrinkle in the Jet Age" according to a *New York Times* reporter.[92]

Both Dowty Rotol and Hamilton Standard found continued markets. Dowty introduced the six-blade R391 Advanced Propeller System for the new Rolls-Royce AE2100 turboprop engine in 1999. The main application was the Lockheed-Martin C-130J and Lockheed-Martin/Alenia Aeronautica C-27J Spartan transports, which went into service with armed forces around the world, including Australia, Canada, Denmark, Italy, Norway, Qatar, the United Kingdom, and the United States, but also included the Saab 2000 airliner for regional airlines (Figure 25). The Smiths Aerospace Group acquired Dowty in September 2000 and subsequently sold the propeller business to GE Aviation Systems in May 2007.[93]

Hamilton Standard continued to be the world's largest propeller manufacturer even though it continued to diversify. The military and regional airline market for the 54H60 and commuter propellers remained active.

[90] Bowles and Dawson, "The Advanced Turboprop Project," 342; Johnson, "Something New for Airliners: Propellers"; Hager and Vrabel, *Advanced Turboprop Project*, vii. There has been a recent revival of propfan and unducted fan technology, now called the open rotor, by NASA, GE, and Rolls-Royce to reflect ongoing concerns over fuel efficiency and new priorities based on reducing the environmental impact of engine emissions. See John Croft, "Open Rotor Noise Not a Barrier to Entry," *Flightglobal*, July 5, 2012, www.flightglobal.com/news/articles/open-rotor-noise-not-a-barrier-to-entry-ge-373817/ (Accessed July 20, 2012).

[91] Megan S. Ryerson and Mark Hansen, "The Potential of Turboprops for Reducing Aviation Fuel Consumption," *Transportation Research Part D: Transportation and Environment* 15 (2010): 305, 313–314.

[92] Bryant, "New Wrinkle in the Jet Age," p. 1.

[93] "Dowty R391 Advanced Propeller System," File A20070022000, RO, NASM; Jamison Roseberry, e-mail to the author, October 11, 2012.

FIGURE 25 The updated Lockheed-Martin C-130J Hercules transport, which includes four six-blade Dowty R391 propellers, indicates the continuing viability of high performance propeller-driven aircraft into the twenty-first century. US Air Force photo/1st Lt. Stephani Schafer.

Hamilton Standard acquired French propeller manufacturer Ratier Figeac in 1998 after cultivating a long relationship that began in the 1990s with the joint development of the highly successful 568F composite six-blade propeller for regional airliners. At the corporate level, UTC merged Hamilton Standard with the newly acquired Sundstrand Corporation, another specialty manufacturer, to form Hamilton Sundstrand in June 1999. Within Hamilton Sundstrand, the Ratier Figeac team introduced a new series of eight-blade propellers. The NP2000 replaced the 54H60 on the C-130, E-2, and C-2 aircraft. The FH 385/386 designed for the Airbus A400M Atlas military transport is currently the world's largest with a seventeen-foot-six-inch diameter and capable of generating an impressive 90 percent efficiency.[94] Hamilton Sundstrand merged with Goodrich to

[94] Hamilton Sundstrand Corporation, "Hamilton Standard History," 1999, www.hamiltonsundstrandcorp.com (Accessed February 26, 2002); Ratier Figeac, "Propellers," and "History," n.d. [2012], www.ratier-figeac.com/ (Accessed October 9, 2012); Hamilton Sundstrand, "Hamilton Sundstrand A400M propeller system certified by European Aviation Safety Agency," April 5, 2012, www.hamiltonsundstrand.com/vgn-ext-templating/v/index.jsp?vgnextoid=16eaaec96b99110VgnVCM1000007301000aRCRD&hsct=hs_news&ciid=b657a045a9f76310VgnVCM1000004f62529fRCRD (Accessed October 9, 2012).

form UTC Aerospace Systems (UTCAS) in July 2012 with 95 percent of all design and production taking place in France.[95]

The propellers produced by GE Aviation Systems and UTCAS feature new, more efficient, and quieter swept-back blade shapes, materials, and control systems. Underneath their spinners are two legacies of the Aeronautical Revolution. The hydraulic constant-speed mechanisms of both the R391 and the NP2000 are modern-day versions of the Hele-Shaw Beacham design of 1924 shaped by Pop Milner and his associates at Gloster, Bristol, Rolls-Royce, and Rotol and the Hydromatic propeller created by Frank Caldwell, Erle Martin, and their colleagues at Hamilton Standard in 1937.[96]

There was also legacy of propeller technology in tilt rotor aircraft, which combined the vertical capabilities of conventional helicopters with the faster, long-range cruise performance of turboprop aircraft. Years of experimentation with the idea of a "tilting propeller" began with Henry Berliner's fixed-wing biplane and its large diameter fixed-pitch propeller mounted on a pivoting vertical shaft near the tip of each wing in 1924. After World War II, the US Army and Air Force collaborated with Bell Aircraft on the piston-engine powered XV-3 convertiplane, which made a series of successful transitions from helicopter to airplane during the period 1958 to 1962 despite several design deficiencies. The follow-on NASA/Army/ Bell XV-15 tilt rotor research aircraft project began in 1971 and culminated in the first full in-flight conversion from helicopter-to-airplane mode on July 24, 1979. Building upon that success, the US Navy approved the Joint-service Vertical take-off/landing Experimental, or JVX, program in 1981. Despite a series of program cancellations and crashes, the first flights of the resultant Bell Boeing V-22 Osprey occurred in 1989 with delivery of the first production model to the US Marine Corps ten years later.[97] Capable of carrying thirty-two troops at over

[95] Daniel Coulom, telephone interview by author, September 18, 2012; UTC Aerospace Systems, "Welcome," 2012, http://utcaerospacesystems.com/ (Accessed September 11, 2012).

[96] "Dowty R391 Advanced Propeller System"; Daniel Coulom and Robert Scheckman, telephone interview by author, February 16, 2010. The American Society of Mechanical Engineers recognized the importance of the Hydromatic propeller series when it named the design an international historic technology in 1990. American Society of Mechanical Engineers, *Hamilton Standard Hydromatic Propeller: International Historic Engineering Landmark*, Book No. HH 10 90 (November 8, 1990).

[97] Martin D. Maisel, Demo J. Giulianetti, and Daniel C. Dugan, *The History of the XV-15 Tilt Rotor Research Aircraft: From Concept to Flight*, NASA SP-2000–4517 (Washington, DC: NASA, 2000), 140–152.

275 mph, Ospreys and their two thirty-eight-foot diameter propellers, or "rotors," went into operational service as transport and medical evacuation aircraft with the Marine Corps in 2007 and with the Air Force in 2009.

In the early twenty-first century, the propeller-driven airplane persists as a niche technology that offered something the jet engine could not. Turboprop airliners carry passengers to and from regional and metropolitan airports. Military forces use them for transporting soldiers and weapons as close to the battlefield as possible over all types of terrain and through various forms of restricted airspace. In the world of general aviation, recreational fliers enjoy themselves on a Sunday afternoon, bush pilots access remote areas unreachable by any other means, and daredevils put on air shows to entertain crowds. Whether anyone wanted to call the technology thrusting them forward through the air propellers or rotors or accept them at all, the technology was a vital part of aviation.

12

Conclusion

The Triumph and Decline of the Propeller

The propeller specialists contributed to the development of the airplane during a vibrant period of innovation in North America and Europe from World War I to the end of World War II. The themes of culture, community, specialty, reinvention, and use take the reader from the mind's eye, the drawing boards, workshops, research and development facilities, and factories populated by the specialists to the spectacular aerial pathways, commercial air routes, and battlefields in the sky that the aeronautical community used to change the world. The propeller specialists weave in and out of this overarching Aeronautical Revolution, a period that witnessed the intense technical development of the airplane and the rise of modern aviation.

A propeller community existed since the early days of the airplane. During the years preceding and following the Wrights' first flight at Kitty Hawk, aeronautical enthusiasts experimented with a variety of propeller designs and materials. They focused on improved versions of the Wrights' fixed-pitch propeller with its permanently set blade angle and made from layers of wood. It was cheap, easy to manufacture, fit well into the established design paradigm of keeping everything in an airplane as light and simple as possible, and, most important, it worked for the thousands of airplanes produced during World War I and into the 1920s. The search for more performance led propeller specialists to reevaluate the status quo regarding the technology. As a result, they pursued simultaneous and interwoven technical trends centered on infrastructure, design, and materials as they began to question their place in the aeronautical community during the war and the immediate postwar period.

American and European military organizations established new aeronautical research and development facilities and workshops during World War I to place propeller technology on a practical and operational footing. The central role of the US government in developing propeller testing equipment and techniques provided the fundamental infrastructure for immediate wartime production needs and future advances in the postwar era. Engineers reacted to dynamic design requirements by using their experience from different disciplines to pioneer their specialty within the new field of aeronautical engineering. With the staff, equipment, and procedures in place, development-based organizations like the US Army Air Service's Engineering Division fostered the latest points of departure in propeller technology.

As the necessary infrastructure took shape, individuals, companies, and governmental agencies set about reinventing the wood fixed-pitch propeller during World War I and continued to pursue this quest through the 1920s. They envisioned a propeller capable of changing the angles of its blades as they rotated through the air. This new variable-pitch propeller maximized the power of an aircraft engine at different speeds, operating conditions, and altitudes. The specialists called it the "gearshift of the air" because it offered the same performance-enhancing benefits that a multi-gear transmission provided an automobile on the ground. The first steps toward that development offered the straightforward challenge of perfecting a reliable and durable mechanism connected to blades made from materials strong enough to withstand the rigors of flight. The myriad of designs and materials led to technical failures and brought into focus tensions related to the appropriate level of government involvement in the development of new technologies in the United States. In contrast, designs from Great Britain and Canada proved to be waypoints on a long journey of development.

As the struggle to reinvent the propeller continued, engineers simultaneously experimented with metal fixed-pitch propellers. They incorporated fundamental technical characteristics that contributed to future advances. A crucial intermediate step was the ground-adjustable-pitch propeller that consisted of a steel hub joined to aluminum alloy blades and capable of having its pitch changed by mechanics and pilots on the ground. The design propelled a new generation of pioneering aircraft in the late 1920s, most notably Charles Lindbergh's *Spirit of St. Louis*. While Lindbergh and others captured headlines, the first commercial airliners and new military aircraft all relied upon the same propellers. This stage in propeller development benefitted from a relationship where government specialists collaborated with and provided technical and financial assistance to private propeller firms. The US Army Air Service's work with Standard Steel of Pittsburgh was crucial in finding a design and a construction material that facilitated the growth of aviation in the late 1920s and early 1930s.

Not all aircraft in the 1920s relied upon Standard Steel propellers. America's largest aircraft and engine manufacturer, the Curtiss Aeroplane and Motor Company, introduced the rival Reed fixed-pitch metal propeller that, to the broader aeronautical community and the airminded public, satisfied the need for a modern propeller. The use of the Reed propeller on American and European air racers, the fastest airplanes in the world, made it a spectacular symbol of "Progress" even though its structural design was flawed. The failure of the design in everyday use illustrated that what constituted innovation for the propeller specialists and the aeronautical community was as messy and ambiguous as the process of technological development itself.

Corporate America took notice as the aeronautical community defined a larger role for the airplane in everyday life in the latter half of the 1920s. Encouraged by government legislation, electrified by an increasing number of spectacular flights like Lindbergh's, and enthusiastic for future business success, entrepreneurs consolidated the pioneer aviation companies through merger and acquisition. They created the large holding companies that represented a comprehensive approach to manufacturing, selling, and transportation within the entire aviation industry. Their goal was to emulate the classic model of a vertically integrated corporation. Part of this consolidation involved a scramble to acquire the latest airframe, engine, and propeller technology. The industrial propeller community, consisting of dedicated companies and specialist engineers, became a valuable commodity to the modern aviation corporation. Thus, the successful reinvention of the propeller went beyond technical prowess and a cultural expression of "higher, faster, and farther;" it was rooted in the business of aviation.

As the specialists created and introduced practical variable-pitch propellers, a new generation of aircraft began to appear in North America and Europe. These newly-christened "modern" airplanes faced a significant problem. With ground-adjustable-pitch propellers, long runways were suddenly too short for takeoff or mountain ranges were too high to clear with an acceptable margin of safety. Air transport companies faced the challenge of running profitable and safe routes across the vast distances of the North American continent. Both airlines and manufacturers learned in 1933 that their new Boeing and Douglas airliners would not be successful until the manufacturers incorporated reinvented propellers. Their inclusion in the synergy of the airplane was not inevitable, but it was necessary if aircraft were to continue to evolve and improve. Due to the efforts of propeller specialists in the United States, aircraft manufacturers and commercial and military operators almost universally adopted these new propellers in their efforts to realize their visions of air travel and airpower in the early 1930s. The result was a spectacular introduction of American aeronautical technology in international events like the 11,300-mile England to Australia MacRobertson Air Race in October 1934.

The national aeronautical cultures of the United States, Great Britain, and Germany reacted to reinvented propeller designs differently for the remainder of the 1930s to the outbreak of World War II. The American aeronautical community's acceptance of variable-pitch propellers from 1932 to 1934 marked the end of a revolutionary phase of aeronautical development and design. Manufacturers and their corporate leadership in the United States continued to innovate and refine them. German and British companies produced and sold licensed American designs and struggled to create their own with varying degrees of success. The dramatic introduction of Nazi aeronautical technology at the 1937 International Aviation Meet in Zurich and the aerial campaigns of the Spanish Civil War indicated a growing German presence in propeller innovation. On both sides of the Atlantic, improved design features facilitated the growth of international commercial and military aviation. They incorporated constant-speed, or automatic pitch-changing mechanisms, and blade feathering that ensured safe flying when an engine failed. These thoroughly modern propellers increased performance and efficiency and enabled flights like those of George Grogan's United Air Lines *Mainliner* over New York in 1938, which made aviation safer overall.

World War II war ushered in the final acceptance of modern propellers among the international aeronautical community. Great Britain made that transition during the first year of its struggle against Nazi Germany. Government-issued technical specifications, the cultural factor of national pride, and the practicalities involved in logistics and the needs of fighting a war with equipment on hand ensured that the Spitfire and Hurricane pilots of the Royal Air Force faced the Luftwaffe in deadly aerial combat without the latest propeller designs. The acceptance, implementation, and use of the most up-to-date performance-enhancing propeller technology during the spring and summer of 1940 quickly became a matter of life, death, and national survival.

As the first global aerial conflict raged on, the air forces of the major combatants relied upon the aeronautical technology of the late 1930s to conduct strategic and tactical bombing campaigns, maintain air superiority, and operate worldwide supply networks with modern airplanes. Propeller manufacturers became vital components of the national war machines that churned out hundreds of thousands of military aircraft. Aerial operations during World War II created a vast network that opened the world to air travel and the decade and a half that followed witnessed the onset of an Air Age characterized by both peace and war. The propeller community and its piston-engine contemporaries helped realize an intercontinental aerial system based on ever-more sophisticated propulsion systems. These refined aircraft dominated international air warfare and travel and revolutionized society's perceptions of space and time to make the world feel smaller.

Gas turbine-powered commercial and military aircraft expanded the boundaries of flying higher, faster, and farther in the newly emergent Jet Age of the 1950s and 1960s. The propeller specialists scrambled to find a continued place for their technology in high performance aeronautics. After a period of initial turmoil, they found it with the turboprop, a combination of a gas turbine engine with an even more refined modern propeller. Turboprop aircraft filled niches, such as regional commuter transportation and tactical airlift, which less efficient jet-powered aircraft could not. International economic turmoil and high fuel prices stimulated work toward another reinvention of the propeller for increased fuel efficiency in the 1970s and 1980s, which due to non-technical cultural, economic, and political circumstances never materialized. Popular resistance to the idea of the propeller-driven airplane, and what it represented in terms of modernity, casts a shadow that prevails to this day.

Propellers are perhaps the most unappreciated component of one of humankind's most important creations, the airplane, even though they have been thrusting those wonders of the modern age forward through the sky for over a century. They persist as a niche technology and they do what they do better than anything else. Regional airlines, military air forces, and general aviation users operate propeller-driven aircraft powered by both piston and gas turbine engines to fly and transport people and cargo all over the globe.

Essay on Sources

In comparison to airframes and engines, there has been little or no discussion of propellers in the vast sea of literature on the history of flight. As a result, the main body of evidence for this project consisted of archival and published primary sources. Secondary materials addressing the technical development of propellers and flight and broader issues in the history of technology had their place as well. The following essay discusses the sources central to this work.

Primary Sources

The primary sources used in this book reflect the interplay between airframe, engine, and propeller specialists and their clients and collaborators, the airline industry and national governments. Delving into the private and public records of all the historical actors revealed a broader context of innovation while understanding the role of the propeller. That research methodology also made up for a paucity of resources in critical areas while still facilitating a book whose sum was greater than its individual parts. In other words, while a propeller manufacturer's records were not available regarding a particular design, that information would appear in the files of a government research organization or an aircraft manufacturer that would be the ultimate user of that product.

Archives

Boeing Company Archives, Seattle, Washington, Long Beach, California, and St. Louis, Missouri

The Boeing Company maintains the world's premier corporate aerospace archive for itself and its legacy companies, which include Douglas, McDonnell, and North American. The collections include the personal papers of executives and engineers, subject files for each aircraft model, and general marketing materials. Boeing was a member company of the United Aircraft and Transport Corporation during the crucial years of 1929–1934. The papers of engineer, Claire Egtvedt, include the verbatim transcripts of the corporation's Technical Advisory Committee meetings, which discuss in detail the technical direction of its member companies. In the case of the variable-pitch propeller, aircraft and engine designers debated the merits of the new technology with the specialists dedicated to their development. The files on the Boeing Model 200, Model 221A, and Model 247, and Douglas DC-2 generated the documentary evidence for how the variable-pitch propeller became an integral part of those modern airplanes.

National Air and Space Museum, Smithsonian Institution, Washington, DC

The National Air and Space Museum maintains several collections vital to understanding the development of the airplane propeller. The Clement M. Keys Papers and Curtiss-Wright Corporation Records documented the acquisition, development, and marketing of the respective inventions of Albert Reed and Wallace R. Turnbull. The Wright Field Technical Documents Library and Propeller Test Reports added to the story of the propeller unit at McCook and Wright Fields. The Fred Weick papers included a 1981 oral interview conducted by United Technologies corporate historian Harvey H. Lippincott that helped recreate the story of Hamilton Aero in Milwaukee. The German/Japanese Captured Air Technical Documents consist of engineering data and reports collected American technical teams at the end of World War II, which included a January 1943 report by Hans Ebert tracing the history of the VDM propeller.

Two other collections proved important to this study of material objects. The museum's vertical files included correspondence, reports, brochures, manuals and spare parts catalogs, press releases, journal and newspaper clippings, and photographs for the myriad of propeller manufacturers and their individual designs. The records of the Registrar's Office yielded vital information about the history of specific propeller and aircraft artifacts in the museum collection that are discussed in the book. While they were hit or miss in terms of depth of detail for some artifacts, they often yielded information found nowhere else.

National Archives and Records Administration, Washington, DC, and College Park, Maryland

Both the US Army's Engineering Division and the Navy's BuAer played vital roles in the synergistic reinvention of the airplane propeller during the interwar period. The correspondence, reports, memoranda, and photographs documenting that work, now held by the National Archives in record groups 342 and 72, respectively, were central to this study. Special mention must be made of the Army's Engineering Division Records that cover the period 1917 to 1950. They are known as the Sarah Clark Files in honor of an unassuming government clerk that maintained those records for nearly forty years between 1919 and 1956, which ensured their efficient use by later generations of historians. Scholars wishing to conduct their own case studies of other component technologies in the United States, which include airframes, armament, engines, instruments, and landing gear, should make them their first point of departure. The patent litigation related to Albert Reed's duralumin propeller kept in the Records of the United States Court of Claims, Record Group 123, was an indispensable source for this project as well. The final Reed patent case served as a detailed summary of the development of the aerial propeller in the United States and Europe through the voices of the people involved and an analysis of the most influential designs.

National Archives of the United Kingdom, London, England

Formerly known as the Public Record Office, or PRO, the National Archives maintains the operational and technical records of the Royal Air Force (AIR) and the Ministry of Aviation (AVIA). For students of the variable-pitch propeller in Great Britain, they contain correspondence, memoranda, reports, and photographs related to the Hele-Shaw design and its Gloster, Bristol, and Rotol iterations, the de Havilland propeller, and the VDM propeller. The operational record books of several Royal Air Force squadrons combined with a series of letters to Fighter Command reconstructed the dramatic days of the constant-speed conversion program. The Ministry of Supply's internal report, "Propellers: Development and Production, 1934–1946," provided a valuable and detailed context for the overall discussion of British propellers.

National Aeronautics and Space Administration, Washington, DC
NASA's legacy organization, the NACA, primarily focused on solving the fundamental aerodynamic challenges facing American aviation during the 1920s and 1930s. While its records document primarily that story in depth, the History Office's NACA and History Subject files contain valuable information relevant to propeller design. The propeller research projects documented in the Langley Research Authorization collection are an important organizational and technical complement to the Army and Navy records at NARA.

United States Air Force Historical Research Agency, Maxwell Air Force Base, Alabama
The miscellaneous organizational histories, papers, oral interviews, and files held at Maxwell document the history of the US Air Force. Two collections were crucial to understanding the Army's fundamental role in propeller development at McCook and Wright fields. The papers and oral interview of D. Adam Dickey, the long-standing civilian researcher present at the beginning in 1917 through the 1950s, and the propeller memoranda of the Bureau of Aircraft Production during World War I revealed the complex infrastructure needed for continued achievement in propeller design and construction.

Other Repositories

University and museum archives, historic trusts, and libraries also held materials essential to the study of the variable-pitch propeller. Regarding Aeroproducts, the Charles Kettering Collection at the Kettering University Archives, Flint, Michigan, and the Allison Branch Archives of the Rolls-Royce Heritage Trust in Indianapolis, Indiana, included materials related to the history of the company. The Resource Center of the Museum of Flight in Seattle, Washington, maintains a file of notable aviation entrepreneurs from the area, including Tom Hamilton. The Ryan NYP *Spirit of St. Louis* Notebooks maintained by the San Diego Aerospace Museum Library and Archives, San Diego, California, helped recreate the process of designing and building Charles Lindbergh's transatlantic monoplane. The Chattanooga Albums and the Paul A. Hiener Collection at the Chattanooga-Hamilton County Bicentennial Library, Chattanooga, Tennessee, and miscellaneous file material held by the Hamilton Sundstrand Community Relations Division, Windsor Locks, Connecticut, offered biographical information on Frank W. Caldwell.

Missed Opportunities

This book did not benefit from access to the archives of the propeller manufacturers, Hamilton Standard, de Havilland, or Rotol. Hamilton Standard's records were part of the extensive United Technologies Archives in Hartford, Connecticut, that documented each of the member companies and equaled the Boeing Company Archives in breadth and detail. Those records included the papers of Frank W. Caldwell and Erle Martin. The archives closed in the mid-1990s, just before the initiation of this project, with each member company taking back its relevant records. Hamilton Standard did not make those records available to the public and their current disposition is unknown. The de Havilland Aircraft Heritage Centre in Hertfordshire, England, maintains the archival legacy of the British company, but all attempts by the author to gain access were ignored. Rotol's records, including board meeting minutes and business ledgers, are held by the Gloucestershire Archives, Cheltenham, England. The author did not access those records, nor the materials held by the Deutsches Museum in Munich, Germany, on VDM due to the apparent duplication of materials found in other collections. Researchers delving deeper into the history of the propeller in Great Britain and Germany should not overlook them. Moreover, the extent of archives documenting the history of propellers in the Soviet Union and Japan are unknown.

Published Primary

The technical development of flight is well-documented in published primary sources in the form of articles and books. For the Aeronautical Revolution, the NACA *Bibliography of Aeronautics* edited by Smithsonian Institution librarian Paul Brockett between 1909 and 1932 should be consulted first by all researchers. Charles Fayette Taylor collaborated with R. K. Smith on a dedicated propulsion bibliography featured in Taylor's *Aircraft Propulsion.*[1] The encyclopedic *Aircraft Yearbook*, *Jane's All the World's Aircraft*, and *Who's Who in American Aeronautics* provided useful reference material on both the people and technology related to the reinvention of the airplane propeller.

Articles discussing the technical development of the airplane propeller appeared in wide range of publications. The editors of the periodicals *The Ace*, *Aerial Age Weekly*, *Aero Digest*, *Aeronautical Journal*, *Aeronautics*,

[1] C. Fayette Taylor, *Aircraft Propulsion: A Review of the Evolution of Aircraft Piston Engines* (Washington, DC: Smithsonian Institution Press, 1971).

The Aeroplane, *Aviation*, *Flight*, *Industrial Aviation*, *Interavia*, *The Slipstream*, and *U.S. Air Services* featured the propeller community and commented on the significance of their designs at critical moments. Professional engineering societies, including the ASME, IAS, RAeS, and the SAE, published papers in their respective journals and transactions that documented new developments in aviation.[2] Manufacturer publications included *Boeing News*, *The Bristol Review*, and *Curtiss-Wright Review*. The most important corporate voice for this study was the UATC's *The Bee-Hive* – the title is in homage to Pratt & Whitney's line of engines named after flying insects – that featured important articles on Hamilton Standard. Trade journals from the broader world of industry, primarily *American Machinist*, *Automotive Industries*, *Manufacturers' Record*, *The Metal Industry*, and the *New England Automobile Journal*, also focused on aeronautical technology.

Government institutions generated reports concerning their work in aeronautical technology, including the propeller. The early efforts to develop a practical variable-pitch propeller from World War I to the late 1920s appeared in the British Aeronautical Research Council *Reports and Memoranda* and the US Army Air Service's *Bulletin of the Experimental Department, Airplane Engineering Division* and *Air Service Information Circular*. The Air Service's 1924 promotional booklet, *A Little Journey to the Home of the Engineering Division*, was also useful in gauging the overall extent of the activities taking place there.[3] While the NACA is known best for its self-generated annual and technical reports, special mention must be made of the Technical Memorandum series, which include translations of foreign articles and items of specific interest to the organization's researchers regarding propellers.

The voluminous records of national patent offices proved an invaluable resource for tracing the history of invention and innovation. As a historical document, a patent contains a concise history of a specific technology in the words and diagrams of the applicant and their patent attorneys. The abstract, description, and drawings present the operation of a new technical solution while the claims section justifies its novelty. The presence of filing and issue dates, often separated by many years, provided timelines when no others were available. Finally, it is important

[2] See *Transactions of the ASME*, *Journal of the Aeronautical Sciences*, *Journal of the Royal Aeronautical Society*, and *SAE Transactions*.

[3] *A Little Journey to the Home of the Engineering Division* (Dayton, OH: Thompson Print Company, n.d. [1924]).

to see whether or not the applicant was alone in his endeavor or if they assigned the rights to their new idea to a commercial concern or national government.

Popular publications that featured articles on the people and organizations that made propellers their specialty and their overall role in modern aviation also shaped this work. Magazines included *The Century Magazine*, *Collier's*, *Fortune*, *Popular Mechanics*, *Popular Science*, and *Time*. Large national newspapers such as the *Atlanta Constitution*, *Baltimore Sun*, *Chicago Daily Tribune*, *Christian Science Monitor*, *Daily Boston Globe*, *Los Angeles Times*, *Wall Street Journal*, and the *Washington Post* covered the introduction of new innovations. The aviation columns of Lauren D. Lyman, Reginald M. Cleveland, and James V. Piersol in the *New York Times*, proved crucial in gauging the role of the propeller and the airplane in everyday life. Regional newspapers like the *Hartford Courant*, *Homestead Messenger*, *Milwaukee Sentinel*, *Pittsburgh Post-Gazette*, and *West Hartford News* provided important background information about manufacturers and users of propellers in the locales they covered.

Advertisements, brochures, and marketing material that appeared in both trade and public journals also have their place in documenting the history of specialist communities and their products. In general, they provide a gauge for evaluating a company's perception of the importance of its own innovations. Specifically, those materials were especially helpful in tracing the progress-oriented rhetoric celebrating both the Standard Steel and Reed propellers in the 1920s. The Aeroproducts publication, *Blades for Victory*, places the role of the company in the context of the World War II production program.[4]

The propeller appears in contemporary books as well. Textbooks related to aircraft propellers by Wilbur Nelson, Henry Watts, and Fred Weick and aviation technology in general by Charles B. Hayward and Grover C. Loening presented the knowledge of the state-of-the-art in aeronautics.[5] Autobiographical narratives written by Charles Lindbergh, Grover C. Loening, William B. Stout, and Eugene Wilson added critical

[4] Aeroproducts Division, General Motors Corporation, *Blades for Victory: The Story of the Aeroproducts Propeller and the Men and Women Who Build It* (Dayton, OH: Aeroproducts Division, General Motors Corporation, 1944).

[5] Wilbur C. Nelson, *Airplane Propeller Principles* (London: John Wiley and Sons, 1944); Henry C. Watts, *The Design of Screw Propellers with Special Reference to Their Adaptation for Aircraft* (London: Longmans, Green, and Company, 1920); Fred E.

information to key moments in the history of the propeller.[6] Of particular note are the first-person narratives of RAF Fighter Command pilots David M. Crook, Alan C. Deere, Gordon Olive, Ian Gleed, and Paul Richey.[7] While the remembrances of famous aviators, engineers, and executives should always be tempered by solid archival research, their discussions of the technology at their fingertips, which included the propeller, were central to this study.

Secondary Sources

Despite the excess of secondary works that overlook the propeller, there are, however, notable exceptions. Synthetic articles started to appear in England and Canada beginning in the 1960s.[8] The first book-length study was George Rosen's *Thrusting Forward: A History of the Propeller*. Rosen worked as a propeller aerodynamicist for Hamilton Standard from 1937 to 1977. Over time, he cultivated an interest in the history of the propeller, which led to his informal status as company historian. Rosen's work is significant and an excellent first point-of-departure for enthusiast and general readers.[9] Unfortunately, Rosen gave his voluminous source material to the United Technology Archives and they are unavailable to

Weick, *Aircraft Propeller Design* (New York: McGraw-Hill, 1930); Charles B. Hayward, *Practical Aeronautics: An Understandable Presentation of Interesting and Essential Facts in Aeronautical Science* (Chicago, IL: American School of Correspondence, 1912); and Grover C. Loening, *Military Airplanes: An Explanatory Consideration of Their Characteristics, Performances, Construction, Maintenance and Operation*, 2nd ed. (Boston, MA: W. S. Best Printing Company, 1918).

[6] Charles A. Lindbergh, *The Spirit of St. Louis* (New York: Scribner, 1953; reprint, St. Paul: Minnesota Historical Society Press, 1993), and *WE* (Putnam: New York, 1927); Grover Loening, *Our Wings Grow Faster* (Garden City, NY: Doubleday, Doran, 1935); William B. Stout, *So Away I Went!* (New York: Bobbs-Merrill Company, 1951); and Eugene Wilson, *Slipstream: An Autobiography of an Air Craftsman* (Palm Beach, FL: Literary Investment Guild, 1967).

[7] David M. Crook, *Spitfire Pilot* (London: Faber and Faber, 1942); Alan C. Deere, *Nine Lives* (London: Hodder and Stoughton, 1959); Ian Gleed, *Arise To Conquer* (London: Victor Gollancz Ltd., 1942); Gordon Olive and Dennis Newton, *The Devil at 6 O'Clock: An Australian Ace in the Battle of Britain* (Loftus, New South Wales: Australian Military History Publications, 2001); and Paul Richey, *Fighter Pilot* (London: B. T. Batsford Ltd., 1941).

[8] K. M. Molson, "Some Historical Notes on the Development of the Variable Pitch Propeller," *Canadian Aeronautics and Space Journal* 11 (June 1965): 177–183; Robert M. Bass, "An Historical Review of Propeller Developments," *Aeronautical Journal* 87 (August/September 1983): 255–267.

[9] George Rosen, *Thrusting Forward: A History of the Propeller* (Windsor Locks, CT: United Technologies Corporation, 1984).

researchers. Dedicated histories of the Airscrew Company, de Havilland, and Rotol that benefitted from access to company records also exist, but their lack of source documentation encourages access to the original documents instead.[10]

Overall, this work reflects the growing body of literature on innovation and flight during the first half of the twentieth century. The Aeronautical Revolution encompassed cultural, economic, military, political, and social changes that created technological communities that took their place as part of world society. At the core of that transformation was the reinvention of the airplane and its most important parts. The individual works of Walter Vincenti, John Anderson, James R. Hansen, Eric Schatzberg, and contributors to anthologies edited by Roger Launius, Peter Galison, and Alex Roland have reshaped historical perceptions of aeronautical innovation during the interwar period.[11]

[10] Tony Deeson, *The Airscrew Story, 1923–1998* (n.p.: Airscrew Howden, n.d. [1999]); C. Martin Sharp, *D.H.: A History of de Havilland* (Shrewsbury, England: Airlife, 1982); and Bruce Stait, *Rotol: The History of An Airscrew Company, 1937–1960* (Stroud, Gloster, England: Alan Sutton Publishing, 1990).

[11] Walter G. Vincenti, *What Engineers Know and How They Know It: Analytical Studies from Aeronautical History* (Baltimore, MD: Johns Hopkins University Press, 1990); John D. Anderson, Jr., *The Airplane: A History of Its Technology* (Reston, VA: AIAA Press, 2002); James R. Hansen, *Engineer in Charge: A History of the Langley Aeronautical Laboratory, 1917-1958* (Washington, DC: National Aeronautics and Space Administration, 1987); Eric M. Schatzberg, *Wings of Wood, Wings of Metal: Culture and Technical Choice in American Airplane Materials, 1914–1945* (Princeton, NJ: Princeton University Press, 1998); Roger D. Launius, ed., *Innovation and the Development of Flight* (College Station: Texas A&M University Press, 1999); and Peter Galison and Alex Roland, ed., *Atmospheric Flight in the Twentieth Century* (Boston, MA: Kluwer Academic Publishers, 2000).

Index

Adjustable and Reversible Propeller Corporation, 83, 88, 91, 103
Advanced Turboprop Project, 339–341
Aerial Age Weekly, 31n38, 75n1
aerial spectacle
  air racing, 147, 167, 178, 224
  government-sponsored, 154, 163, 259, 268
Aerial Steamer, 18
Aeromarine Plane and Motor Company, 29
Aeromatic Propellers, 325
aeronautical community, 3, 4, 5, 6–7, 12, 13–14, 51, 76, 139, 234, 272, 306, 328
  American, 42, 120, 214, 231, 244, 348
  British, 208, 210, 261–262, 265, 268, 272, 281, 286, 304
  engineering logic, 206, 215, 234, 235
  German, 255
  and metal propeller, 26, 126, 134, 143, 176
  and Reed propeller, 154, 155, 158, 178–179
  and variable-pitch propeller, 87, 98, 198, 246
  and wood propeller, 32, 38
aeronautical infrastructure, 8, 45, 73, 181, 345, 346
Aeronautical Revolution, 2, 2n3, 9, 73, 206, 234, 345
  and ground-adjustable propeller, 116, 126, 144
  and modern airplane, 203, 212, 217, 233, 277, 306, 308
  and Reed propeller, 167, 169, 178
  and variable-pitch propeller, 104, 204, 211, 220, 305
aeronautical specialty, 3–7, 9–11, 103, 196f15, 329, 346
Aeroproducts Division of General Motors, 313–314
  Model 606 propeller, 333, 334
  postwar developments, 324, 328, 333, 334
  Unimatic propeller, 252, 313–314
  and World War II, 316, 319
Air Age, 4, 236, 273, 306, 324, 349
Air Ministry
  and de Havilland Aircraft Company, 265
  and Fighter Command conversion program, 294
  and Hele-Shaw Beacham propeller, 108–110, 207
  and rearmament, 282–284, 317
  and Rotol, 267
Air Propellers, Inc., 191
Aircraft Research Association, 337
airmindedness, 4, 6, 7, 9, 15, 135, 146, 324, 347
Airscrew Company Ltd., 264, 277, 279f21, 287, 318
Allison Division of General Motors, 252, 334, 339, 340
  T56/501 engine, 331, 333
aluminum, aluminum alloy. *See* duralumin

American Propeller and Manufacturing Company, 23–25, 26, 27, 103
  Universal propeller, 91–94, 101, 102
  and World War I, 39, 40
Army Air Forces, 308
  509th Composite Group, 313
  56th Fighter Group, 315–316
  39th Bomb Group, 310
Army Air Service
  Engineering Division, 36
Arnold, Henry H., 83, 216

Bacon, David L., 92
Bakelite Micarta, 98, 100, 118–120, 192
Balfour, Harold, 281
Bane, Thurman H., 82n20, 82n21, 82n24, 83n26, 84n29, 84n30, 86n34, 87, 87n40, 88, 88n42, 88n43, 88n44, 90n48, 91n52, 92n57, 95n65, 95n66, 98n78, 99n80, 100, 100n84, 113n117, 119n5, 119n7, 122n16, 122n17, 123, 123n19, 151, 152n15, 164n55, 165n58
Bastow, Stuart, 28, 29, 176
Bastow-Pagé propeller, 28–29, 176
Bazley, Halsey R., 198
Beacham, Thomas Edward, 109, 207, 208, 236, 243, 264
Blanchard, Jean-Pierre Francois, 17
Blanchard, Werner J. "Pete"
  at Aeroproducts, 313–314, 333
  at Curtiss, 202, 203, 227
  at Engineering Projects, 250–252, 272
Boeing Airplane Company
  B-17, 239, 308, 309f23
  B-29, 308–310, 313, 315
  Monomail, 212–214, 247, 217–221, 228, 233, 236
Boggs, G. Ray, 80
Bonson, E.W., 23
Breech, Ernest, 252
Bristol Aeroplane Company Ltd., 210, 262, 266
Brookes, Donald S., 279, 304
Brow, Harold J., 155, 166
Bureau of Aeronautics (BuAer), 168n68, 173n81, 186n16, 228n78
  on commercial sales, 228, 239
  and metal propeller, 134, 152–153, 168, 173, 248
Bureau of Standards, 168
Burgess Company, 25
Burroughes, Hugh, 110, 207, 209

Caldwell, Frank W., 32–33, 44, 196f15, 230f18, 322, 343
  awards, 229, 240
  at Curtiss, 33–35
  at Hamilton Standard, 188, 197–201, 242
  at McCook Field, 37, 42
  patents, 128, 144, 188, 197
  and propeller whirl testing, 47–48, 53–54, 73
  and S.A. Reed, 163–167, 170, 176, 179
  at Standard Steel, 144, 187
  at United Aircraft, 310, 331
  and variable-pitch propeller, 211, 214, 219, 220, 237–238, 245
Camp Borden, Royal Canadian Air Force, 112, 114f9
Canadian Propellers Ltd., 308
Caproni, Giovanni, 29
Cayley, George, 18
Chadwick, Roy, 264
Chatfield, Charles H., 195, 196f15, 213
Chauviére, Lucien, 22
Collier Trophy
  1925, 162, 178
  1933, 229, 230f18
  1938, 235
  1987, 340
composites, 277, 318–319, 337, 342
Constant Speed Airscrews Ltd., 270–272
Cooke, Desmond, 284, 289, 292, 294, 299–300, 304
Cozens, Henry Illife, 278, 281, 285
Croce-Spinelli, Joseph, 19
Crook, David M., 302–303
culture of performance, 3–6, 13
Curry, John F., 170, 172
Curtiss Aeroplane and Motor Company, 34, 39, 161, 181, 201
  and Reed propeller, 149, 153, 161, 165, 166, 347
Curtiss aircraft
  Condor, 242
  CR-3, 146, 147f13, 157
  JN-3, 35
  JN-4, 80, 84, 88, 98
  P-36, 250
  P-40, 249

Curtiss Electric propeller, 203, 228, 240, 250
and World War II, 311–313, 316
Curtiss-Wright Corporation, 181, 211, 240, 272, 332, *See* Propeller Division

Damon, Ralph S., 250
Darley, H. S. "George", 298
Daugherty, Earl S., 75, 76f7, 79–80
Davenport, Arthur, 264
de Havilland Aircraft Company Ltd,
and Fighter Command conversion program, 292–299, 301–303
Hamilton Standard license, 229, 231–232, 262, 268
and market dominance, 265–266
and rearmament, 266, 277, 282–283
and World War II, 317–318
de Havilland Propellers Ltd., 331
de Havilland, Geoffrey, 27
Deere, Alan C., 280, 290, 303
Deutsche Versuchsanstalt für Luftfahrt (DVL), 253, 255
DH-4 "Liberty Plane", 73, 119
and Reed propeller, 154, 160, 168
and Standard Steel propeller, 132, 137
and variable-pitch propeller, 82, 90
and wood propeller, 50, 121
Dickey, D. Adam, 55, 67, 68, 73
background and education, 60
evaluation of German propellers, 322
head of propeller unit, 72–73
Materiel Division propeller, 215
Dicks Aeronautical Corporation, 144, 248
Dicks, Thomas A,
collaboration with Frank Caldwell, 126–128, 129, 132, 133
at Dicks Aeronautical, 144, 248–249
at Standard Steel, 94–95
Dicks-Luttrell Propeller Company, 94
Dienstbach, Carl, 16, 21, 22n17
Doolittle, James H. "Jimmy", 159, 167, 224, 226
Dornier Do 17, 257
Douglas Company
DC-1, 221–223
DC-2, 205f17, 223–224, 233
DC-3, 1, 241, 243–244, 246
DC-series, 7
Sleeper Transport (DST), 1, 2f 1, 245
Dowding, Hugh, 262, 276, 294
Dowty Rotol Ltd., 332, 336
R391 Advanced Propeller System, 341, 342f25, 343
duralumin, 29, 30, 31, 134, 255
Army-Standard Steel blades, 126–127
drop-forging, 124
Hamilton Aero blades, 186
Reed propeller, 148–150, 162, 163, 165, 167
Durand, William F., 43, 51, 52
Dwight, R. D., 30

Earle, Robert L., 250
Ebert, Hans, 253, 254, 261, 273, 320, 321
Ehinger, C. E., 333
engineering
aeronautical, 33, 45, 72–73, 100, 211, 346
Army style, 46, 53, 72–73, 115, 118
electrical, 60, 73
mechanical, 33, 47, 73, 95, 104, 211
Engineering Division, Army Air Service
and metal propeller, 117–134
and propeller testing, 53–66
Propeller Testing Laboratory, 59, 61f5
and Reed propeller, 163–171
and variable-pitch propeller, 81–91
engineering logic
conservative, 88, 99, 111, 118, 205, 263
performance-based, 136, 206, 222, 233, 234
Engineering Projects, Inc., 251–252, 313
Esnault-Pelterie, Robert, 26
Evans, Stanley H., 264

Faehrmann, Hermann, 122
Fairey Aviation Company Ltd,
Battle bomber, 271, 279, 287
Fairey-Reed propeller, 157, 160, 209
Fales, Elisha N., 125
Fechet, James E., 88, 173
Fedden, Roy, 210–211, 211n17, 231, 262n80, 266n96, 267, 267n98, 267n99, 268, 268n102, 268n104, 283, 283n24, 284n28
fixed-pitch propeller, 22, 22f2, 31, 37, 38f3, 44, 47
compared to variable-pitch propeller, 206, 208, 213, 222, 231
in Germany, 260

fixed-pitch propeller (*cont.*)
in Great Britain, 263, 264, 277–280, 279f21, 287–288
ground-adjustable, 117f10, 129, 130f11, 131f12
Reed propeller, 146, 163, 169–170
and Wright brothers, 20–22, 345
Ford Motor Company, 119, 169, 249
Franklin, William H., 299
Freedman Aircraft Engineering, 325
Freeman, Wilfrid R., 267, 268

Gallaudet, Edson F., 30
Gallup, David L., 33
Gardner, Lester D., 177
Gazely, Richard C., 70
General Electric (GE), 48, 62, 98, 340, 341, 343
Unducted Fan (UDF), 339–340
Gibson, Hugo C., 23
Giffard, Henri, 17
Gillmore, William E., 69, 70n74, 83n26, 88n44, 94n60, 102, 102n89, 102n90, 173n84
Gleed, Ian, 288, 303
Gloster Aircraft Company, 108, 110, 111, 206–207, 207n3, 207n4, 209, 210, 210n15, 210n16, 267n99
Gouge, Arthur, 264
Graf, John E., 177
Great Aerodrome, 20
Grey, C. G., 6, 157, 158, 177
Grogan, George, 1, 2f 1, 245, 348

Halahan, P. J. H., 287
Hale, W. A., 56, 57, 68, 73
Hall, Donald, 135–138
Hamilton Aero Manufacturing Company, 184, 187, 188–189, 190, 193, 210
and metal propeller, 185–187
Hamilton Standard Propeller Corporation
constant-speed counterweight, 237–240, 242, 292–296, 322
controllable-counterweight, 180, 199–201, 200f16, 220, 262
23E50 propeller, 308
45C2 propeller, 223
54H60 propeller, 331, 341
568F propeller, 342
formation, 188–189
Hydromatic propeller, 242–244, 243f 19, 330, 331
and World War II, 307–308
Hamilton Sundstrand Corporation, 342
Hamilton, Thomas F. "Tom", 39, 190, 195, 196f15, 236–237
at Hamilton Aero Manufacturing, 184, 185, 187, 210
at Hamilton Standard, 188, 189
Hart 1920 Model propeller, 82–88, 89f8
Hart and Eustis propeller, 76f7, 77–82, 91
Hart, Seth, 78
Hartzell Propellers, 175, 325, 336
Hathorn, Charles, 149, 150, 151f14
Hawker Hurricane, 275, 285, 288, 291, 304
Hayward, Charles B., 31
Hearle, Frank T., 27, 232
Heath, Spencer, 23–24, 27, 91–94, 100–102, 103, 176
Heddernheimer Metal Company, 253
Hegy, Ray, 185–187
Helbig, W., 253
Hele-Shaw Beacham propeller, 109–111, 208, 262
Bristol development, 262, 266–267
Gloster development, 111, 206, 210, 262, 268
Rolls-Royce development, 262, 267
Hele-Shaw, Henry Selby, 108–109, 207, 208, 236, 243, 264
Hicks, Harold A., 120
higher, faster, and farther, 5, 5n9, 13, 14, 272, 334
and the Jet Age, 306, 326, 338, 349
and propeller development, 143, 179, 198, 327
Hilliker, George, 150, 151f14
Hives, Ernest W., 267, 284
Howard, Clinton W., 175, 214
Hughes, Howard R., 10, 235, 235n1, 246, 246n30
Huston, Elmo F., 310

Ide, John Jay, 105
incorporation of aviation, 180–181
indeterminate synergy of invention, innovation, and use, 13, 14, 123, 143, 178
Ingells steel propeller, 58
Ingells, James, 122
Institute of the Aeronautical Sciences (IAS), 177, 211, 240, 311
International Air Meeting, 256–259
inventive institutions, 10
Iso-Rev Propellers, 325

J. A. Fay and Egan Company, 175, 194
Jane, Fred T., 25, 31
Jet Age, 14, 326–329, 335, 338, 341, 349
Johnson, Robert S., 316
Jones, Casey, 151, 151f14
Jones, Ivor, 295
Junkers, Hugo, 104
Junkers-Flugzeugwerk, 229
Juvenal, Chester G., 310

Kilner, Walter G., 173
Knowles, Henry, 264
Kraeling, Harry A., 95, 174, 188–189

Lampton, Glen T., 93–94, 226
Lane-Burslem, Eric, 292, 297
Langley, Samuel P., 20
Le Peré LUSAC-11, 82, 119
Leighton, Bruce G., 152
Leitner, Henry, 110
Leitner-Watts propeller, 110, 124
Lesley, Everett P., 51, 51n19, 52
Lindbergh, Charles A., 116, 140, 245
  and Standard Steel propeller, 117f10, 140, 145, 346
  as technical adviser, 222, 223
Lockheed/Lockheed-Martin aircraft
  C-130J, 341, 342f25
  L-188 Electra, 333
  P-38, 249
  Super Electra, 235, 246
  Vega, 214, 225
Loening, Grover C., 44
Lund and Dwight propeller, 30
Lund, James B., 30
Luttrell, James B., 94
Lynam, E. J. H., 105–106, 110, 208

MacDill, Leslie, 65n60, 68n71, 68n72, 70, 71n78, 92n57, 94, 94n60, 94n61, 101n85, 101n86, 102, 155n29, 162n54, 164n57, 168n66, 169n69, 169n71, 173n80
MacNeil, Charles S. J., Jr., 251–252, 272, 313–314
MacRobertson Race, 204, 231–234, 235, 236, 262
Mairesse, Robert, 80–81
Mannella, Alexander F., 137, 193
Manning, Leroy, 119
Mantz, Paul, 103
Manufacturer's Aircraft Association (MAA), 153, 178
Martens, Arthur, 253
Martin, Erle, 191–192, 230f18
  at Hamilton Standard, 238, 242, 311, 330, 343
Martin, Harold S., 98, 169
Massachusetts Institute of Technology (MIT), 33, 37, 251, 323
materials, 306, 325, *See* Bakelite Micarta, composites, duralumin, steel, wood
Materiel Division, Army Air Corps, 70
  competition with industry, 100–102
  Propeller Laboratory, 66, 69f6
  and propeller testing, 66–70
  and Reed propeller, 173–174
  and variable-pitch propeller, 212–217, 247, 250
Matthews Brothers Manufacturing Company, 39, 184, 185
Maughan, Russell J., 156
McCauley, Ernest G.
  collaboration with Frank Caldwell, 96–97, 102, 127, 128
  McCauley Aviation Corporation, 144
McCook Field. *See* Engineering Division, Army Air Service
McGlasson, Robert, 271
McIntosh, Lawrence W., 166
McStay, Carl E., 81
Mead, George J., 182, 195–197, 196f15, 218, 244
Messerschmitt Bf 109, 257, 260, 287, 293, 320
metal. *See* duralumin, steel
  as design choice, 124
  limitations, 31–32
  and modernity, 12, 127
Metal Airscrew Company Ltd., 108
Metal Propellers Ltd., 110, 124
Metz, Arthur W. F., 292, 297, 298
Meusnier, Jean Baptiste, 17
Micarta. *See* Bakelite Micarta
Military Aeroplanes, 44
Miller, Frank, 152
Miller, Walter, 155, 166
Milner, Harry Lawley "Pop", 110–111, 209, 211, 231, 267, 343
Mitchell, William "Billy", 83, 87, 119
modern aircraft, 2, 6–7, 13, 221, 243, 249, 276, 347
Moffett, William A., 152, 169
Monteith, Charles N., 90, 195, 196f15, 218
Moss, Sanford A., 98

Moy, Thomas, 18
MRC Ltd., 296

National Advisory Committee for Aeronautics (NACA), 35, 42–43, 92, 217
  engine cowling, 205, 221
  research programs, 51–52, 73, 125, 159, 315, 328–329
National Aeronautic Association (NAA), 150n13, 162, 229, 235, 340
National Aeronautics and Space Administration (NASA), 329, 337, 339, 340, 343
National Air Races, 132, 155, 158, 162, 224, 326
national pride, 268, 317, 348
Navire Aerien, 19
Nelson, Arvid, 186, 187, 190, 198, 230f18
New York Times, 35, 72, 87, 225, 228, 326, 341
Nippon Gakki Ltd., 323
Nutt, Arthur, 158

Ogburn, William Fielding, 8
Olive, Gordon, 274, 293, 300, 301, 303

Pagé, Victor W., 28–29, 176
patents, 178, 184
Patrick, Mason M., 92, 100–102, 159–160, 173
Paucton, Alexis-Jean-Pierre, 18
Pénaud, Alphonse, 19–20
Pierson, Rex K., 264
Pittsburgh Screw and Bolt Corporation, 248–249
Planophore, 19
plastic. *See* Bakelite Micarta
Popular Science, 316, 340
Post, Wiley, 225, 225n65, 238
Poulsen, Carl M., 262–263
Praeger, Otto, 87
Pratt & Whitney Aircraft Company, 181–182, 199, 223, 340
  Hornet engine, 212, 216, 224, 226
  Wasp engine, 175, 201, 216, 218, 225, 236, 327
Progress, 4, 11, 146, 161, 179, 233, 347
propeller community, 10, 12, 13, 44, 104, 144, 234, 345, 349
  hierarchy, 108, 165, 206, 208, 234
  industrial, 180, 211, 347
  and the Jet Age, 328, 330
  and metal propeller, 123, 124, 144, 197
  and rearmament, 306
  and variable-pitch propeller, 27, 75, 102, 145, 212, 273
propeller components
  cuffs, 315–316
  governor, 109, 237, 240, 243f 19, 251, 295
  paddle blades, 315–316
  split hub, 128, 130f11, 186
Propeller Division, Curtiss-Wright Corporation, 312f24, 319t3, 324
  establishment and innovation, 248–250
  postwar decline, 332–333
  and World War II, 311–312
propellers. *See* fixed-pitch propeller, variable-pitch propeller
  Canada, 111–114
  France, 19, 27, 229, 232, 343
  general aviation, 325
  Japan, 110, 207, 229, 265, 322–323
  multipiece, 77
  Soviet Union, 319, 339
propfan, 339–340
propulsion innovations, 9
  emergency power, 286
  fuel, 264, 269, 286
  supercharging, 77, 87, 98, 107, 109
  water injection, 316
Pulitzer Trophy Race, 132, 156, 157, 158, 159, 166
pursuit of practicality, 12

Ratier propellers, 232–233, 319, 342
Rawdon, Henry S., 168
Reed Propeller Company v. United States, 176–177
Reed Propeller Company, Inc., 153, 171, 174, 187, 201
Reed, S. Albert, 147–149, 151f14
  D-type propeller, 150–151, 151f14, 154, 163, 171, 172–173
  patents and litigation, 149–150, 174t2, 177, See *Reed Propeller Company v. United States*
  R-type propeller, 158–159, 163, 168, 173
  Sylvanus Albert Reed Award, 177, 240
Reichsluftfahrtministerium (RLM), 255, 261, 271
reinvention, 3, 6, 15, 73, 75, 339, 345–349

Reissner, Hans, 104–105
Rentschler, Frederick B., 184, 187, 189, 193, 198
Republic P-47, 315–316
Requa-Gibson Company, 23
Rhode Island (airship), 28
Richey, Paul, 287–288, 303
Rittenhouse, David, 147f13, 157
Rolls-Royce Aero Engines, 262, 267
Roma (airship), 98
Rotol Airscrews Ltd. *See* Hele-Shaw Beacham propeller
  and aerial spectacle, 268–269
  formation, 267–268
  postwar, 330, 332
  and rearmament, 283–285
  Rotol Hurricane, 287
  Rotol Spitfire, 285, 289, 291
  and World War II, 317, 318–319
Rowe, Brian, 340
Royal Aeronautical Society (RAeS), 207–210
Royal Air Force (RAF)
  Advanced Air Striking Force, 282, 287, 288
  Air Component, British Expeditionary Force (BEF), 282, 288
  Bomber Command, 265, 283
  87 Squadron, 288
  54 Squadron, 285, 289–290, 291, 294
  Fighter Command, 274–281, 285, 290–295, 296, 302
  41 Squadron, 285
  Hornchurch Fighter Station, 289–290
  Long-Range Development Unit (LRDU), 268–269, 269f20
  19 Squadron, 279, 279f21, 281, 285
  92 Squadron, 298
  1 Squadron, 287
  74 Squadron, 279, 299
  65 Squadron, 284–285, 290, 292–294, 299–301, 301f22
  609 Squadron, 298, 302–303
  611 Squadron, 298
  33 Maintenance Unit, 298
Royal Aircraft Establishment (RAE), 105–107, 110, 208, 270
Royal Canadian Air Force (RCAF), 112
Russell, Frank H., 162, 170, 171
Ryan NYP *Spirit of St. Louis*, 73, 116, 117f10, 140, 346
Saunders, Gerald A. W., 299
Schneider Trophy Race, 147f13, 157, 159, 209
Schory, Carl, 189, 201
Schroeder, Rudolph W. "Rudy", 87, 119, 169, 244
Schwarz propeller, 260, 277, 318, 321
Sheldon, John, 17
Sidgreaves, Arthur F., 267
Simmons, James Lee, 23
Skeel, Bert E. "Buck", 157
Skinner, C. E., 58
Smith Engineering Company propeller, 215–216, 224–227
Smith, John W., 120–121, 122, 123–124
Smith, Mascom A., Jr., 60, 67, 72
Smith, S. Harold, 215
Smiths Aerospace Group, 341
Sōchirō, Honda, 323
Spanish Civil War, 261
Squier, George Owen, 35, 53, 151
Standard Steel Propeller Company
  ground-adjustable propeller, 137, 173, 174
  organized as corporation, 140
  origins, 94–95
steel
  and culture, 94
  hollow-steel blades, 95, 122, 144, 207, 249, 251
  Smith steel propeller, 120
Stengel, Henry Ivan, 29, 176
Stout, William B., 120, 123
Sugimoto, Osamu, 323
Sumitomo Metal Industries Ltd., 322–323
Supermarine Spitfire, 274, 275, 280, 291, 298

Taylor, C. Fayette, 64
test and evaluation
  flight, 50
  whirl, 45, 47, 74
  wind tunnel, 21, 50, 68, 125, 159, 328
The Spirit of St. Louis (book), 139
Theorie de la vis d'Archimede, 18
Tibbets, Paul, 313
tilt rotor aircraft, 343
Tupolev Tu-95 bomber, 339
turboprop, 329–330, 337, 338–339, 341, 344. *See* Advanced Turboprop Project
Turnbull, Wallace R., 111–113, 114f9, 211, 242, 312
  at Curtiss, 201–203
  No. 2 propeller, 113–114

US Air Mail Service
  interest in new propellers, 87, 119, 132
  and Reed propeller, 154, 160, 168, 174
Unholtz, Bernard, 253, 270, 320
United Aircraft and Transport Corporation (UATC), 181–183, 194
  Technical Advisory Committee (TAC), 194–198, 196f15
United Aircraft Corporation (UAC), 236–237, 244, 311, 331
United Technologies Corporation (UTC), 331, 342
Uppercu, Inglis M., 29, 31
user innovation, 13, 275, 304
UTC Aerospace Systems (UTCAS), 343

Valk, William E., Jr., 152, 175
Van Zandt, J. Parker, 156
variable-pitch propeller, 18
  constant-speed, 107, 109–110, 237–240, 251, 266–267, 292–293
  controllable-pitch, 76
  electric, 113, 114f9, 201–202, 226, 227–228
  feathering, 241–247, 308, 309f23, 348
  hydraulic, 107–109, 197–200, 207–208, 223, 254–255
  mechanical, 85, 95–97, 115, 215, 224
  reversible, 77, 248, 313, 330
Vaughan, Guy W., 249
Vereinigte Deutsche Metallwerke (VDM), 252–256, 261
  comparison to American and British designs, 270
  in Great Britain, 270–272
  and public spectacle, 256–259
  and World War II, 320–322
Vickers Wellesley, 269, 269f20
Vincent, Jesse G., 58, 82, 122, 123
von Steiger-Kirchofer, Carl, 20

Waite, William, 149
Walker, Charles C., 264
Wall Street Journal, 41, 193, 307
Walsh, Raycroft, 190–191, 198, 216, 224, 228, 230f18
Washington Aeroplane Company, 23
Watts, A. G., 298
Watts, Henry C., 110, 208, 264, 277, 278, 318
Weick, Fred E., 51, 195, 196f15
Westinghouse Electric and Manufacturing Company
  Bakelite Micarta, 98, 118–120, 192
  whirl testing, 53–58, 54f4, 81, 94
Williams, Alford J. "Al", 155
Wilson, Eugene E,
  at BuAer, 134, 168
  at Hamilton Standard, 180, 189–190, 192, 193, 198, 214
  at Standard Steel, 188
wood
  advantages, 26, 31, 37
  construction, 25–26, 40–41
  and culture, 25
  as design choice, 16, 44
  limitations, 27, 34–36, 37, 41–42
  and modernity, 12
  and Wright brothers, 22
Woodward Governor Company, 237–238
Woodward, Elmer E., 237–238
World War I, effect on propeller development, 46, 75, 115, 345–346
  Germany, 104–105
  Great Britain, 105–108
  United States, 36–42, 45, 72, 99, 123, 170
World War II, 306, 319, 348
  Axis propeller industry, 323
  British production, 318
  United States production, 316
Wright Aeronautical Corporation, 181
  Cyclone engine, 202, 221, 223, 226, 246, 327
  Whirlwind engine, 135, 137, 174, 186, 202
Wright Field. *See* Materiel Division, Army Air Corps
Wright, Theodore P., 211
Wright, Wilbur and Orville, 4, 16, 22f2
  invention of propeller, 20–22

Yates, William W., 305
Yount, Barton K., 87